The Mechanization of Nature: Matter, Body, and Motion in Blasius of Parma's *Physics*

Studien und Texte zur Geistesgeschichte des Mittelalters

Begründet von

Josef Koch

Weitergeführt von

Paul Wilpert, Albert Zimmermann und
Jan A. Aertsen

Herausgegeben von

Andreas Speer

In Zusammenarbeit mit

Wouter Goris, Kent Emery, Jr.
und Georgi Kapriev

BAND 140

The titles published in this series are listed at *brill.com/stgm*

The Mechanization of Nature

Matter, Body, and Motion in Blasius of Parma's Physics

By

Henrik Lagerlund
Sylvain Roudaut
Erik Åkerlund

Critical edition by

Robert Andrews

BRILL

LEIDEN | BOSTON

The Library of Congress Cataloging-in-Publication Data is available online at https://catalog.loc.gov
LC record available at https://lccn.loc.gov/2025025443

Typeface for the Latin, Greek, and Cyrillic scripts: "Brill". See and download: brill.com/brill-typeface.

ISSN 0169-8028
ISBN 978-90-04-73427-2 (hardback)
ISBN 978-90-04-73496-8 (e-book)
DOI 10.1163/9789004734968

This book is printed on acid-free paper and produced in a sustainable manner.

Contents

List of Figures VIII
Preface VII

Introduction 1

Interpretive Essays

1 Scales and Weights: Introducing Mechanics in Natural Philosophy 13
Sylvain Roudaut

2 Causality and Causal Powers in Blasius of Parma's *Physics* 45
Joël Biard

3 Blasius of Parma on Motion in the Sublunary World 72
Henrik Lagerlund, Sylvain Roudaut and Erik Åkerlund

4 Rarefaction and Condensation in Blasius' *Physics* 93
Erik Åkerlund

5 Blasius of Parma's Account of Artifacts in the *Physics* 102
Henrik Lagerlund

6 Building a Unified Theory of Nature: The Functions of the 'New Analytical Languages' in Blasius of Parma's *Questions on the Physics* 115
Sylvain Roudaut

Edition: Blasius Pelacanus de Parma Quaestiones super VIII libros Physicorum

Description of the Manuscripts 163
Robert Andrews

1 Liber II q. 1: Utrum domificator in faciendo domum faciat aliquid rebus naturalibus condistinctum 168

2 Liber II q. 2: Utrum quodlibet res naturalis habeat in se principium motus et quietis 178

3 Liber II q. 7 Utrum ad cuiuscumque rei naturalis generationem praeter particulare agens requiratur influxus causae universalis, quae causa universalis dicitur sol 201

4 Liber II q. 8 Utrum inter agentia particularia, supposita semper generali influentia superiorum, possit qualitas una vel plures formam substantialem producere 216

5 Liber IV q. 2 Utrum entia naturalia distantia ab eorum locis naturalibus moveantur ad illa impedimentis subtractis 228

6 Liber IV q. 3 Utrum corpora naturalia appetant moveri ad eorum loca naturalia per lineas breviores 246

7 Liber IV q. 7 Utrum rarefactio sit possibilis 258

8 Liber V, q. 1 Utrum agens naturale, hora qua agit in passum, agat in ipsum secundum se totum et quamlibet eius partem 272

Appendix: Blasius Pelacanus de Parma, *Quaestiones super VIII libros Physicorum* (redactio secunda) 285

Bibliography 290

Index 311

Preface

This book is the result of the research project "The Mechanization of Philosophy 1300–1700" financed by Vetenskapsrådet (Swedish Research Council) grant agreement no. 2019–02777. It studied the medieval background to the mechanical philosophy of the seventeenth century. One of the major results of the project is presented by these new essays on Blasius of Parma as well as the edition of eight questions from his question commentary on Aristotle's *Physics*. The present book brings to the foreground, for the first time to an English audience, a neglected, but highly interesting and novel thinker. We would especially like to thank Joël Biard, who has pioneered the work on Blasius, for agreeing to be involved in this project and for contributing one of the six essays included. He has also generously agreed to let us use some of his transcriptions of Blasius' questions. The work has been conducted at the Philosophy Department at Stockholm University and many of these essays have been presented at the *Stockholm Workshops in the History of Philosophy*-series. We would like to thank the participants of these seminars for their valuable comments. As always, we are indebted to our families for their patience and support.

Stockholm, May 2024

Figures

Interpretive Essays

1.1 Equal-armed balance with equal weights, one of which is at the end of a fixed arm, the other hanging freely 14

1.2 Illustration of a body (A) having a greater positional heaviness than another body (B) 20

1.3 A body (B) in equilibrium with a heavier one (A) due to its positional heaviness 23

1.4 Two vases connected by a pipe, illustrating the effect of weights in fluid bodies 30

6.1 A body divided into an infinity of proportional parts 135

6.2 Representation of the inverse relations between heat and coldness in a subject 150

6.3 Reconstruction of a case of uniformly difform distribution of heat, as described by Blasius of Parma 154

Edition

II Q1

1a Vat. lat. 2159 f. 85ra, Biblioteca Apostolica Vaticana (Vatican Library) 171

1b Vat. lat. 3012 f. 27va, Biblioteca Apostolica Vaticana (Vatican Library) 171

II Q2

1a *in ima col. f. 86va* V1, Biblioteca Apostolica Vaticana (Vatican Library) 183

2 *in ima pag. f. 88vb* V1, Biblioteca Apostolica Vaticana (Vatican Library) 198

IV Q2

1 *in ima col. fol.* 115ra V1, Biblioteca Apostolica Vaticana (Vatican Library) 232

Introduction

Blasius of Parma, born between 1350 and 1354 near Parma, and passing away on April 23rd, 1416, in the same city, is a figure often overlooked in the annals of late medieval philosophy, particularly compared to figures such as William Ockham or John Buridan. Yet, Blasius enjoyed significant acclaim during his lifetime and was recognized as one of the era's most innovative and daring thinkers. His provocative ideas, often characterized as a form of medieval materialism, earned him the epithet the "Diabolic Doctor" (*doctor diabolicus*).

Blasius' educational journey began with studies in logic and medicine at Pavia, where he later embarked on a teaching career around 1377.[1] His scholarly pursuits flourished across several universities of Northern Italy at the end of the fourteenth and beginning of the fifteenth century. Blasius taught in the 1380s at the universities of Bologna and Padua, which he left in 1388 for a brief stay in Florence, before returning to Pavia where he taught between 1388 and 1407. It seems as though Blasius' influence in Pavia gradually declined due to the growing reputation of a new generation of philosophers, that of Paul of Venice in particular. In 1412, Blasius was called to teach in the *studium* of Parma, where he remained until his death in 1416.

During the early 1380s, Blasius spent a brief period in Paris, which by then was the most significant hub of intellectual activity, where he likely immersed himself in the works of prominent Parisian thinkers such as John Buridan, Albert of Saxony, Nicole Oresme, and Marsilius of Inghen. The exposure to this diverse intellectual climate undoubtedly influenced Blasius's philosophical development.

Despite his acclaim, Blasius encountered controversy when, in 1396, the Bishop of Pavia, the Franciscan William Centueri, condemned several of his teachings, particularly those concerning materialist interpretations of soul and nature. This ecclesiastical censure, however, did little to deter Blasius' academic pursuits, as he retained his position and continued to refine his theories. It is

1 Detailed bibliographical information on Blasius of Parma may be found in G. Federici Vescovini, "Pelacani, Biagio, detto Biagio da Parma", in *Dizionario biografico degli Italiani*, vol. 82 (Rome: Istituto dell'Enciclopedia Italiana, 2015), 95–100; G. Federici Vescovini, "Blaise Pelacani de Parme. Portrait en '*doctor diabolicus*'," in C. Angotti, M. Brinzei, and M. Teeuwen (eds.), *Portraits de maîtres offerts à Olga Weijers* (Turnhout: Brepols, 2013), 117–126; G. Federici Vescovini, *Astrologia e scienza. La crisi dell'aristotelismo sul cadere del Trecento e Biagio Pelacani da Parma* (Florence: Enrico Vallecchi, 1979), 21–39; J. Biard and A. Robert, "Introduction," in Joël Biard and Aurélien Robert (eds.), *La philosophie de Blaise de Parme: physique, psychologie, éthique* (Florence: SISMEL—Edizioni del Galluzzo, 2019), 3–12, esp. 3–6.

 | DOI:10.1163/9789004734968_002

likely, however, that this condemnation prompted Blasius to temper some of his theses, or to present them as mere hypotheses or "possible determinations" of a question, a method through which Blasius often conveyed his ideas, as can be seen in the works presented below.

Over the course of his career, Blasius had a significant influence in Italy and taught many students. Vittorino da Feltre (1378–1446) is recognized as one of Blasius' most prominent students.[2] He taught natural philosophy, moral philosophy, astrology, and mathematics, this final discipline covering both what we would call today pure mathematics and applied mathematics (more precisely, astronomy and optics in his case). Blasius wrote commentaries on several works of Aristotle, namely his *Quaestiones de anima, Quaestiones de generatione et corruptione, Quaestiones super libros Meteororum, Quaestiones de caelo*, and his *Quaestiones super libros Physicorum*. This last work, which constitutes the subject of the present volume, has come down to us in two different redactions, between which the condemnation of 1396 took place. While the first redaction of Blasius' *Questions on the Physics* was finished between 1385 and 1388, the second redaction was composed in 1397.

Besides his commentaries on Aristotle's works, Blasius also produced other kinds of writings, demonstrating his interest in a wide range of topics and his familiarity with nearly all the scientific disciplines of his time. He wrote two major mathematical works, namely an influential treatise on Thomas Bradwardine's *Treatise on Proportions* and some questions on John Peckham's *Perspective*. Blasius' fame as a mathematician was well known throughout Europe, since his works continued to be quoted up to the sixteenth century. Among many others, Leonardo da Vinci, for instance, quoted Blasius' works.[3]

In the field of logic, Blasius authored a commentary on the logic of Peter of Spain. Among various shorter works, Blasius composed two texts (one in the form of a treatise, the other in the form of a question) on the science of weights, which correspond to the discipline we would today call 'statics,' as well as two texts on the problem of the intensity of forms (one devoted to the geometric representation of intensive properties, the other on the explanation and measure of intensive variations).

Through this broad intellectual repertoire, Blasius demonstrates his acquaintance with a variety of philosophical and scientific traditions. As a philosopher trained in Italy in the fourteenth century, where the curriculum in philosophy was closely connected to that of medicine, Blasius was acquainted with the medical theories of the time, which actually play a central role in the

2 It has been conjectured that Giovanni Marliani was also a student of Blasius, but his estimated date of birth (ca. 1420) appears to make this impossible.

3 Federici Vescovini, *Astrologia e scienza*, 295.

philosophical theories he developed on the nature of bodies and souls.[4] The medical theory of qualitative complexions, from which physicians developed explanations of bodily, biological, and psychological functions, is central to the articulation of the reductionist stance that Blasius often adopts while discussing what he calls 'problems' in natural philosophy. During Blasius' time, the medical theories developed in Italy by figures such as Peter of Abano were intertwined with astrological concepts concerning the influence of celestial spheres on natural phenomena in the sublunary realm. This special status granted to astrology held significant importance in shaping Blasius' approach to physics.

Regarding his philosophical influences, two primary traditions significantly influenced Blasius' philosophical positions and methods: the Parisian tradition of the fourteenth century and the Oxfordian tradition. The Parisian tradition, propagated by philosophers such as John Buridan, Albert of Saxony, Nicole Oresme, and Marsilius of Inghen, played a pivotal role in Blasius' philosophical development. His approaches to logic, physics, and psychology exhibit a distinct similarity in method, characterized by a logical and semantic framework inherited from the Parisian masters. Blasius' positions often gravitate towards forms of reductionism, aligning with the nominalist tendencies of these scholars whom he frequently references, either implicitly or explicitly.

In addition to the Parisian tradition, Blasius also drew inspiration from the Oxfordian tradition, particularly the lineage of the so-called Oxford Calculators including Thomas Bradwardine, William Heytesbury, John Dumbleton, Richard Kilvington, and Richard Swineshead. This influence is notably evident in Blasius' works, particularly in his *Questions on the Physics*. Here, one of Blasius' distinctive features is the employment of quantifying methods, attributing numerical values to various physical magnitudes. Moreover, Blasius

4 On the Italian learning context in the fourteenth century, see J. Chandelier and A. Tabarroni, "Philosophie, médecine et frères mendiants à Bologne dans la première moitié du XIVe siècle," in J. Chandelier and A. Robert (eds.), *Savoirs profanes dans les ordres mendiants en Italie, XIIIe–XVe siècle* (Rome: Publications de l'École française de Rome, 2023), 199–233; A. Tabarroni, "La nascita dello Studio di Medicina e Arti a Bologna," in C. Casagrande and G. Fioravanti (eds.), *La filosofia in Italia al tempo di Dante* (Bologna: Il Mulino, 2016), 25–36; A. Tabarroni, "Ambienti culturali prossimi a Dante nell'esilio: lo Studio bolognese di Arti e Medicina," in E. Malato and A. Mazzucchi (eds.), *Dante fra il settecentocinquantenario della nascita (2015) e il settecentenario della morte (2021)* (Rome: Antenore, 2016), 327–349; D. Buzzetti, R. Lambertini, and A. Tabarroni, "Tradizione testuale e insegnamento nell'Università di Medicina e Arti di Bologna dei secoli XIV e XV," in *Annali di storia dell'università italiane*, vol. 1 (Bologna: CLUEB, 1997), 77–96; G. Fioravanti, "I filosofi e la medicina: una progressiva autonomia," in C. Casagrande and G. Fioravanti (eds.), *La filosofia in Italia al tempo di Dante*, 65–76; C. Crisciani and G. Fioravanti, "I filosofi e i medici come gruppo: autorappresentazione e autopromozione," in C. Casagrande and G. Fioravanti (eds.), *La filosofia in Italia al tempo di Dante*, 77–90.

systematically employs thought-experiments involving counterfactual or even impossible scenarios, reminiscent of the 'imaginary' physics (or physics *secundum imaginationem*) developed by the Oxford Calculators to assess the validity of rules governing physical processes.

However, unlike the Oxford Calculators, whose thought-experiments were often exercised in a logical context, Blasius' thought-experiments are primarily motivated by an empirical stance, driven by the desire to identify *causal factors* operating in nature. In this regard, his use of thought-experiments aligns more closely with Galileo's, serving as a speculative device for conducting physical investigations into the causes of natural phenomena.[5]

The amalgamation of diverse influences in Blasius's physics justifies our inclination to view his natural philosophy as a bridge between the main trends of the fourteenth century and the emergence of early modern physics. The new ideas he advocates in his physics can be encapsulated by the term 'mechanization,' which appears in the title of the present volume, for several reasons.

Firstly, Blasius assigns central importance to statics, or what was termed in medieval times the 'science of weights,' in his natural philosophy. Statics, a branch of mechanics, is integrated into Blasius' views of local motion, which diverges significantly from the standard Aristotelian theory of motion.

Secondly, closely linked to the preceding point, Blasius emphasizes the importance of heaviness and lightness in his explanation of motion. Unlike early modern accounts of 'mechanical philosophy,' exemplified in Descartes, Blasius maintains a qualitative ontology wherein primary qualities such as heat, coldness, wetness, dryness, as well as heaviness and lightness are fundamental and irreducible features of nature. However, akin to later mechanical philosophy, the properties of weight hold special importance in Blasius' explanations of natural phenomena.

Thirdly, Blasius leans towards a reductionist view of nature akin to mechanical philosophy, as he seeks in his physics to diminish the role of substantial forms. While Blasius never fully embraces a completely materialist ontology in which all substantial forms and accidents are reduced to dispositions and configurations of matter, he considers the reducibility of substantial forms as a possible hypothesis.[6]

---

5 On thought experiments in Galileo, see C.R. Palmerino, "Galileo's Use of Medieval Thought Experiments," in Katerina Ierodiakonou and Sophie Roux (eds.), *Experiments in Methodological and Historical Contexts* (Leiden / Boston: Brill, 2011), 101–125.

6 For more on this see J. Biard, *Hypothèse matérialiste et pensée radicale. La philosophie de la nature de Blaise de Parme* (Paris: Vrin, 2022), 13–39.

Fourthly, Blasius presents a novel theory of motion that fundamentally revises the traditional opposition between violent and natural motion central to Aristotelianism, mirroring developments in modern physics where this opposition also collapses. In his *Questions on the Physics*, Blasius aims to demonstrate that the conventional understanding of this opposition is flawed, as many motions described as 'natural' are, in fact, violent. Blasius contends that motions can be indifferently labeled as violent or natural, depending on the perspective, leading him to view this opposition as relative rather than absolute.

It is, foremost, for these reasons that we see Blasius as an early mechanical philosopher, but his commentary is important for another reason as well. It represents, we think, a new way of doing physics in the Aristotelian tradition. Perhaps the clearest way to show this is to illustrate how it breaks down the traditional barrier between what was called speculative and productive philosophy or science, or, in Robert Kilwardby's famous division from *De ortu scientiarum*, between the speculative and the practical (that which aims towards activity) sciences.[7] In his division, physics and mathematics belong, together with metaphysics, to the speculative sciences while the practical sciences involve those activities captured by languages (logic and grammar), ethics, and mechanics. The mechanical sciences refer primarily to the human body, and aim to aid the body in specific ways. The kinds of sciences he mentions are, hence, those having to do with nutrition, armament, building, and, of course, medicine.

Kilwardby makes an effort to defend the mechanical sciences as real sciences belonging to philosophy. Some have argued, he notes, that they belong to art and not to science, since they only deal with the individual and contingent. Kilwardby argues that philosophy and science need to be taken in a broad sense. Furthermore, when Aristotle says that science is of the necessary and universal he speaks only of the speculative sciences, but the mechanical sciences can be said to incorporate aspects of both, since they make use of the speculative sciences, which means that they touch upon the necessary and universal, while at the same time aim towards the practical and action on the individual and contingent. The sciences are on this view primarily divided by their end; the speculative aim towards truth while the practical aim towards action primarily in relation to bodies.

7 See Robert Kilwardby, *De ortu scientiarum*, ed. Albert G. Judy (Oxford University Press: Oxford, 1976) and the study in Alfonso Maierú, "Robert Kilwardby on the Division of the Sciences", H. Lagerlund and Paul Thom (eds.) *A Companion to the Philosophy of Robert Kilwardby* (Brill: Leiden, 2013), 353–389.

Even though the distinction is clear it is, as indicated, not absolute and there seems to be room for what is practical to be speculative and vice versa. However, the aim of physics never mixes with the aim of mechanics, and traditionally physics, of course, makes no use of mechanics. Some parts of mechanics on the other hand make use of mathematics and physics in their aim. In this way, mechanics in Kilwardby's view has similarities with what later (in the seventeenth century) was to become called experimental science, that is, in its practical aim.[8]

The medieval use of the term "experimental science" can, foremost, be found in Roger Bacon. He seems also to set a practical goal for experimental science, which, as he says, aims to reveal the trickery of magic in the same way as logic reveals sophistical thinking.[9] He, furthermore, seems to see sciences like optics and the study of phenomena like rainbows as particularly useful experimental sciences.[10]

An interesting feature of Blasius' *Physics*-commentary is, as noted above, the explicit use it makes of statics or what in the Middle Ages was called the science of weights (*scientia de ponderibus*). It would in Kilwardby's division primarily sort under the mechanical sciences, since its aim is activity and the individual, but it clearly also involves aspects of the speculative sciences. It is what came to be called a 'middle' science like optics and on Kilwardby's classification it straddles both the practical and the speculative. In Bacon's view, it would be similar to optics a particularly useful experimental science.

It has to be said that this is an original aspect of Blasius' work. Mechanics in this sense played no part in Aristotle's *Physics* and little, if any, part in commentaries on *Physics* by medieval thinkers before Blasius. He was able to do this primarily because of his analysis of the distinction between artifacts and nature.[11] He seems to have realized that natural things can equally well be seen

8 See Anstey, P.R. and Vanzo, A., *The Origins of the Experimental/Speculative Distinction* (Cambridge University Press: Cambridge, 2023), 19–42.

9 He writes: "This science [experimental science] alone, therefore, knows how to test perfectly what can be done by nature, what by the effort of art, what by trickery, what the incantations, conjurations, invocations, deprecations, sacrifices, that belong to magic, mean and dream of, and what is in them, so that all falsity may be removed and the truth alone of art and nature may be retained. This science alone teaches us how to view the mad acts of magicians, that they may be not ratified but shunned, just as logic considers sophistical reasoning." (Roger Bacon, *Opus maius*, translated by Robert Burke, New York: Russell and Russell, 1962, vol. II, Part VI, Chapter 2, 587.)

10 For more on this see Jeremiah Hackett, "Roger Bacon on *scientia experimentalis*", in J. Hackett (ed.) *Roger Bacon and the Sciences* (Brill: Leiden, 1997), 277–315.

11 Blasius' view of the distinction between art and nature is discussed in essay 5. By rejecting an ontological distinction between art and nature Blasius is able to freely move between

as artifacts or as machines, and, hence, such machines as are used in the mechanical sciences, that is, the lever, capstone, pulley, screw, and stamp, can be used to study nature as well as the principles that govern it. Blasius' preferred machine is the scale, which was also used by Jordanus de Nemore already in the thirteenth century to combine classical mechanics with Aristotelian science. The scale is by him seen as the primary tool to study aspects of nature, and he proves that all the other machines can be reduced to it—a result to which also Galileo refers.[12]

The science of weights as it was approached by Jordanus also incorporates references to mathematical results by Archimedes. Blasius refers to Archimedes several times in his *Physics*-commentary,[13] but this use of mathematics is also supplemented with mathematical results found in the Oxford Calculators tradition. As mentioned above already, Blasius frequently incorporates results from the Oxford philosophers into his *Physics*, supplementing his Aristotelian demonstrations with a mathematical element, which greatly broadens the usefulness of these demonstrations. By including all these elements in his commentary, Blasius gives physics new tools as well as undermines the traditional view of physics as a purely speculative science, since it incorporates practical aims similar to Bacon's experimental sciences.

Following these reflections, we would like to highlight five elements that make Blasius *Physics*-commentary stand out in the medieval tradition. He unites aspects that, to a large extent, previously were unrelated, namely (I) an Aristotelian method, (II) the mechanical sciences (or statics), and (III) the mathematical approach to physics developed by the Oxford Calculators. There are two further aspects to Blasius' approach to physics which we would like to mention, and those are (IV) his use of thought-experiments and (V) his probabilism. As is shown by Joël Biard in his essay included in this volume (Essay

---

the two and there is no philosophical problem for him to use art to study nature. The practical aim of art is used in physics as a tool or an artifact would be used in everyday life.

12 In *Delle machine* from c. 1593, Galileo, for example, writes: "We, desiring now to discuss this matter, in order to proceed in an orderly way, shall begin by considering the nature of the first and simpler machines, to which the others reduce or from which they are compounded. These simple machines are five in number: the lever, the capstan, the pulley, the screw, and the stamp or force of impact. All of these again reduce in a certain way to a single one, that is to the balance." For the translation quoted here, see W.R. Laird, "The Scope of Renaissance Mechanics" *Osiris* 2, 1986, 43–68.

13 See Essay 1 for further discussion of the use Blasius makes of Archimedes. Many of his references are actually to the pseudo-Archimedean *De insidentibus in humidum* or *De insidentibus in liquido*. It is, however, unusual to find references to Archimedes in commentaries on *Physics*, but there was a growing trend for this at the end of the fourteenth century.

2), Blasius clearly develops a probabilism similar to that of the ancient Academic skeptic Carneades and presented by Cicero in the *Academica*. It is an unusual position in the Middle Ages, but similar views were held by John of Salisbury in the twelfth century and Nicholas of Autrecourt in the fourteenth. It also became a standard position among "scientific" philosophers in the seventeenth century, developed by Pierre Gassendi and later mentioned by Pierre Bayle as well.[14]

Blasius integrates his probabilism with his use of thought-experiments. On several occasions, he offers alternate explanations of physical phenomena, presenting them as probable; thus they become thought experiments or alternate possible explanations. His materialism is one such alternate physical explanation.[15] In this thought experiment, all substantial forms are rejected.

It is also possible to understand the side issues he discusses throughout his work in a similar fashion. In several of his questions, Blasius inserts what he calls 'problems' into the discussion. They sometimes seem out of place and may introduce issues from such a wide variety of fields as mechanics, medicine, astrology, physiognomy, mathematics, and various other curiosities of the natural world; but it seems reasonable to view them as thought experiments about various problems associated with physics. They often also have a mathematical nature and introduce the reader to more 'applied' problems in physics. Perhaps they can also be seen as sharing the aim of the experimental sciences in Bacon's sense, since they to some extent are meant to replace 'magic' with science.

Taken together, (I)-(V) introduce, we think, a new approach to physics which enriches the Aristotelian foundation, but ultimately also goes beyond it and invigorates the commentary tradition on the *Physics*.

The interpretive essays in this book study six carefully selected aspects of Blasius' commentary. They also present primary topics manifested in the eight questions, edited by Robert Andrews, at the end of the book. These six aspects are Blasius' (1) use of mechanics, (2) analysis of causality, (3) account of motion, (4) analysis of rarefaction and condensation, (5) view of artifacts, and (6) the use of a new analytical language in physics.

The introduction of mechanics into natural philosophy is one of the most prominent features of Blasius' commentary, as already noted. The way in which this is done is presented and explained in the first essay, written by Sylvain Roudaut. The role of Blasius' examples from statics, or the "science of weights,"

---

14 See H. Lagerlund, *Skepticism in Philosophy: A Comprehensive, Historical Introduction* (New York: Routledge, 2020), Chapter 7.

15 See Biard, *Hypothèse matérialiste*.

in his *Questions on the Physics* remains to some degree enigmatic. However, it is clear that his fusion of the mechanistic tradition into the context of Aristotelian natural philosophy expands the explanatory scope of mechanics, and transforms natural philosophy into something that looks more like a kind of mechanistic natural science. Blasius' treatment of mechanics in natural philosophy goes hand-in-hand with more general features of his philosophy, and ultimately constitutes what can be considered an anti-Aristotelian stance, even though it is presented as part of a commentary on Aristotle.

Included in natural philosophy, these changes also have repercussions when it comes to Blasius' understanding of the nature of causation. The second essay, written by Joël Biard, treats Blasius' role in the transformation of conceptions of causation and power in the Late Medieval period. Blasius does not so much try to unify the conception of causation as to determine the different powers that are at work in each natural change. Though quite traditionally Aristotelian in some parts, and from some perspectives, he is eclectic and innovative in others. Ultimately, the patchwork of different powers at work in different sorts of change requires us to not try to explain everything in a fixed scheme, but rather investigate what powers and causes might be at work in any particular change.

A notion closely connected to that of causation is motion. The third essay presents Blasius's treatment of motion and change in the sublunary world, from a background in the Aristotelian tradition as well from Ockham's and Buridan's treatments of this issue. Though to a large extent employing quite traditional terminology, Blasius arrives at some quite surprising and unusual positions. From the traditional position that everything seeks to be and remain in its natural place, Blasius arrives at the unorthodox position that all local motion in the sublunary world is *violent*. In other words, in the sublunary world there is no such thing as "natural change" or "natural motion". All change, including death, is something that is imposed on a thing from the outside, and thus is not natural to it. Seen from the background of the history of this question, Blasius' move undermines the whole division between that which is natural and that which is not, and opens up the possibility of a quite different conception of "nature" with new vistas for its study.

Buridan and Ockham, together with the broader Aristotelian tradition, figure prominently in Blasius' treatment of motion. This is the case also for his treatment of condensation and rarefaction, presented in the fourth essay, written by Erik Åkerlund. In philosophical traditions (such as the one that Blasius is in) where vacuum is not considered possible, condensation and rarefaction are baffling phenomena in need of an explanation. In comparison to the views of Ockham and Buridan, Blasius ultimately seems to come closest to Ockham's position, where condensation is understood to be a thing's parts coming closer together "all the way down", but without rejecting a more "Buridanean" way of speaking about it, referring to change of quantity. As in other questions, Blasius is eclectic in his treatment of the phenomena of condensation and rarefaction, as well as inventive in his use of examples and counter-examples.

As was seen above, Blasius treats all motion in the sublunary world as violent, as opposed to natural. What does this do to his understanding of the relation between the natural and the artificial? This issue is dealt with in the fifth essay, written by Henrik Lagerlund, which treats of Blasius' understanding of artifacts. Once again, Blasius takes his cue from Ockham and Buridan on this issue. In general, Blasius follows a general trend of the time, based on a nominalist metaphysics, to blur the line between the natural, on the one hand, and the artificial, on the other. More specifically, although he does not deny that something is introduced by the building of a house, say, Blasius does deny that that which has come about is something over and above the natural, that is, that out of which it is made. In this radical nominalism with respect to artifacts, Blasius equates the artificial with the natural when it comes to its ontological aspect, instead focusing on the intentional aspect in the making of the artifact. An artist leaving his mark on an artifact is not fundamentally different from the mark left in nature by God. There is nothing more to an artifact, from an ontological perspective, than nature.

Addressing, finally, a more general view of Blasius' philosophy once again, in the sixth essay, written by Sylvain Roudaut, Blasius' overall use of languages and methodologies is treated. Given the background of John Murdoch's suggestion that "new analytical languages" are a distinctive feature of the unification of the sciences in the fourteenth century, Blasius's language, methods, and theories are presented and analyzed in three different areas: the theory of supposition, the theory of continuity and infinity, and the theory of proportions and intension of forms. In all of these areas, Murdoch's hypothesis regarding languages, as far as Blasius is concerned, is judged to be vindicated. The "new analytical languages," incorporating also theories and methodologies, play a great role in Blasius' *Questions on the Physics*, at different levels of generality. They ground Blasius' approach to physics, incorporating empiricist observations as well as mathematical analysis, in his eclectic unification of the sciences, as presented in the earlier essays.

Interpretive Essays

CHAPTER 1

Scales and Weights: Introducing Mechanics into Natural Philosophy

Sylvain Roudaut

1 An Intriguing Problem

The first question Blasius asks about Book II of Aristotle's *Physics* (in the second redaction) concerns the reality of artifacts such as houses. Are artifacts some real things somehow distinct from or existing over the natural things of which they are made? (See Essay 5 for a separate study of this question.) In the first article comprising this question, Blasius examines what he calls a 'problem,' in line with the general method he follows in his questions.[1] The problem does not match what one might expect when it comes to discussing the ontological nature of artifacts. As a matter of fact, the problem concerns a scale having two arms of equal length at the end of which two equal weights are suspended. On one side, the weight is suspended at the end of a bar rigidly fixed to the arm so that the bar and the arm make a right angle, whereas on the other side the weight is suspended at the end of a freely moving string (see Fig. 1.1). How can we explain that, in this case, the weight suspended at the end of the freely moving string will be heavier than the other one?

The presence of such a problem in Blasius' *Questions on the Physics* might seem at first glance quite mysterious. The real function of what Blasius calls "problems" in this work is an enigma in itself, which exceeds the scope of this study. Suffice it here to say that, beyond the amusing pedagogical break that such 'problems' could offer to someone trying to catch students' attention while introducing them to the complexity of Aristotle's *Physics*, these problems are certainly part of Blasius' own intellectual project. Through these problems, involving mechanics, medicine, astrology, physiognomy, mathematics, and various curiosities of the natural world, Blasius offers a new approach to physics. He integrates into physics new scientific disciplines and theoretical materials that enable him to construct a consistent worldview synthesizing most of the scientific knowledge available at his time.

1 See text 1 (II, q. 1) in the present volume, nn. 14–15.

 | DOI:10.1163/9789004734968_003

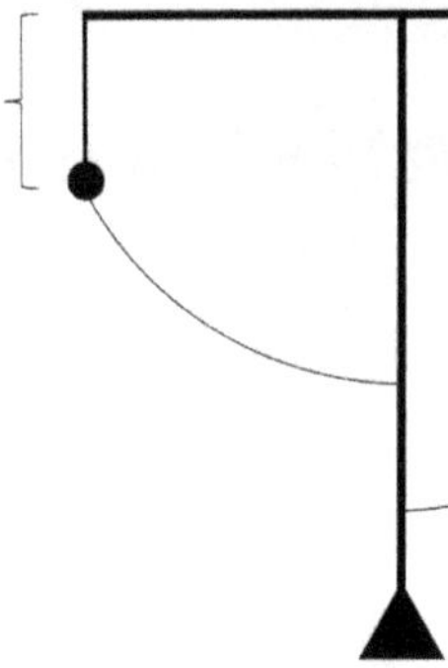

FIGURE 1.1 Equal-armed balance with equal weights, one of which is at the end of a fixed arm, the other hanging freely (Let us note the presence of diagrams corresponding to this problem in the margins of the two manuscripts containing Blasius' second redaction of his questions on the *Physics*)

Statics or, according to the terminology of the time, the 'science of weights,' belongs to those disciplines that Blasius undertakes to integrate into his natural philosophy. What is important in the case of the science of weights is that, more than other disciplines, this field of medieval knowledge plays a crucial role in Blasius' philosophical system, being central to his anti-Aristotelian stances and, more precisely, to the new theory of motion he advocates in his physics.

The *Questions on the Physics* is not the only place where Blasius deals with the science of weights. Blasius wrote two works inspired by works on statics belonging to the Jordanian tradition, i.e., writings left by Jordanus de Nemore, a significant scientific figure of the thirteenth century whose exact identity remains partially undetermined.[2] The scientific value of these two other writings has been deemed relatively low by their respective editors, namely Joseph Brown and Marshall Clagett.[3] Admittedly, these works do not match the depth, consistency, and complexity of Jordanus de Nemore. However, Brown and Clagett have not done justice to Blasius of Parma's contribution to the science of weights. His *Treatise* stands out from most works of the period due to its combination of theorems of pure statics and developments in hydrostatics—a

2 Blasius of Parma, *Tractatus de ponderibus*, ed. and trans. M. Clagett, in E.A. Moody and M. Clagett, *The Medieval Science of Weights* (Scientia de Ponderibus). *Treatises Ascribed to Euclid, Archimedes, Thabit ibn Qurra, Jordanus de Nemore and Blasius of Parma* (Madison: The University of Wisconsin Press, 1952), 229–279; Blasius of Parma, *Questiones super Tractatum de Ponderibus*, ed. J.E. Brown in ead., *The* Scientia de ponderibus *in the Later Middle Ages*, PhD Dissertation (Madison: University of Wisconsin, 1968), 348–569.

3 See, e.g., Brown, *The* Scientia de ponderibus, 126, 136; cf. Clagett in *The Medieval Science of Weights* 233–234.

combination absent in the usual treatises on the science of weights of that time. We will, however, explore the decisive role that certain axioms related to hydrostatics play in Blasius' general physics. Furthermore, the *Questions* on Jordanus of Nemore distinguish themselves from the rest of this literature through a stylistic approach that was, again, rarely used: Blasius offers a critical analysis of the coherence of the axioms proper to the science of weights in the form of disputed questions, seeking to test the consistency of these axioms with the principles of the physics within which they are meant to fit.

Most of all, Brown and Clagett did not analyze Blasius' *Questions on the Physics* and, therefore, the way Blasius put into use the scientific material inherited from the science of weights in the context of this work. Despite not reaching the same level of sophistication as Jordanus de Nemore's works, Blasius' unique contribution lies in his application of the science of weights to address various questions arising from the interpretation of Aristotelian physics. One of the great merits of Graziella Federici Vescovini's pioneering study was to identify the influence of the Archimedean tradition in Blasius' general philosophical system. Federici Vescovini highlighted the connection between Blasius' reflections on weights and some other anti-Aristotelian theses in physics.[4]

The recent editions of several important texts of Blasius, including the ones offered in this volume, make it possible to reassess Federici Vescovini's analysis and evaluate more accurately the general function of the science of weights in his physics. This essay aims to provide new insights into the centrality of the Archimedean tradition in Blasius' physics and to demonstrate the systematic connections that this distinct scientific model, separate from Aristotle's, maintains with Blasius' major theses on the nature of matter, bodies, and motive forces in nature. In this way, it will give the reader a better sense of the peculiar character of Blasius' philosophical worldview. The next section (Section 2) introduces the basic elements of the medieval science of weights as it was accessible to Blasius. We shall then see how Blasius draws on these influences to extend the epistemological scope of explanations based on weights, thus incorporating ahead of his time mechanistic models of explanation in a novel way (Section 3), before focusing on the anti-Aristotelian features of his conception of weights (Section 4).

4 G. Federici Vescovini, *Astrologia e scienza. La crisi dell'aristotelismo sul cadere del Trecento e Biagio Pelacani da Parma* (Florence: Enrico Vallecchi, 1979), 279–281, 291–296.

2 Blasius and the Medieval Science of Weights

The medieval science of weights already had quite a long history when Blasius was teaching in Northern Italy in the late fourteenth century. Coming from ancient sources, in particular the Archimedean corpus as well as other influential texts such as the pseudo-Aristotelian *Problemata mechanica* (which seems to have had a merely indirect influence on the Latin medieval world), the science of weights had enjoyed considerable developments thanks to Arabic thinkers before reaching the Latin world.[5] By the time of Blasius, the 'science of weights' was the term of art designating what we would call today 'statics,' i.e., the branch of mechanics studying the equilibrium of bodies. The discipline that Blasius and his contemporaries called the 'science of weights' was named so due to its intricate history, by which it had come to acquire a special status.

The mechanics (*mēkhanikē* / μηχανική) inherited from the Greeks originally meant the study of machines or devices and comprised both a theoretical aspect (the knowledge of general physical laws) and a practical one (the concrete engineering of mechanical devices). Through its reception by Arabic philosophers and scientists, who sought to clarify the relations among the different types of knowledge received from the Greeks and to achieve a better classification of their arts and sciences, the practical arts of making machines became sharply distinguished from the theoretical study of the rules governing the equilibrium between bodies depending on their weights and the

5 After Duhem's pioneer work (P. Duhem, *Les origines de la statique*, 2 vols. (Paris: Hermann, 1905–1906)), the classic study of the long transmission of the Archimedean corpus in the Latin world remains Clagett's monumental *Archimedes in the Middle Ages*; cf. M. Clagett, *Archimedes in the Middle Ages*, 5 vols. (Madison: The University of Wisconsin Press, 1964–1984). See also Moody and Clagett, *The Medieval Science of Weights*; Brown, *The* Scientia de ponderibus. For a recent overview of Archimedes' influence (not limited to the Middle Ages, however), see M. Ceccarellia and S.A. Paipetis (eds.), *The Genius of Archimedes—23 Centuries of Influence on Mathematics, Science and Engineering. Proceedings of an International Conference Held at Syracuse, Italy, June 8–10, 2010* (Dordrecht: Springer, 2010), and the bibliography therein. The question of the transmission of Archimedean sources and the tradition of the *Problemata mechanica* through the Arab world has made significant progress in recent years; see, in particular, M. Abattouy, "The Aristotelian Foundations of Arabic Mechanics: From the Ninth to the Twelfth Century," in C.H. Leijenhorst, C.H. Lüthy, and J.M.M.H. Thijssen (eds.), *The Dynamics of Aristotelian Natural Philosophy from Antiquity to the Seventeenth Century* (Leiden / Boston / Cologne: Brill, 2002), 109–140; M. Abattouy, "Greek Mechanics in Arabic Context: Thābit ibn Qurra, al-Isfizārī and the Arabic Traditions of Aristotelian and Euclidean Mechanics," *Science in Context* 14 (1/2), 2001, 179–247, and, for further references, the bibliography in M. Abattouy and S. Al-Hassani, *The Corpus of Al-Isfizārī in the Sciences of Weights and Mechanical Devices* (London: Al-Furqān Islamic Heritage Foundation, 2015).

forces exerted upon them. The Latin medieval world inherited this distinction between these two distinct sets of knowledge, i.e., a practical knowledge belonging to what was designated as the 'mechanical arts,' on the one hand, and a theoretical knowledge possessing the status of a true science, namely the science of weights (*scientia de ponderibus*), on the other hand.

To grasp the specific way in which Blasius utilizes the 'science of weights,' it is thus important to keep in mind that, in his view and that of his contemporaries, this discipline was defined as a science.[6] During Blasius' time, this characterization posed several questions, particularly regarding the precise nature of this science and its relationships to other scientific disciplines. Indeed, the science of weights as it was conceived in the Latin tradition comprised both a mathematical dimension and a physical one. To this extent, the science of weights belonged to the so-called 'middle sciences,' that is, sciences like optics, astronomy, or theoretical music, which contain elements coming from physics but also from mathematics. Aristotle himself seemed to grant 'mechanics' such a status but only made scattered references to this particular discipline, leaving almost completely unspecified its exact nature.[7] Blasius' perspective on the nature of the science of weights aligns with the general view held by his contemporaries, who regarded it as a discipline bridging physics and mathematics. As a skilled mathematician who authored writings on the theory of proportions and on perspective,[8] Blasius recognized the potential of a discipline that, more

6 The influential classification of the sciences inherited from al-Fārābī's *De scientiis* and disseminated through Gundissalinus' own *De scientiis* presents the *scientia de ponderibus* as a science comprising two parts, and distinct from the science of devices or machines (*scientia de ingeniis*), which itself represents the pratical application of the science of weights. See al-Fārābī, *De scientiis, "translatus a Girardo Cremonensi."* In al-Fārābī, *Über die Wissenschaften / De scientiis*, ed. F. Schupp (Hamburg: Felix Meiner, 2005), 84: "Verum scientia ponderum comprehendit de re scientie gravium duas res, aut speculationem in ponderosis inquantum mensurantur aut mensuratur cum eis, et est inquisitio de radicibus sermonis in ponderibus, aut considerationem in ponderosis que moventur, aut cum quibus movetur, et est inquisitio de radicibus instrumentorum quibus elevantur res graves, et super que permutantur de loco ad locum. Ingeniorum vero scientia est scientia modi preparationis ad faciendum convenire omnia quorum modi demonstrantur in doctrinis quarum narratio preteriit cum sermone et demonstratione, super corpora naturalia, et in acceptione eorum et situ ipsorum in eis actu." Cf. Dominicus Gundissalinus, *De scientiis*, ed. M.A. Alonso (Madrid/Granada: Consejo Superior de Investigaciones Científicas, 1954), 108–112; Dominicus Gundissalinus, *De divisione philosophiae*, ed. L. Baur (Münster: Aschendorff, 1903), 121–124.

7 Aristotle, *Physics* II 2, 193b23; *Posterior Analytics* I 9, 76a24; I 13, 78b37.

8 Blasius of Parma, *Questiones circa Tractatum proportionum Magistri Thome Braduardini*, eds. J. Biard and S. Rommevaux (Paris: Vrin, 2005). Blasius of Parma, *Questiones super perspectiva communi*, eds. G. Federici Vescovini and Joël Biard (Paris: Vrin, 2009). On this text, see also

than optics, music, or astronomy, could provide novel insights into key aspects of natural processes. Blasius' mathematical expertise likely instilled in him the confidence to see the value of this interdisciplinary science, which could be applied to various physics-related inquiries.[9]

Another question connected to the status of the science of weights was its relations to other scientific disciplines, most specially physics and mathematics. Acknowledging the hybrid nature of the science of weights raised the question of the exact place of this discipline within the traditional classification of sciences inherited from the Aristotelian tradition. Is it primarily subordinated to physics? Or should it rather be defined as a branch of mathematics? Blasius does not dwell on this issue, being obviously more interested in the applications of statics than in methodological reflections over its status. However, Blasius' revolutionary move lies precisely in this integration. By embedding the science of weights within a commentary on the *Physics*, Blasius assimilates it into the scientific framework of Aristotelian natural philosophy—a method oriented toward seeking explanations and causes in nature.

Before showing some of these applications and highlighting the theoretical function of the science of weights in Blasius' theory of nature, a few words on Blasius' most immediate source for this discipline, i.e., Jordanus de Nemore, will help us better appreciate the truly innovative character of his own approach to statics. The distinctive character of Jordanus de Nemore's works was the will to unify two different approaches to weights and their properties, namely the Aristotelian one, on the one hand, and the Archimedean one, on the other.[10] The Aristotelian approach to weights favored explanations of equilibrium and disequilibrium among rigid bodies owing to a dynamic point of view. In this approach, the notions of heaviness and lightness were analyzed through the concept of *motion*. In Aristotle's theory of motion, the speed of some object's fall is taken (erroneously) to be proportional to its heaviness, which makes it possible to study the respective weights of objects in equilibrium by taking into consideration the speed with which they fall downwards when a state of equilibrium is disturbed. The *Problemata mechanica* was for this tradition an important text that indirectly influenced the medieval developments of statics, eventually becoming directly available in the West in the

---

G. Federici Vescovini, "Le questioni di 'perspectiva' di Biagio Pelcani de Parma," *Rinascimento* 2/1, 1961, 163–243.

9 To this extent, it is no surprise to find Blasius quoting his own mathematical works in his writings on the science of weights; cf. Brown, *The* Scientia de ponderibus, 127–128.

10 See Moody and Clagett, "General Introduction," in Moody and Clagett (eds.), *The Medieval Science of Weights*, 3–20, here at 7.

sixteenth century. This pseudo-Aristotelian treatise, likely composed in a Peripatetic milieu, established several important theorems about weights by considering the forces exerted upon objects and the speed of their fall, taking the circular trajectories of bodies suspended on an equal-armed balance as a way to measure their speed.

By contrast, the Archimedean approach to statics, presented in *On the Equilibrium of Planes* (Περὶ ἐπιπέδων ἰσορροπιῶν), favored a purely geometric approach, leaving aside considerations about speed and forces. Archimedes drew on the concept of 'center of gravity'—a concept notably absent from the *Problemata mechanica*—to demonstrate in an elegant way results already established in the *Problemata* through the concepts of force and speed. The more mathematical approach of Archimedes was also reflected in the rigorous deductive style of his treatise.

Jordanus de Nemore, probably active between 1230 and 1260, made what can easily be considered the most important contribution to statics in the Latin world, alongside other scientific achievements.[11] Jordanus utilized the circular trajectories of weights placed on a balance to demonstrate his most significant results. This particular method results in connecting the heaviness of an object and the degree of curvature of its fall. When placed on a scale with another object, the heavier object will be defined as the one having the straighter trajectory to the ground. Explained on this basis is the way in which the length of the arm on which an object is suspended affects its weight. As we shall see, Blasius generally follows the same method for measuring the heaviness of an object:

> The longer the arm on which a suspended heavy body will be, the more it will weigh and the faster it will move downward. This is obvious, because the longer the arm is, the less curved will be the suspended [body's] fall; since etc.[12]

Jordanus employed this approach to weight to describe physical systems in equilibrium in terms of different forces interacting in a particular way, according to what we would call today 'component forces.' The concepts he employed, broadly speaking, stemmed from an Aristotelian-based dynamics, even in his most creative ideas. For instance, a new concept of considerable importance

11 A good synthetic introduction to Jordanus' works may be found in J. Høyrup, "Jordanus de Nemore, 13th Century Mathematical Innovator: An Essay on Intellectual Context, Achievement, and Failure," *Archive for History of Exact Sciences* 38/4, 1988, 307–363.

12 See text 6 (IV, q. 3) in the present volume, n. 30.

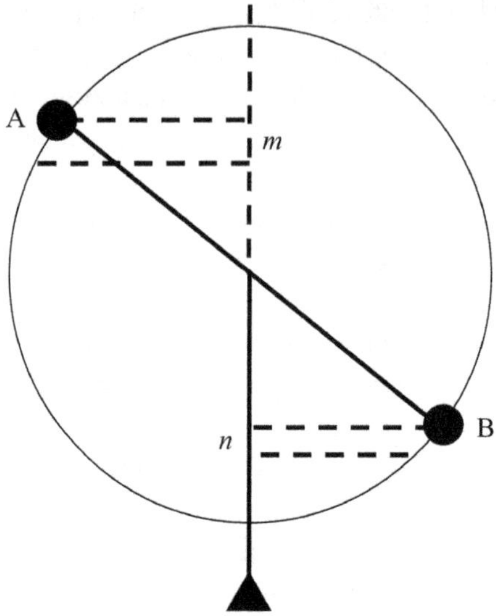

FIGURE 1.2
Illustration of a body (A) having a greater positional heaviness than another body (B)

for the study of weights forged by Jordanus was the notion of 'positional heaviness' (*gravitas situalis*), or 'heaviness according to the place' (*gravitas secundum situm*). In Jordanus' science of weights and the later commentaries written on it, this technical notion referred to the heaviness that an object comes to acquire when put in a particular position in a physical system, like an object suspended at the end of an arm connected to a fulcrum. The effect of this particular position is that, in such a case, the object apparently weighs more or less proportionally to its distance to the fulcrum, as recalled by Blasius in the above quotation.

This notion of positional heaviness or heaviness according to place, as opposed to heaviness absolutely speaking (*simpliciter*), was likely a transposition to the case of weights of the distinction between being absolutely speaking (*esse simpliciter*) and being in a certain way (*esse secundum quid*) central to Aristotelianism. Jordanus defined the way to measure the positional heaviness of a body by taking a part of its circular trajectory, as small as one wishes, and by projecting it on a vertical line (that of the fulcrum in Fig. 1.2 below). A body is said to have more positional heaviness than another when the projected distance on the vertical line is greater. Following Jordanus' theory, the body A in the figure above (1.2), despite having the same intrinsic weight as the body B, has a greater positional heaviness than B (the segment m being greater than n) and, thus, should move downward and lead the system toward a state of equilibrium.

Jordanus harmonized the conceptual framework about weights proper to Aristotelianism with the Archimedean method. Indeed, despite his reliance on a 'dynamic' approach to statics typical of Aristotelian-inspired works, he presented his results in the rigorous deductive form favored by Archimedes. Thus, in Jordanus' treatises and the writings inspired by his works, results are presented as theorems strictly deduced from a set of initial postulates and already demonstrated theorems like in a mathematical work following an axiomatic method. One century before Blasius' period of activity, Jordanus' synthesis between the Aristotelian and the Archimedean traditions on weights represented a very elegant and highly consistent system of demonstrations for complex problems involving the equilibrium of rigid bodies.

That said, Jordanus' statics was limited to the study of weights and their properties in an abstract way, avoiding any concrete application of the laws it established to explain individual phenomena of the natural world, except for the special case of the balance at the center of his analyses. Jordanus did not venture into considering different types of bodies (such as liquid ones) or considering qualities other than heaviness and lightness in the problems he analyzed. In short, Jordanus restricted himself to the internal development of statics, becoming the most influential authority on the science of weights in the Latin world before the modern period, but did not seem to have envisioned (at least not in any of his writings at our disposal today) its integration into a new system of natural philosophy.[13] Despite the inferior intrinsic quality of Blasius' writings on weights compared to Jordanus' achievements, this integration is at the center of Blasius' physics.

3 The Science of Weights in Blasius' *Questions on the Physics*

3.1 *Basic Principles and Applications*

The science of weights plays an important role in Blasius' natural philosophy. Problems and theorems coming from this discipline appear in many places of his *Questions on the Physics*. As we have seen, a problem of this type appears in his question about artifacts (II, q. 1) but other questions pose a similar enigma about weights to be solved as their initial 'problem.' In Book VII, Blasius even chooses to ask as his fifth question: "Whether two weights, one of which is suspended at the end of a balanced arm, and the other is released to its own nature (*suae naturae dimissum*), can move locally in this disposition equally fast?"[14]

13 On the influence of the Jordanian writings for the late medieval Latin tradition of the science of weights, see See Moody and Clagett, "General Introduction," in Moody and Clagett (eds.), *The Medieval Science of Weights*, esp. 9–10, 14–20.

14 Blasius of Parma, *Quaestiones super VIII libros Physicorum* (secunda redactio), VII, q. 5

Posing a question of statics as a main question in a commentary on Aristotle's *Physics* seems to be an unprecedented case in the late medieval commentary tradition, and is certainly original.[15]

The most interesting application of the science of weights, however, is to be found within the body of Blasius' *Questions*, where it serves to address the main philosophical issues around which Blasius organized his reading of Aristotle's *Physics*. Blasius assigns a special role to heaviness and lightness compared to other qualities such as heat and coldness, which do not have according to him the same status, as we shall see. The status granted to the science of weights leads Blasius to a thorough revision of the Aristotelian conception of motion, to which he opposes an 'Archimedean' view that is according to him preferable in some respects. Through these applications, it will appear that the science of weights directly contributes to the development of mechanistic theses within Blasius' natural philosophy.

To appreciate this point, let us first give some examples of problems related to the science of weights in Blasius' *Questions on the Physics*. In his third question on Book IV of the *Physics*, drawing on some remarks of Aristotle,[16] Blasius tries for instance to understand whether bodies reach their natural place following the shortest possible lines, i.e., straight lines. Blasius inserts different examples of bodies placed on a balance in different configurations to establish the rules governing the behavior of falling bodies. Following Jordanus' statics, Blasius uses the circular trajectories of bodies placed on the arms of a balance to evaluate their respective weights. In such a system, a body is said to be heavier than another when its descending trajectory follows a circle of a greater diameter than does the other.

The solution to a problem examined at the beginning of Blasius' second question on Book II of the *Physics* (II, q. 2) follows directly from this way of measuring a thing's proportional heaviness. In this context, Blasius asks why a weight placed at the end of one arm of a two-armed balance will be in equilibrium with a weight four times lighter placed at the end of the opposite arm, which is four times longer than the other. Blasius' intention here is not to demonstrate this fact, as it trivially follows from the law of the lever, which states that weights are in equilibrium when placed at distances inversely

('Utrum aequaliter gravia, quorum unum in brachio aequilibris est appensum, reliquum vero simpliciter suae naturae dimissum, possint aeque velociter in hac dispositione localiter moveri'), ms. Vat. Lat. 2159, 193va–194vb.

15 None of Blasius' main sources (John Buridan, Albert of Saxony, Nicole Oresme, Marsilius of Inghen) for question-commentaries on Aristotle's *Physics* contains any such question.

16 Aristotle, *Physics*, IV.1, 208b8–12.

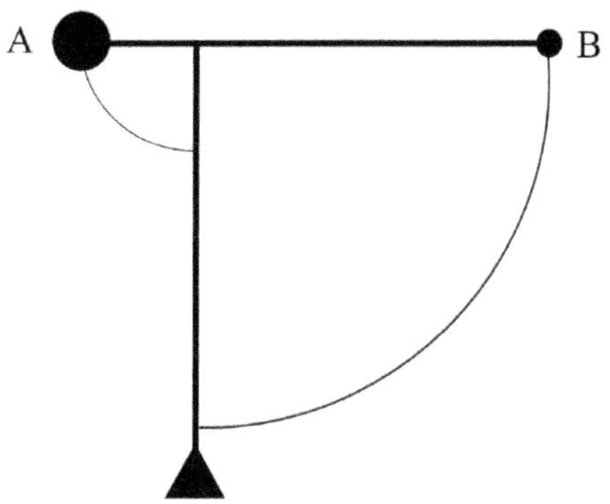

FIGURE 1.3
A body (*B*) in equilibrium with a heavier one (*A*) due to its positional heaviness

proportional to their weights. Blasius is rather interested in explaining this very law. In order to do so, he draws on the properties of the circle, showing that the quarter of a circle traveled by the lighter weight placed on the longer arm will be greater than the other, thus proportionally increasing its 'positional heaviness' (see Fig. 1.3 above). The positional heaviness of *B* (in the example above), being four times more distant from the fulcrum than *A*, increases proportionally its intrinsic weight, which is four times smaller than *A*'s.

With this method at hand, we are now in a position to understand the problem mentioned at the beginning of this study, which involved two equal weights placed on an equal-armed balance with a fixed hanger, on one side, and a freely hanging one, on the other.[17] Blasius' explanation rests on the same property. Here, the body hanging at the end of a freely moving string is positionally heavier than the other body because the trajectory that it will follow when falling down is straighter than the other has (see Fig. 1.1).

This geometric way of measuring the positional weight of objects placed in a system in equilibrium depending on their virtual trajectory (i.e., their trajectories if they were to fall downward) underlies Blasius' distinction between absolute weight and what he often calls 'positional heaviness' or 'heaviness according to the place' following Jordanus de Nemore. When a body is placed on a balance, its distance from the center of equilibrium (i.e., from the fulcrum) increases its positional heaviness since, when it gets farther from the fulcrum, the trajectory of a body becomes "less curved," "straighter," or "closer to the straight line," in Blasius' terms.

17 This case is taken up from a theorem established by Jordanus in his *Elementa*. See Jordanus de Nemore, *Elementa supra demonstrationem ponderum*, ed. E.A. Moody in Clagett and Moody (eds.), *The Medieval Science of Weights*, 136–138. Cf. also *Liber Jordani de ponderibus*, ed. E.A. Moody in Clagett and Moody (eds.), *The Medieval Science of Weights*, 158–160. This proposition, however, is absent from Jordanus' *Liber de ratione ponderis*.

A heavy body not placed in a system of physical constraints will always fall down to the ground following a straight line. When connected to a physical system involving different forces acting on it as a resistance to its fall, the laws of proportionality between positional heaviness and distance from the center of equilibrium will apply. In such a physical system, like in a two-armed balance, where a circular trajectory necessarily results from the combination of two (or more) forces, the falling trajectory will similarly be straighter when the body is positionally heavier. Blasius deduces from this the reason why carrying an object upon one's head makes it lighter than carrying it with one's extended arm. He suggests that in the former case, the object's potential fall follows a more curved path, since the constraint of the head aligned with the spine forces it to take a more indirect trajectory toward the ground:

> Fourth corollary: A heavy [body] in equilibrium can weigh much more in equilibrium than where it would rest out of its place following a straight line. This is clear because not only can it weigh in equilibrium, but it can weigh more by a greater distance from the equilibrium center; however, this cannot occur in a straight [line]; therefore, etc.
>
> Fifth corollary: It does not happen that a heavy [body] in equilibrium moves, all other things being equal, as fast as where it would move by a straight line. This is obvious, because it does not happen that, all other things being equal, a straight line resists [the thing] moving downward more than a curved line would, however close to straightness it may get.
>
> From these [corollaries] the cause may also be deduced why a weight [placed] on your head weighs less than if you were carrying it with your arm, since then it would move faster than by a curve.[18]

This application of a general theorem about weights to the empirical situation of carrying a heavy object on one's head is not the only case where Blasius strays from the paradigmatic analysis of scales. He also wants to shed light on apparently puzzling properties of heavy bodies in nature, like the fact that an animate body (such as a strong person) can lift bodies heavier than its own weight, while an animate being, when it does not use its strength, does not elevate another body heavier than itself when they are placed on either side of a balance:

> And we put forward this conclusion: A human being cannot lift with equilibrium a greater weight than without equilibrium. It is declared that

18 See text 6 (IV, q. 3) in the present volume, nn. 32–34.

> without equilibrium a human being can lift a stone heavier than she is, as is immediately obvious. You see that one human being lifts and carries two other humans simultaneously without equilibrium; however, the same human cannot lift a weight with equilibrium unless [this weight] is less heavy than she is. Therefore, if a human being were heavy like ten pounds, and a stone *A* weighs ten pounds, if this human were placed on one arm, and the stone on the other, there would be no motion, as you know; because there is no motion from a proportion of equality, and even less will [one thing] lift something heavier than itself.[19]

3.2 *The Geometric Use of Lines: Some Methodological Clarifications*

These few examples of Blasius' approach to studying weights shows that, due to its focus on the trajectories of falling bodies, the comparison between degrees of curvature in trajectories constitutes the methodological criterion for differentiating heavier and lighter bodies. In this context, to be 'more or less curved' can mean two things: (1) a trajectory in space following a greater circle is 'less curved' in the sense that, from a geometric point of view, it has a smaller degree of curvature, or (2) a trajectory is 'less curved' than another if it follows a circle of the same diameter but represents a smaller arc of that circle.[20] According to Blasius, this geometric point of view also has a physical meaning. Ultimately, all properties of weights amount to the fact that a straight line expresses *less resistance* to a body's trajectory than a curved one. This explains for instance that a heavy body moving downward when placed on a balance initially in equilibrium moves slower and slower while the same body moves faster and faster in free fall motion:

> The first [corollary] is: Some truths being collated, the more curved the line is, the more it resists and the less effectively the mobile is applied to it. This is self-evident. Because the more a line is curved, the more it differs from the straight line by which heavy [bodies] move faster and naturally toward their natural place.
>
> Second corollary: On the same circle, a greater portion is more curved than a minor [one], the truth of which, I say, is obvious from the second conclusion of the first section of the *Almagest*.

19 See text 6 (IV, q. 3) in the present volume, n. 42.

20 Blasius of Parma, *Tractatus de ponderibus*, part 1, 'Suppositiones et petitiones,' 240, ll. 42–43: 4. Minoris circuli cirumferentia plus curvatur; mariorisque minus; 240, ll. 61–62: 5. Eiusdem circuli et maior arcus plus curvatur minore.

> These words or truths admitted in this way, the first corollary is: The more a heavy [body] in equilibrium moves downward, the slower and slower it moves. This is obvious, because the more it moves downward, the greater will be the arc of the same circle it describes. But by the second truth of what has been admitted, a greater portion is more curved than a minor [one] on the same circle. Thus by the first truth, it moves slower and slower, contrary to what happens for a heavy [body] moving downward by a straight line, whose motion is always faster and faster.[21]

As evidenced by the second corollary in the passage just quoted, a 'more curved' trajectory can refer not only to a degree of curvature in the usual sense of the term but also to a portion of a circle that is greater than a shorter portion of the same circle. Usually, however, Blasius employs this expression to refer to a circle of smaller diameter, 'more curved' being equivalent, in contemporary terms, to the degree of curvature as proportionally inverse to the radius of a circle. The appeal to degrees of curvature for comparing circles of different radiuses allows Blasius to explain phenomena like the one described in the opening problem of his second question on Book II of the *Physics*, or ones similar to the following:

> An oblong weight [placed] at the same distance from the horizon as the arm to which it is fixed, weighs more than if it were freely hanging [*ad circumvolubile verteretur*] at the end of the arm. This is obvious, because the whole oblong [body] fixed with the arm and any of its parts describes a quarter circle that the hanging [body] would cover. And as a consequence, it would weigh more according to that place than according to this one.[22]

Methodologically speaking, the 'circle' approach followed by Blasius relies on comparisons between trajectories and, thus, between *types* of lines. As a commentator on Aristotle's *Physics*, Blasius is well-versed on problems stemming from comparing lines of different sorts and, correlatively, motions of different types. In Book VII of the *Physics*, Aristotle had expressed objections against the possibility of comparing circular motions to rectilinear ones due to difficulties arising from comparing straight and curved lines. Blasius is not impressed by these objections. Despite being of different kinds, he says, straight and curved lines are still comparable:

21 See text 6 (IV, q. 3) in the present volume, nn. 27–29.

22 See text 6 (IV, q. 3) in the present volume, n. 35.

> The first reply could be this one, that "straight and curved motions are not mutually comparable," as it is written in the seventh book of the *Physics*, referring to the text.
>
> This reply is worthless, because although such motions are not mutually comparable with respect to their space, for a straight line cannot be proportionally compared [*commensurare*] to a curve and the other way round since one can only touch the other on a point, they are however mutually comparable in fastness and slowness, as it is said in the fourth book of this [work].[23]

As can be seen, Blasius underlines that despite pertaining to different genera, the straight and curved lines corresponding to trajectories of moving bodies can be compared with respect to *speed*. Speed is indeed predicated univocally of these two types of trajectories, and therefore represents a 'common measure' (expressed by the Latin *commensurare*) allowing their comparison thanks to the mathematical theory of proportions. While Blasius is not absolutely original in this claim, this point offers a good example of the way fourteenth-century natural philosophers sought to quantify different types of physical magnitudes by circumventing principles of Aristotelian physics, most importantly the non-comparability of distinct genera.

3.3 *Extending the Science of Weights in Aristotle's* Physics

The integration of elements originating in the science of weights into a question-commentary on Aristotle's *Physics* enables Blasius to go well beyond the typical case studies involving balances. Blasius intends to extend these case studies in two main ways, namely by reflecting on cases where balances have unusual properties or dimensions, and on cases involving the internal structure of bodies.

The first way in which Blasius goes beyond the problems raised about ordinary balances has to do with what could be called 'imaginary physics.' Blasius conceives imaginary scenarios in which scales reach gigantic sizes, leading to the investigation of how weights exceeding by far the dimensions of earthly objects relate to the center of the universe, i.e., the ultimate place toward which heavy bodies tend. In question two on the second book of the *Physics*, for instance, Blasius shows that what is true for two weights placed on a two-armed balance does not always apply when extreme cases are considered. The normal behavior of a balance in equilibrium is such that a body heavier than

23 See text 2 (II, q. 2) in the present volume, nn. 20–21.

another, when placed on an equal-armed balance, will fall downward following a straighter path that reflects its greater weight. However, when one imagines a balance of 'cosmic' dimension extended over an important part of the universe—which Blasius thinks as finite, as do all his Latin contemporaries—this rule does not apply in the same way:

> [...] Conceive an upright beam perpendicular to the surface of earth. Then if to this beam is applied an arm equal to the diameter of the earth, at the end of which a weight is suspended, in this case this heavy [body] will move downward by a curve until it is in direct contact with the center of the world. But if this arm were made ten times longer so that it equals half the diameter of the world, it would not move in such a way that its motion would describe a quarter of a circle. Since, in this way, it would rise in such a motion and the arm would by no means equal the diameter of the world, being here on the contrary only equal to the diameter of the earth. Nor in this case would it move describing a quarter of a circle. Therefore, from what has been said, it is obvious that the proposition has to be modified in such a way that, although a weight is made heavier in a longer arm and moves downward faster, it does not move faster downward in an arm of any length whatsoever, as [anyone] clever can see clearly.[24]

This way of forging thought-experiments about the center of the world and gigantic balances greater than the earth's diameter is striking. It demonstrates Blasius' interest not only in properties of weights as observable in a terrestrial reference frame but more generally for the laws of statics as structuring the cosmos and applicable to all bodies and parts of the universe without exception.

Another observation will confirm this point. In a question on the natural place of bodies raised in Book II of the *Physics*, Blasius takes from Richard Swineshead (active ca. 1340s), one of the so-called Oxford Calculators, an interesting thought-experiment in which a column-shaped body falls down vertically toward the center of the earth and, thus, toward the center of the world—the Earth being imagined in this thought-experiment as perforated.[25] Like Richard Swineshead before him, Blasius is interested in the way that the different parts of this column-shaped body contribute to the object's fall. Granted that the different parts of the body do not behave as a completely unified whole

24 See text 2 (II, q. 2) in the present volume, n. 24.
25 See text 5 (IV, q. 2) in the present volume, n. 7.

but retain their own individual properties and operate as such, the body should start decreasing in speed when some parts of it reach the center of the universe, while other parts are still moving toward it. Since the parts already located at the center of the universe will not be moved toward it any longer, they will act as a resistance against the parts still falling downward (even though all these parts are rigidly connected to one another in the body). As the body keeps moving toward the center of the universe, the resistance increases, since more parts reach the center, and the speed of the body proportionally decreases. From this, Blasius draws the conclusion that the body will need an infinite amount of time to reach the universe's center and, in fact, will never reach it.

Blasius' treatment of this thought-experiment is perhaps not very original, given Richard Swineshead's previous analysis of it in his *Liber calculationum*. However, Blasius' treatment presents something absent from Swineshead's *Liber calculationum*, namely a serious reflection on the way the total weight of a body or physical system is related to its parts and their dispositions. Blasius is not merely interested in the logical paradox stemming from the case just described. He also seeks to analyze the structure of bodies possessing weights, i.e., the way the shape of a body may affect the equilibrium of its parts or, put differently, how the configuration of its parts influences their positional heaviness and the forces they exert.

This interest may shed light on the reason why Blasius chose to open his question on Book II of the *Physics*, concerned with the ontology of artifacts, with a problem involving scales. The possibility of altering the natural behavior of bodies by reconfiguring the internal disposition of its parts may constitute a good argument in favor of the reality of artifacts.[26] This interpretation should be taken with caution, however. The fact that the 'problems' placed at the beginning of Blasius' questions are generally disconnected from the content of these questions should invite us to consider this as a mere hypothesis.

Be that as it may, Blasius' interest in understanding how the internal structure of parts affects the weights of objects also concerns liquid bodies and not only rigid ones. In question four on Book IV of the *Physics*, Blasius studies the case of a system made of water in which a void is created, this case involving the respective weights of the elements comprising the system.[27] Blasius describes a system equivalent to a water siphon made of one vase filled with water connec-

26 On this point, see Henrik Lagerlund's essay of Blasius' theory of artifacts in this volume.

27 J. Biard and S. Rommevaux, "Blaise de Parme, Questions sur la Physique, IV, qu. 4 et qu. 5, édition critique et traduction," in J. Biard and S. Rommevaux (eds.), *La nature et le vide dans la physique médiévale. Études dédiées à Edward Grant* (Turnhout: Brepols, 2012), 317–378.

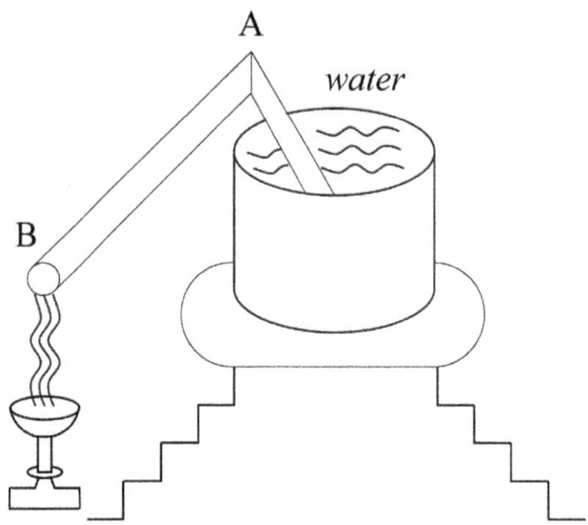

FIGURE 1.4
Two vases connected by a pipe, illustrating the effect of weights in fluid bodies

ted to another one by a pipe comprising two parts meeting at one point (point *A* in Fig. 1.4 above).

Blasius in this case reflects on the nature of the void and the principle that 'nature abhors a vacuum,' given that, by emptying the air contained in the pipe, the vase filled with water will transfer its water to the other. Interestingly, his explanation of why the water will move out of the water-filled vase is that if the part of the pipe that is not directly plunged into water (corresponding to the segment *BA* in Fig. 1.4) is long enough, i.e., longer than the other part, the weight of the water contained in this part will be greater than that of the water contained in the other part of the pipe. Combined with the impossibility of the void, this fact explains that the longer part of the pipe will move the water contained in the other part, just like the heavier body on a two-armed balance will lift the lighter one. According to Blasius, therefore, the situation instantiated in the siphoning of the water, in accordance with the principle of communicating vessels, not only obeys the same laws as those of statics but is *structurally equivalent* to an unequal-armed balance.

In the second question on the same book (IV, q. 2), Blasius mentions a case where a heavy body made of earth (thus heavier than water) is placed into a receptacle containing water, and starts to sink. This receptacle has the shape of a reverse pyramid and contains all the water of the universe. The fall of the object in the medium, consequently, has the effect of moving the entire mass of water which, taken as a whole, weighs more than the solid body. Possessing no clear concept of pressure and, thus, no good explanation to offer regarding why the body could not reach the bottom of the receptacle, Blasius considers this case as a paradox violating a rule usually associated with the Aristotelian theory of motion, namely that the force of the agent must always be superior to (or, in medieval terms, must always respect a 'proportion of greater inequality') the resistance of the patient:

> A small piece of earth existing in air, released to its own nature [*demissa suae naturae*], would by its falling down to water's inferior place displace all the water of the universe out of its natural place. In this way a body could not reach its place naturally without another body being removed out of its place. And the consequence is proved: Let *A* be some piece of earth of one or two pounds having the shape of a sphere. And imagine if you will that all the water of the universe is in a pyramid-shaped vase with the sharp edge downward and its base upwards. Now let this [piece of] earth fall into the water. And since you posit that its natural place is under all the water, it will move downward to the bottom of the vase; and consequently it makes all the water at the bottom of the vase rise, lifting it. And unless it is conceded that this [consequence] is true, it follows that an action is caused by a proportion of minor inequality, and indeed the whole water of the universe is much heavier than the piece of earth mentioned, therefore, etc.[28]

The case of hydrostatics, which draws Blasius' interest as we will see shortly, underscores the difficult problem of understanding the respective role of the material parts of a body in its total weight and equilibrium. It also reflects the then-current state of the science of weights, which Blasius mostly inherits from the Jordanian tradition. Besides texts coming from the Jordanian tradition, Blasius had access to texts coming more directly from Archimedes, such as the *De insidentibus aquae*, in which the roles of the center of equilibrium with respect to the parts of a body plays a greater role than in the dynamic approach of the Jordanian tradition. However, the small role played by the notion of the center of gravity in Blasius' theory of weights, as well as the absence of a proper concept of pressure, explains to some extent why he is has difficulties in offering satisfactory solutions to some of the physical puzzles with which he engages.

The reader familiar with the famous thought-experiments through which Galileo elaborated his theory of gravity and motion will probably find a similar echo in Blasius' own problems of imaginary physics. This similarity testifies to Blasius' innovative approach to the science of weights. He employs methods

28 See text 5 (IV, q. 2) in the present volume, n. 10.

mainly inherited from the so-called Oxford Calculators relying on the use of the void, gigantic dimensions, and other imaginary scenarios. But whereas his English predecessors mostly focused on finding rules for measuring motions and qualities, Blasius' emphasis on weights in his *Questions on Aristotle's Physics* serves to explain structural features of the natural world, including the ultimate explanation of the configuration of the universe and the motion of bodies within it.

3.4 *Materialist and Mechanistic Hypotheses*

The centrality of the science of weights for Blasius' theory of nature may again be noticed in the different hypotheses that he formulates in his writings. As is known, the radicality of Blasius' ideas, which earned him a formal condemnation by the Bishop of Pavia in 1396, was such that he liked to present them by offering different possible solutions to a given problem, and to formulate the most extreme ones in the form of hypotheses. One fascinating hypothesis imagined by Blasius in the first redaction of his questions on the *Physics* as well as in the questions he dedicated to the *On Generation and Corruption* is the following.[29] It could be the case, says Blasius, that the world is made of matter only, and that this matter is divided and configured in different ways only because of qualities, which include heaviness and density. The different species of beings we see around us, in this hypothesis, would be just different types of configurations of matter, in which various figures and operations correspond to what we call 'species.'[30]

This hypothesis is an extreme form of materialist reductionism, for two reasons.[31] First, because in this hypothesis, the duality between matter and form

---

29 Blasius of Parma, *Quaestiones super VIII libros Physicorum* (prima redactio), ms. Vat. Chig. O IV 41, I, q. 7, 237^{ra-rb}. The same 'fantasy' is taken up in Blasius' questions on the *De generatione et corruptione* where the role of weights in this thought-experiment is emphasized; cf. Blasius of Parma, *Quaestiones circa libros de generatione et corruptione*, ms. Vat. Chig. O IV 41, I, q. 1, f. 1rb: Modus autem ponendi quattuor elementa erit iste quia secundum veritatem tota massa materie divisa est in quatuor partes difformes in raritate, et densitate, et in qualitatibus primis et aliis, unde est una pars rarissima et levissima locum supremum obtinens, quae dicta est ignis usque in hodiernum diem.

30 On this hypothesis, see J. Biard, *Hypothèse matérialiste et pensée radicale. La philosophie de la nature de Blaise de Parme* (Paris: Vrin, 2022), 16–19. On Blasius' ontological hypotheses about matter, see also J. Biard, "Matière, forme, qualité. Blaise de Parme et le statut de la matière," in T. Suarez-Nani, A. Paravicini Bagliani (eds.), *Materia. Nouvelles perspectives de recherche dans la pensée et la culture médiévales (XIIe–XVIe siècles)* (Florence: SISMEL—Edizioni del Galluzzo, 2017), 191–215.

31 For useful comments on the extent to which one may qualify Blasius' doctrine as a materialist system, see Biard, *Hypothèse matérialiste*, 9–10, who recalls that medieval 'material-

proper to the Aristotelian theory of bodies is simply eliminated: according to this hypothesis, only matter would be needed to explain the sensible world. Second, this hypothesis is a clear form of materialism because it seems to attribute all motions and states of affairs in the world to heaviness and lightness. Matter and its configuration (i.e., the figures present in it) are sufficient to account for natural phenomena. Only properties most closely connected to the essence of matter, namely density and rarity, as well as heaviness and lightness, are needed in this hypothesis.

In the first redaction of his questions on the *Physics*, just like in the questions on the *On Generation and Corruption*, Blasius eventually rejects this hypothesis. The distinction between matter and form is, according to him, necessary to explain several natural phenomena, like the fact that boiling water naturally reverts to its natural state of coldness when removed from its source of heat.[32] To believe that only matter exists prevents one from giving a satisfactory account of this fundamental fact of nature. Blasius goes back, for this reason, to the good old hylomorphic doctrine of Aristotle. For similar reasons, Blasius believes it necessary to postulate a real distinction between matter and the qualities inhering in it. Even if such never happens in the normal course of nature, God could for instance separate the heat existing in a hot body from the body in which it exists. This possibility shows that, ontologically speaking, qualities are really distinct from matter, in which they 'inhere' but from which they can always be separated.

The view that qualities are really distinct from their subject was common by Blasius' time.[33] However, this rather consensual view generated a problem related to the cause of motion. If accidents are things really distinct from the substances in which they inhere, what is their exact causal role in natural processes? Are substances the causes of natural processes, or rather accidents, or perhaps a mix of them? This problem became particularly important in the fourteenth century when it came time to investigate the causes of natural motion such as the fall of heavy bodies.

In line with the analyses of John Buridan, Albert of Saxony, and other 'Nominalists' connected to the Parisian milieu, Blasius holds that the qualities of heaviness and lightness play a primary role as efficient causes of the natural

---

ism,' which is a term that does not refer to a philosophical doctrine in the Middle Ages, should be carefully distinguished from 17th-century materialism, for instance.

32 Cf. Biard, *Hypothèse matérialiste*, 29.

33 On the late medieval doctrine of 'real qualities', see R. Pasnau, *Metaphysical Themes 1274–1671* (Oxford: Oxford University Press, 2011), 179–199, and the bibliography therein.

motion of bodies.[34] Buridan and Albert both believed that the substantial form of a stone was the principal agent of the stone's fall, while its heaviness was its instrumental cause.[35] Albert, in particular, denied in his *Questions on the De caelo*—a text which Blasius knew well, having commented on it[36]—that some heaviness deprived of its substantial form and matter would fall downward.[37]

Unlike his predecessors, Blasius seems to grant a very limited role to the substantial form when it comes to explaining the fall of a heavy body. Indeed, he claims that a consecrated host which, according to Christian faith, has no longer any substantial form, moves because it retains its accidents and, so, its heaviness:

> […] A consecrated host, placed in the sphere of air, moves downward. Now it is asked: Does this [host] move naturally or violently? And it is argued that it moves in neither of these ways. Because there is no substantial form in it that is said to be a principle of motion and rest […]. Not violently [either], because there is no violence in the case proposed.
>
> To this argument it is replied that this motion is natural, and so that [the host] is moved downward naturally. Indeed, it is enough in the case proposed that there be heaviness. And in the same way I have said that some matter, deprived of any substantial form, when it is informed however by heaviness, this matter with its heaviness moves downward naturally. And this applies to a consecrated host, although there is in this case no matter.[38]

---

34 On the Parisian Nominalists' positions concerning this issue, see E.D. Sylla, "Aristotelian Commentaries and Scientific Change: The Parisian Nominalists on the Cause of the Natural Motion of Inanimate Bodies," *Vivarium* 31/1, 1993, 37–83.

35 John Buridan, *Quaestiones super libros Physicorum*, II, q. 4, 267; Albert of Saxony, *Questiones super libros Physicorum*, VIII, q. 6, 1035, ll. 30–32; cf. Albert of Saxony, *Quæstiones in Aristotelis De cælo*, ed. B. Patar (Louvain-la-Neuve / Louvain—Paris—Dudley, MA: Éditions de l'Institut supérieur de philosophie / Peeters, 200), III–IV, q. 9, 500–504.

36 On Blasius' commentary on Albert's *Questions*, see Graziella Federici Vescovini, "Notes sur la circulation en Italie du commentaire d'Albert de Saxe sur le De caelo," in J. Biard (ed.), *Itinéraires d'Albert de Saxe: Paris-Vienne au XIVe siècle*, Paris, Vrin, 1991, 235–251, here 240–246.

37 Albert of Saxony, *Quæstiones in Aristotelis De cælo*, III–IV, q. 10, 508, ll. 98–00: Secunda conclusio. Gravitas separata a materia et a forma substantiali existens sursum non descenderet nisi formae substantiales vices supplerentur aliunde.

38 See text 2 (II, q. 2) in the present volume, nn. 78–79. Let us note that this argument is already found for instance in Buridan's own questions on the *Physics*, II, q. 5, 282, ll. 6–13.

According to Blasius, matter informed by heaviness without any substantial form would still move downward, as made clear in this passage. Elsewhere, Blasius specifies that it is not matter as such that produces motion but only the qualities inhering in it, such as lightness and heaviness. He writes that "considering a piece of matter *A* deprived of all forms, this matter in such a disposition neither moves nor is at rest"[39] and, elsewhere, that "concerning the case of some earth deprived of all heaviness [...] this earth deprived [of heaviness] does not fall downward since it does not have what is required for its fall."[40] Heaviness and lightness are therefore the two principles necessary and sufficient for explaining local motions occurring in nature, as opposed to substantial forms, which determine the specific nature of bodies but do not immediately cause any type of motion.

This thesis concerning the exclusive role of heaviness and lightness in local motion highlights what could be called the 'mechanistic' orientation of Blasius' physics. The hypothesis described above reveals indeed the epistemological value that Blasius attributes to material density and weight in explaining the behavior of natural bodies. It is noteworthy that, even if he never fully endorses a radically materialist ontology where *only matter* exists, Blasius still considers material density and weight as the two core properties of matter, which play the main explanatory roles in physics and *could* suffice, in such a hypothesis, to explain nearly all motions occurring in nature.

4 Archimedes vs Aristotle: Some Anti-Aristotelian Theses

As we have seen so far, the role of the science of weights in Blasius' natural philosophy is to complement the conceptual framework of Aristotelianism so as to achieve a more comprehensive theory of nature, while orientating accounts of natural phenomena toward more mechanistic explanations. Nonetheless, Blasius' will to bring elements related to the science of weights into an Aristotelian-based framework does not always follow a conciliatory approach. At times, the way Blasius applies concepts pertaining to the science of weights to problems of physics departs from what Aristotle would have thought about weights. His questions on the *Physics* contains theses revealing an explicit conflict between the two authorities he tries elsewhere to reconcile, namely Aristotle and Archimedes.

39 See text 2 (II, q. 2) in the present volume, n. 26.

40 See text 5 (IV, q. 2) in the present volume, n. 69.

4.1 *The Ontology of Heaviness and Lightness*

First, let us give one example of how Blasius draws on the science of weights to propose analyses that could not be acknowledged by an orthodox Aristotelian. This first example has to do with Blasius' ontological analysis of heaviness and lightness, a telling indicator of the special importance that Blasius grants to these qualities. He analyzes the nature of heaviness and lightness most thoroughly in his questions on Aristotle's *On Generation and Corruption*, a work worth mentioning here for appreciating his views on the issue. In short, Blasius believes that each of the four elements (water, air, fire, and earth) has its own heaviness (in the case of water and earth) or lightness (in the case of fire and air). In other words, two elements seeming to share the same heaviness or lightness do not, in fact, share the same specific property. The lightness or air, for instance, is specifically different from the lightness of fire, in the sense that these two properties do not belong to the same species.

This point as well as its philosophical implications need some explanation. Asking whether the primary qualities (heat, cold, wet, dry) that elements seem to share are of the same species was not uncommon in the context of questions on the Aristotelian corpus. This particular question arose from passages—the most explicit of which were found in *On the Generation of Animals*—where the Stagirite posited a distinction between types of heat.[41] There, Aristotle explained that celestial heat (heat coming from the sun) must be distinguished from elemental heat (fire's heat) and that these types of heat, in turn, should not be confused with animal heat (heat as the vehicle of the soul in animate beings). Aristotle's interpreters were then left with a conundrum to solve: Is it always the case that the same predicate (e.g., 'hot') really refers to the same specific property, in the sense of the exact same species of property?

When applied to the case of the elements, this question was particularly important, since Aristotle's theory of the transformations of the elements into one another as well as his theory of mixtures both directly relied on the existence of qualities shared between them (air and fire, for instance, share heat).

By the time of Blasius, this problem had already a long history in the tradition of commentaries on Aristotle's natural philosophical texts, and different answers had been advocated. In his questions on Aristotle's *On Generation and Corruption*, Blasius claims that the four basic elemental qualities (heat, cold, wet, and dry) found among the elements are of the same species. The heat of

41 Aristotle, *On the Generation of Animals* II 3, 736b30–737a5.

air, in other words, really belongs to the same species as the heat of fire. Blasius' justification of this point is straightforward. Two properties belong to the same species when they cannot be distinguished through the effects they produce. Now, the heat in air and the heat in fire produce the same effect. Therefore, they must belong to the same species.

An interesting aspect of Blasius' treatment of this question is that, in a doubt about whether the same analysis equally applies to heaviness and lightness, he rejects that the effects of heaviness and lightness in two elements are the same, leading him to deny that these properties belong to the same species. For instance, if lightness were the same in air and fire, these two elements would tend to the same place, which is not true. When placed within the sphere of fire, some air would move downward. However, its lightness would still be responsible for this motion, proving therefore that the lightness of air is different from that of fire. Blasius is led to conclude, in other terms, that heaviness and lightness are *relative* properties, in the sense that the lightness of air is heavier than the lightness of fire. When an element reaches its natural place, ceasing to move upward or downward, it loses its proper weight.[42]

This view, clearly explained in Blasius' questions on Aristotle's *On Generation and Corruption*, is taken up in his questions on the *Physics*.[43] Here, Blasius quotes the pseudo-Archimedean *De insidentibus in humidum* or *De insidentibus in liquido*. Being one of the rare source materials for hydrostatics available in the medieval Latin tradition, the *De insidentibus in humidum* was not written by Archimedes himself but nonetheless gathered elements coming from genuine Archimedean works. According to the initial definitions opening this treatise, an object is neither heavy nor light when it is located in its natural place, if by 'heaviness' one means 'specific heaviness' (and similarly for lightness), i.e., the weight proper to an object compared to another object of a different species but of the same size.[44] The notion of 'specific weight' was a notion serving to qualify the way in which an object weighs when placed in a given medium. More precisely, the *De insidentibus in humidum* made a distinction between weight according to number (or absolute weight, in modern terms) and specific weight, thus highlighting the property of bodies to weigh more or less in different types of mediums, a property on which Archimedes' hydrostatics was based. But this pseudo-Archimedean text employed this distinction in such

42 Blasius of Parma, *Quaestiones circa libros de generatione et corruptione*, II, q. 4, ms. Vatican Chigi O IV 41, 37va.

43 See text 5 (IV, q. 2) in the present volume, n. 6.

44 (Ps-)Archimedes, *De insidentibus in humidum*, ed. Moody and Clagett in *The Medieval Science of Weights*, 40, def. 7–9.

a way that it challenged an important claim made by Aristotle in *De Caelo*, namely that elements—except fire—keep their weight in their natural place.[45]

In their questions on the *De Caelo*, the Parisian Masters Nicole Oresme and Albert of Saxony drew on the pseudo-Archimedes to back up their views on weights. Both Oresme and Albert drew a distinction between actual and potential weight. They explained that, while a heavy body retains its potential heaviness (due to its substantial form) when it is at rest in its natural place, it does not have any actual heaviness, i.e., any actual tendency to move downward.[46] This distinction allowed the Parisian Masters to reconcile Aristotle's and Archimedes' divergent statements about the intrinsic vs relative character of weight, thereby proposing the hybrid figure of an 'Aristotelian Archimedes'.[47]

Blasius knew the distinction between specific weight and absolute weight, as he employed in his *Questions on Jordanus' Elements* a similar distinction between '*gravis*' and '*ponderosus*.'[48] In the *Questions on Aristotle's Physics*, however, just like in his treatise *De ponderibus*,[49] he takes the position of the Parisian Masters one step further by accepting the principle according to which elements do not have weight (without further specification) in their natural place, making weight *in general* a property essentially relative to its medium.

We can note that this important move, which seems to transform the traditional conception of heaviness and lightness as two positive contraries into a merely relative opposition, echoes other reflections on contrariety found in

45 Aristotle, *De Caelo* 4.4, 311b6–10; on fire, cf. 311a15. Different theories of weights challenging Aristotle's views existed in late Antiquity. On some of them, see M. Rashed, "Contre le mouvement rectiligne naturel: trois adversaires (Xénarque, Ptolémée, Plotin) pour une thèse," in Riccardo Chiaradonna and Franco Trabattoni (eds.), *Physics and Philosophy of Nature in Greek Neoplatonism. Proceedings of the European Science Foundation Exploratory Workshop (Il Ciocco, Castelvecchio Pascoli, June 22–24, 2006)*, (Leiden: Brill, 2009), 17–42. For an overview of ancient conceptions of weights, see D. O'Brien, *Theories of Weight in the Ancient World. Four Essays on Democritus, Plato, and Aristotle. A Study in the Development of Ideas*, 2 vols. (Leiden: Brill, 1981–1984).

46 Albert of Saxony, *Questiones super libros de caelo*, III, q. 3; Nicole Oresme, *Questiones super De celo*, in *The* Questiones super De celo *of Nicole Oresme*, ed. C. Kren, PhD Dissertation (Madison: The University of Wisconsin, 1965), 773–777.

47 On Oresme's references to Archimedes and the pseudo-Archimedes, see C. Kren, *The* Questiones super De celo *of Nicole Oresme*, 1036, 1087–1092.

48 Cf. Brown, *The* Scientia de ponderibus, 136–137.

49 This principle is the first *suppositio* of Blasius' *De ponderibus* (1952), a fact that confirms the source material used by Blasius in hydrostatics since this principle, absent from the genuinely Archimedean *De insidentibus aquae* translated in the thirteenth century by William of Moerbeke, constitutes the first '*petitio*' of the pseudo-Archimedean *De insidentibus in humidum*, which Blasius calls *De insidentibus in liquido* or *De inundantibus in liquido*, ascribing it to Archimedes.

Blasius' writings. In a study devoted to Blasius' theory of intensive properties, Joël Biard has noted that Blasius is attracted by the view that heat and coldness are not positive contraries, cold being a mere privation of heat.[50] This view, leading to abandoning a central tenet of Aristotelian philosophy, based on the existence of couples of primary qualities as positive contraries, is seen as a probable hypothesis by Blasius. Nonetheless, Blasius typically continues to discuss heat and coldness, as well as heaviness and lightness, as real contraries, but it is not completely clear whether and to what extent he really endorses an ontology of positive contraries.

At any rate, Blasius holds that heavy objects cease to be heavy when they reach their natural place, in line with the Archimedean position and with his own criterion associating a nature with a specific effect. Blasius' insistence on this point is striking, all the more so since he tries to ground it on philosophical arguments. It leads to a significant departure from Aristotelian physics in favor of an alternative theory of weights based on hydrostatics. This peculiar approach to the nature of the properties of weight has serious consequences as regards the notions of nature and motion. Since a body reaching the place corresponding to its nature does not weigh, it cannot move, leading to dissociate to some extent the concepts of nature and motion.[51]

At this point, the reader may object that the same reasoning could be used to prove that heat, for instance, is also a relative property, since fire seems to be able to produce through heat effects that even some very hot air cannot produce. Blasius does not deal with this objection, and it is not clear what answers would be available to him here. What is clear, however, is that Blasius' main motivation for holding this view lies in the fact that heaviness and lightness are the primary causes of motion. To this extent, they determine the universe's general configuration in the sense that these properties account for the place that each specific nature has in the world.

These analyses concerning the status and role of heaviness and lightness testify to the centrality of the problem of weights in Blasius' thought. As we have seen, the radically reductionist hypothesis or thought-experiment mentioned above in which only heaviness, likely reduced to degrees of material density, accounts for all motions and processes is rejected by Blasius, who

---

50 See J. Biard, "Blasius of Parma on the Calculation of the Variation of Qualities and Aristotelian Physics," in D.A. Di Liscia and E.D. Sylla (eds.), *Quantifying Aristotle. The Impact, Spread and Decline of the Calculatores Tradition* (Leiden / Boston, Brill, 2022), 232–254, at 242, where Biard suggests that Blasius may be maintaining the traditional conception of contrariety for purely practical reasons despite endorsing a new ontology of qualities.

51 See the essay devoted to Blasius' theory of motion in this volume.

always seems to grant a role to other properties such as heat, cold, dry and wet. But the explanatory power of this thought-experiment explains the appeal it exerted on Blasius who, while not endorsing it completely, believed that the properties of weights (heaviness and lightness) were the essential causes of motions occurring in nature. Heat and cold as well as dry and wet also have physical effects, namely alternative effects on bodies that are subject to them, but the most important type of motion existing in nature, namely local motion, is ultimately due to the properties of weights. This fact explains the eminent role played by the science of weights in his theory of nature and highlights what we have suggested calling a 'mechanistic' orientation proper to Blasius' physics.

4.2 *Weight and Void*

Another point of departure of Blasius' theory of weights from the Aristotelian background that dominated the period in natural philosophy lies in the role of the void. In Book IV of the *Physics*, Aristotle talked about the void, but only to refute its existence, which had been admitted by ancient atomists. In fact, Aristotle denied the very *possibility* of the void. Thus, due to the incompatibility of the void with his physics, Aristotle did not seriously consider cases where the void could have some epistemological value, even as a mere theoretical hypothesis. Blasius' stance on the epistemological role of the void is markedly different. Although he also rejects its existence, he believes that appealing to the void through thought-experiments may be useful for investigating properties and rules of the natural world. One example of this appeal to the void for reflecting on the laws of nature is Blasius' criticism of the then-most dominant rule for determining the speed of motion, i.e., the rule generally called by scholars the 'rule of Bradwardine,' which Blasius seeks to restrict to a better-defined domain of application by pointing out its problematic character when motion in the void is taken into account.[52]

---

52 On this, see S. Rommevaux, "Le vide *secundum naturam* dans les *Questions sur la physique* de Blaise de Parme," in J. Biard and S. Rommevaux (eds.), *La nature et le vide dans la physique médiévale. Études dédiées à Edward Grant* (Turnhout: Brepols, 2012), 293–316. On the role of the void in Blasius' theory of the rules of motion, see S. Rommevaux, "Ajout de conditions dans les règles du mouvement formulées par Blaise de Parme dans la seconde rédaction de ses Questions sur la Physique," in A. Robert and J. Biard (eds.), *La philosophie de Blaise de Parme: physique, psychologie, éthique* (Florence: SISMEL—Edizioni del Galluzzo, 2019), 153–174; S. Rommevaux, "Les règles du mouvement de Blaise de Parme dans les deux versions de ses *Questions sur le traité des rapports de Thomas Bradwardine*," in J. Biard and S. Rommevaux (eds.), *Mathématiques et théorie du mouvement* (*XIV*e–*XVI*e *siècles*) (Villeneuve d'Ascq: Presses Universitaires du Septentrion, 2008), 31–51.

This approach also applies to the science of weights, where Blasius relies on thought-experiments involving the void that could not be tolerated in a standard Aristotelian framework. For instance, in the opening of the third question on Book IV of the *Physics*, Blasius describes a case based on the familiar two-armed balance.[53] Let us consider an equal-armed balance with two suspended pockets at its ends: one containing air along with its inherent heaviness, and the other containing only a vacuum, but which possesses an equivalent quantity of heaviness as the air-filled pocket. This case is problematic because it seems to lead to a paradox. When placed on a balance in this way, the air-filled pocket will be heavier than the one filled with void and will fall down accordingly. However, when detached from the balance and placed in the sphere of air, the pocket filled with air will move slower than the pocket filled with void. Yet, the same heaviness is contained in both pockets.

Blasius does not make it completely clear why the pocket containing air will be heavier in the first case and not in the second, as this argument is simply placed among the usual list of arguments opening a question in favor of a given position. Presumably, in the first scenario (i.e., when placed on a balance), even though the two pockets contain the same quantity of heaviness, the pocket containing air will offer a greater resistance to the downward motion of the void-filled pocket than the other way round, resulting in the fall of the air-filled pocket first, while no resistance is met when the void-filled pocket is detached from the balance and placed freely within the sphere of air. Let us also remember that, according to Blasius, an element is neither heavy nor light when it is in its natural place.

This thought-experiment is obviously inspired by Aristotle's observation that a bladder is heavier when it is filled with air than when it is empty—an observation allowing him to deduce that air must possess some weight even in its own sphere.[54] Blasius' belief is that it makes sense to treat the void as a substrate of inherence for the properties of weight. By Blasius' time, it was not uncommon to reflect upon the ontological status of sensible properties like colors or heat and their causal powers by imagining situations where these properties would be separated from their subjects. Due to the theological importance of the Eucharist, theologians assumed that accidents could be—at least supernaturally—separated from their substrate of inherence. However, assuming that weight properties not only can be separated from concrete substances but can also *inhere* in the void itself, and that the void may consequently have motion, is far more unorthodox.

---

53 See text 6 (IV, q. 3) in the present volume, n. 5.

54 Aristotle, *De Caelo* 4.4, 311b1–15.

Later in the same question, Blasius returns to this puzzle and sketches a way to explain paradoxical properties stemming from attributing properties of weight to the void.[55] When placed within the sphere of air, a pocket filled with vacuum but containing some heaviness would conserve its same shape, which would facilitate the dividing of the medium, thus making it less affected by the medium's resistance. In contrast, a pocket with air would tend to have an elongated shape, making it more subject to the medium's resistance. The argument is interesting, and highlights the reason why Blasius fashions this thought-experiment using pockets presumably made of elastic material rather than rigid bodies. Indeed, the tendency of non-rigid bodies to assume a spherical shape whose center of gravity is the center of the world, explaining the elongated shape of the air-filled pocket, belongs to the axioms of Archimedean statics. It allows Blasius to avoid paradoxical conclusions about weights and to maintain a fundamental theorem of the science of weights, namely that an object cannot be heavier than another when they are placed in a system in equilibrium at equal distance to the center of equilibrium, if the first is not heavier than the other when considered apart from any system in equilibrium.

These examples of how Blasius relies on the void for investigating basic properties of weights are typical of his methodology. They show how thought-experiments, by allowing concepts like the void, make it possible to study limit cases in order to identify the causal factors underlying laws of nature.

5 Concluding Remarks

Did Blasius ever experiment with real balances? The relative simplicity of the devices described in most of his problems related to the science of weights makes that hypothesis plausible. But other cases involving balances with giant arms or void-filled pockets remind us of the mostly speculative nature of the medieval science of weights, which Blasius pushed to the extreme by forging thought-experiments impossible to realize physically.

Despite its speculative character, the science of weights represents for Blasius a tool for understanding the general order of nature that goes well beyond the limited cases of balances in equilibrium. The passages where Blasius suggests that weights could suffice to explain a large spectrum of natural phenomena indicate the predominant role played by statics in his physics and at the same time underscore his deviations from Aristotelian natural philosophy.

55 See text 6 (IV, q. 3) in the present volume, nn. 36–40.

This will to reduce a certain class of natural phenomena to material parts and the properties of weights is in line with several theses of Blasius that tend to restrict the explanatory role of substantial forms and elementary qualities such as heat, coldness, wetness, and dryness.

We have seen that Blasius is inclined to base what may be called in modern terms his general dynamics, i.e., his general theory of the causes of local motion, on the properties of weight rather than on these other primary qualities. A similar tendency may be noticed in different places where Blasius enquires about the cause of more particular phenomena, like when he claims in his *Questions on On Generation and Corruption* that water extinguishes fire not because of its coldness or wetness but because its material parts smother the effect of fire by obstructing the tiny holes of the combustible materials through which fire can intensify.[56] Several cases mentioned above and others present in the questions edited in this volume similarly emphasize the role of material parts and the properties of weight in explaining processes for which one does not need qualities, substantial forms, or other entities of the medieval framework.

Considering these observations, it becomes evident why the assessment made by Brown, Clagett, and Moody regarding the limited value of Blasius' contributions to the science of weights overlooks the true originality of his work. Blasius did not contribute to the progress of the science of weights understood as a science encoded in the form of theorems rigorously deduced from a set of definitions in the Archimedean fashion. His originality lies in the articulation of results belonging to this science applied to a general theory of nature in which it serves to explicate both structural features of the universe as well as more specific phenomena. Blasius contributed, in other words, to the integration of statics into physics and to its promotion as an essential branch of physics.

Admittedly, the science of weights was not the only discipline that Blasius wanted to bring into the framework of physics to build a new vision of nature, but it was certainly one of the most important. In this respect, the orientation of Blasius as a natural philosopher should be qualified from the point of view of mechanics with the same nuances Joël Biard employed in describing Blasius as a 'materialist' philosopher. Blasius was certainly not a materialist philosopher in the modern sense of the term, but a medieval philosopher inclined toward materialism, i.e., formulating materialist hypotheses as convincing alternatives to Aristotelianism while retaining an ontology of substances,

---

56 Blasius of Parma, *Quaestiones circa libros de generatione et corruptione*, II, q. 4, 36va: [...] Aqua ratione sue humiditatis non extinguit ignem. Patet quia humiditas aeris et aque sunt eiusdem speciei et humiditas aeris non corrumpit ignem, quare nec humiditas aque. [...] Secunda propositio est: aqua non corrumpit ignem de per se, ratione sue frigiditatis.

accidents, substantial forms, and qualities. Similarly, Blasius never envisioned a completely mechanical philosophy as some thinkers will come to do in the seventeenth century. But although he did not aim to reduce everything to matter and weights, Blasius did reorient physical explanations toward mechanistic hypotheses and granted a key role to statics in physics.

CHAPTER 2

Causality and Causal Powers in Blasius of Parma's *Physics*

Joël Biard

1 Introduction

Late medieval philosophy reformulates all the main concepts of Peripatetic natural philosophy, among which the concept of cause is central. But it is not a linear evolution that would lead to the concept of causality as it is utilized in modern philosophy, which is focused on efficient causality and the laws of nature. Indeed, some medieval texts anticipate such a transformation, even prepare the way for it; for example, the study of the rules of movement according to cause and effect (that is, *motus quoad causam* and *quoad effectum*, in the terminology of medieval physics) by the comparison of the ratios between the mover and the thing moved, or the increasingly more important emphasis accorded to agent causes and the resulting diminution of final causes, the redefinition of the relations between superlunary and sublunary motions, etc. All these elements are present in some texts of the period, be it in those of the Oxford Calculators or of John Buridan and other Parisian Masters of the fourteenth century, and consequently in all the texts they influenced throughout Europe. But we cannot really understand the transformations and innovations of natural philosophy in this period, if we do not give up a teleological account of the history of science and philosophy. The elements must be located in their context, so that we are dealing with a transformation of the framework of the interpretation of nature that is at once progressive and global.

The physics, and more broadly the natural philosophy, of Blasius of Parma should be considered as an original attempt to understand the dynamics of nature: How many principles are required to understand the existence and the movement of nature? What are the driving forces, the motors, and the powers that account for natural changes in the four genera of motion defined by Aristotle (generation, alteration, increase, and local motion)? Within this agenda, Blasius of Parma does not so much seek to unify the concept of cause, but rather to know to what extent one can determine the various powers at work in natural change.

 | DOI:10.1163/9789004734968_004

2 Causality: A Hyperbolic Doubt

2.1 *Multiple Causes, Various Causes*

Nowhere, to my knowledge, does Blasius provide an explicit general definition of cause. Occasionally, he invokes the following description, basically what we today call a counterfactual approach to causation: if something is posited, another thing follows; and if it is removed, then the other thing is removed as well. But this characterization is limited to a certain type of cause, that is, a cause *per se*. It is a factual description of the relation between cause and effect (based only on our perception) that presents a model of efficiency, and departs from any apprehension in terms of potency and act.[1] But this definition is unusual in commentaries on Aristotle. To my knowledge, it is, for example, not present in the *Questions on the Physics* by John Buridan, Albert of Saxony, or Nicole Oresme.[2] In contrast, we find it used about efficient causality in the *Commentary on the Sentences* by William Ockham.[3]

Actually, what the previous figures typically use to describe the notion of cause is a series of distinctions. Such an approach is indeed very usual, and we find it with few differences in Nicole Oresme and Albert of Saxony.[4] But in those texts the distinctions are applied only to efficient causes.

How are these distinctions presented by Blasius of Parma? First, he endorses the Aristotelian quadripartation of the causes, but supplies few further details. The first redaction of his *Physics* commentary includes the question of whether there are four causes for any natural thing. However, the only surviving manu-

1 Cf. for example Albertus Magnus, *Physica*, Liber II, tr. 2, c. 1, ed. Hossfeld vol. IV,1, 96b: "dicitur causa, ex quo fit res, cum insit ei"; the efficiency itself is presented as a principle of movement (ibid., c. 3, 101b).

2 Nevertheless, Buridan comes near this description in his *Questions on the Metaphysics*, during the discussion of natural necessity. He refuses a proposition attributed to Plato: "positis causis sufficientibus, eo modo quo sunt sufficientes ad aliquid producendum, sequitur necessario productio, ita quod nihil sit impediens" (*In Metaphysicen Aristotelis questiones*, Parisiis, 1518 (repr. Frankfurt a. M., Minerva, 1987), VI, qu. 5, XXXV^{vb}).

3 See among other places, Guillelmus de Ockham, *Ord.*, d. 1, qu. 3, O.T. I, 16; *Reportatio* II, qu. 3–4, O.T. V, 60, about God; *Expositio in libros Physicorum*, lib. VII, c. 3, O.P. V, 629–630, with this small addition: "si enim negetur ista propositio, videtur perire omnis via ad sciendum aliquid esse causam alterius". Ockham admits however a wider definition of cause: "quaecumque res absoluta requiritur in esse reali ad esse alicuius, est causa illius in aliquo genere causae" (O.T. I, 417).

4 Nicholas Oresme, *Questiones super Physicam (books I–VII)*, II, qu. 8, ed. Stefano Caroti, Jean Celeyrette, Stefan Kirschner, Edmond Mazet (Leiden-Boston, Brill, 2013), 219–230; Albert of Saxony, *Quaestiones in Aristotelis Physicam*, II, qu. 10, in *Expositio et quaestiones in Aristotelis Physicam ad Albertum de Saxonia attributae*, ed. B. Patar (Louvain-Paris, Éditions de l'institut supérieur de Philosophie—Éditions Peters, 1999), t. II, 389–394.

script ends after the presentation of the preliminary arguments, so that we do not have the subsequent theses nor the determination.[5] The same question is not to be found in the second redaction.[6] We learn only, during other developments, that Blasius takes up the division between intrinsic causes (matter and form) and extrinsic causes (agent and final).[7]

Another distinction, also well known since Aristotle, is that causes may be described as by themselves (*per se*) or by accident (*per accidens*). "Fifth evidence: since the title of the question includes '*causa*', it is known that the cause of something is twofold: some kinds are said to be a cause by itself, whose effect cannot be produced if this cause does not exist; but a cause by accident is that whose effect can be when the cause does not exist."[8] The cause by itself thus is a cause without which the effect cannot be produced. This may be clarified by examples. One (from the first redaction) is purely Aristotelian: the builder is a cause *per se* of a house, the musician *per accidens*.[9] The other (from the second redaction) concerns a relation in which one term no longer exists: the father is not the cause *per se* of the son, but a cause by accident, since the effect can subsist once the cause no longer exists. We shall return to this distinction, but first we must mention other divisions.[10]

Some causes are principal causes, while others are instrumental causes. This distinction will come to be important when we examine the relation of sub-

---

5 *Questiones super libros Physicorum* [QP], *prima redactio*, II, qu. 7. This division of the causes was evoked in book I, but with a simple reference to book II for more details. The text can be found in manuscript Vat. Chig. O IV 41. Most of the texts of Blasius of Parma quoted in this paper, when they have not yet been edited, can be found in a provisory edition on the website dedicated to this master, URL: https://bdeparme.hypotheses.org/

6 The first redaction of the *Questions on the Physics*, of which we have a single incomplete manuscript, dates from before 1396, the date of the condemnation of Blasius by the Bishop of Pavia, and it was probably written at Padua between 1392 and 1396. The second is posterior to the condemnation, perhaps written in 1397. The text of the second redaction is found in mss. Vat. Lat. 2519 and Vat. Lat. 3012. Here below, I provide the reference to the folios of Vat. Lat. 2519, even when the questions can be found in Blasius's website.

7 QP, 1a red., I, qu. 3, 229va, 2a evidentia.

8 QP, 2a red., I, qu. 2, 63va: "Quinta evidentia, quia in titulo questionis ponitur 'causa', sciendum ⟨est⟩ quod duplex est causa alicuius rei: quedam dicitur causa per se cuius non potest esse effectus quin sit illa causa, sed causa de per accidens est cuius effectus potest esse illa non existente"; cf. QP, 1a red., I, qu. 4, ms. Chigi O IV 41 231ra, 1a evidentia. We read alternatively *per se* or *de per se*, and *per accidens* and *de per accidens*. These expressions are equivalent.

9 See Aristotle, *Physics*, II, 5, 195b25–29; *Post. An.*, I, 4. Jonathan Barnes translates καθ' αὑτό by "in itself", which is perhaps better; but we try to maintain the parallel of by itself / by accident.

10 See QP, 1a red., I, qu. 3, 229^{va-vb}.

stances to accidents, but it also holds among substances. The example Blasius gives is that of a hand that instrumentally causes the movement of a stone, while it is the soul that is the principal cause.

Some causes are universal, others particular. A cause is universal if it indifferently produces a plurality of effects; it is particular if it produces an effect determinately. God and the Intelligences are examples of universal causes, while fire is said to be a particular cause, because it can heat but not cool.

Some causes are total, and others partial. This distinction will become important in describing a cause as a sufficient explanation of an effect.

Blasius also adds, without example or explanation, that some causes are distant, others remote—we may think of father and grandfather, for example. These divisions are not mutually exclusive nor ordered; they may overlap.

Returning to the division between cause by itself and cause by accident, the text of the second redaction specifies that a same thing can, according to different points of view, be called a cause by itself or a cause by accident:

> Note however that the same cause, relatively to the same thing, can be called a cause by itself and a cause by accident, but according to diverse reasons. So the proposition "the father is" is a cause by itself relative to the proposition "the son is;" however, the same thing that is a father is a cause by accident of this thing that is a son.[11]

Here, causality by itself is relegated to the cognitive level, while the real relation is called 'by accident', even in a case where the link seems undisputable. But what matters is to differentiate causality *in essendo* and causality *in inferendo*:

> Something can be called a cause of another thing in four senses: in one sense in inferring only, and so we say that the proposition 'a human being is a donkey' is the cause of the proposition 'a human being is.' In a second sense, we can say that something is the cause of another in being only, and so we say that God is the cause of anything produced or producible [...]. In a third sense we can say that something is the cause of another in being and inferring at the same time, as we would say that this proposition 'a mortal rational animal exists' is the cause in being and in inferring

11 QP, 2a red., I, qu. 2, 62va: "Nota tamen quod eadem causa respectu eiusdem rei potest dici causa de per se et causa de per accidens, secundum tamen diversas rationes. Unde ista 'pater est' est causa de per se respectu huius 'filius est', tamen illa eadem res que est pater est causa de per accidens illius rei que est filius."

> of this proposition 'something capable of laughing exists.' And this cause can be twofold, either remote or immediate.[12]

Throughout question 2 of Book I of the *Physics*,[13] Blasius is cautious about the distinction between the linguistic and cognitive dimensions of the analysis of causality, and the real relations of causal dependency: "in no scientific development enter the principles or causes of natural things, but propositions signifying that it is the case on the part of the principles and causes of natural things."[14] Blasius often refers to and compares real causes and epistemic causes, including the fact that the knowledge of premises either does cause or does not cause knowledge of the conclusion.

It is about the *per se* cause that Blasius utilizes the following description: a cause is a thing such that if it is posited, an effect follows, and if it is removed, the effect is removed. But this criteria is not sufficient, he thinks: "And before all I note that the following argument is not valid: this being posited, that is posited; and this being removed, that is removed; therefore this is by itself the cause of that."[15] The example offered to justify this restriction is that of the eyelid: if it is open, it allows for vision; and if it is shut prevents vision, while by itself it is not the cause of vision.

These formulations suggest that the treatment of the causes is mainly dedicated to the question of whether a cause is knowable. Blasius of Parma does not doubt that some beings cause (in different senses) other beings (that is, that they are their intrinsic constituents or that they have the power to bring them into existence). The main questions are: (1) Are such causes, and partic-

12 *Questiones in tractatus logice magistri Petri Hispani*, I, qu. 27, ed. J. Biard et Gr. Federici Vescovini, collab. V. Sorge et O. Rignani (Paris, Vrin, 2001), 177: "quadrupliciter potest aliquid dici causa alterius: uno modo in inferendo tantum, et sic dicemus quod hec propositio 'homo est asinus' esset causa huius 'homo est', quia sequitur 'homo est asinus, ergo homo est'. Secundo potest dici aliquid esse causa alterius in essendo tantum et sic dicimus quod deus est causa cuiuslibet rei producte vel producibilis [...]. Tertio potest aliquid dici causa alterius in essendo et in inferendo simul, sicut diceremus quod ista propositio 'animal rationale mortale est' est causa istius in essendo et in inferendo, scilicet 'risibile est', et illa causa potest esse duplex, remota et immediata."

13 QP, 2a red., I, qu. 2: "Utrum cognita causa totali unius rei cognoscatur statim ista res et non aliter."

14 QP, 2a red., I, qu 2, V1, 64vb: "in nullo processu scientifico ingrediuntur principia seu cause rerum naturalium sed ingrediuntur propositiones significantes sic esse ex parte principiorum et causarum rerum naturalium."

15 QP, 2a red., I, qu 2, V1, 63vb: "Et ante omnia noto quod non valet hec argumentatio: hoc posito ponitur illud et hoc remoto removetur illud idem, igitur hoc est causa de per se illius effectus."

ularly intrinsic causes, sufficient to account for their effects? And (2) are these relations knowable?

2.2 *Causality and Demonstrability*

In the commentaries on the *Physics*, several questions are dedicated to the knowledge of causality. In the first redaction, there is a series of questions on this subject: question 2 of Book I, "Can the knowledge of something cause or produce (*causare vel efficere*) the knowledge of another thing?"; question 3: "Does in the natural world the knowledge of principles lead to the knowledge of their effects?";[16] and question 4: "Does perfect knowledge of a natural thing presuppose perfect knowledge of all its causes"?[17] His investigation clearly concerns epistemic causality, while presupposing certain real causal relations. In the second redaction, the treatments are principally in question 2 of Book I: "When the total cause of something is known, is this cause immediately known?" and question 6 of Book II "Can it be proved evidently that something is the cause of another thing?"[18]

This last formulation in part takes up the terms of a debate that was conducted in Paris in the middle of fourteenth century, on the occasion of discussions involving Nicholas of Autrecourt.[19] It is well known that, even if Nicholas' texts for the most part have been lost, formulations that are attributed to him were discussed, notably by John Buridan.[20] Blasius' question is phrased in terms of demonstrability and evidence, and a criterion he puts forward is that it is possible to reduce (*reducere*) some propositions to a first principle, which is exactly an issue inherited from Autrecourt.[21]

The second article, which tries to answer the question, is developed in two sections. In the first stage we find a determination specified as "probable"; let us

---

16 Ms. Vat. Chig. O IV 41, 229ra–230vb.

17 Ibid., 230vb–232rb.

18 Blasius of Parma, QP, 2a red., II, qu. 6, 94ra: "utrum possit evidenter probari aliquid esse causa alterius."

19 Nicholas of Autrecourt, *Tractatus utilis ad videndum an sermones peripatheticorum fuerint demonstrativi* [otherwise called *Exigit ordo*], ed. J.R. O'Donnell in "Nicholas of Autrecourt", *Medieval Studies*, 1 (1939), 179–266; Nicholas of Autrecourt, *His Correspondence with Master Giles and Bernard of Arezzo*, ed. L.M. De Rijk (Leiden—New York—Köln, Brill, 1994); see also, besides Grellard (next footnote), Zénon Kaluza, *Nicolas d'Autrécourt, ami de la vérité*, "Histoire littéraire de la France", 42-1 (Paris, Imprimerie nationale, 1995).

20 The texts in which Buridan attacks Nicolas of Autrecourt, without naming him, are listed and commented upon in C. Grellard, *Croire et savoir. Les principes de la connaissance selon Nicolas d'Autrécourt* (Paris, Vrin, 2005), ch. 8 and 9, 229–282.

21 Another indication of the influence of Autrecourt is the recurrence of the term *apparentia* to designate various sorts of cognitive apprehension.

be reminded that this term implies that this position can be rationally argued (proved), though not strictly demonstrable. The doctrine set out in a second stage corresponds to Aristotle's thought, and is called "more profound."[22] We shall examine these in order.

The first answer to the question of knowing whether something is the cause of another involves multiple reasons for doubting the demonstrability of a causal relation, and employs skeptical arguments. The first set of arguments draws attention to changes in sensible perception, whether due to the senses, the medium, or the object itself. Many errors can result from exterior sensation, concerning both objects proper to a sense as well as other objects. But sensation is the basis of all natural knowledge.

Then, if this basis is insecure, the whole edifice of knowledge is undermined, and the internal cognitive powers may make many errors. Consequently, it causes a general doubt about knowledge that I myself may possess. Although the text underlines that "for me and others, scientific knowledge is possible",[23] yet indeed we can never be certain that we have any specific knowledge: "Nevertheless, it is never clearly evident to anyone about some truth that he knows it".[24] So, illusions (or even the possibility of errors of the senses) already lead to a doubt whether there is any knowledge. This series of conclusions is followed by another one that weakens even more the edifice of knowledge.

Even when you comply with the rules of formal validity of an argument, it cannot clearly appear that the argument necessarily concludes. This may seem surprising, even contradictory. It is not a matter of taking the formal validity wrongly—the point of departure is the supposition that it is a formal argument—but we must overcome a simple formalism in order to consider that an argument asserts something true about the world (a scientific syllogism, a material consequence …). Because of the limitations considered above, a thing can appear otherwise than it really is, even in a formally conclusive argument.

It is not the place here to expound in detail Blasius of Parma's theory of knowledge, which is not so much about refusing the possibility of certitude but rather redefining the conditions of human knowledge. Even though perfect knowledge in an absolute sense is reserved to God, a human being can acquire some knowledge, designated in the first redaction of the commentary

22 QP, 2a red., II, qu 6, 94vb: "In primo questionem determinabo probabiliter; in secundo magis profunditus et iuxta intentionem Aristotelis."

23 QP, 2a red, II, qu. 6, 94vb: "licet mihi et alteri sit notitia scientifica possibilis."

24 Ibid.: "nulli tamen est clare evidens de aliqua veritate quod sciat illam."

on the *Physics* as *cognitio in genere*,[25] which does not permit us to know a thing according to all the manners it is knowable, and cannot exclude some degree of error,[26] but involves a discursive process of knowledge that is gradual and therefore perfectible.[27]

What does this imply, more specifically, for knowledge of causes? In the passage cited above, which enumerates reasons for doubting some individual pieces of knowledge, one conclusion still applies these general principles to a particular case of causal relations. A radical consequence may be drawn from the first conclusions, namely: "although it is plausible that the fire *a* is now produced from this straw, however neither to me nor to you (even supposing a meticulous attention) is it evident that some fire is produced."[28] Doubt about the production of fire from straw, accompanied by a new reference to illusions, is even reinforced by an argument from dreams.[29] We can then generalize this doubt to the causation of one thing by another. We cannot strongly refute (*potissime improbari*) that some thing, be it this fire or that stone, exists from eternity, so we cannot demonstrate strongly (*potissime demonstrari*) that some effect is newly produced or producible.[30]

---

25 QP, 1a red., I, qu. 4, 231[rb]: "Cognitio simpliciter dicta est cognitio perfecta quam solus Deus habet; sed per cognitionem in genere intelligo notitiam perfectam humanitus acquisitam"; ibid., 232rb: "cum dicitur quod nullus effectus potest a homine perfecte cognosci, verum est perfecte simpliciter, sed bene perfecte in genere".

26 Blaise de Parme, *Questiones super Perspectiva communi*, I, qu. 6, éd. G. Federici Vescovini et J. Biard, avec la collab. d'Orsola Rignani et Valeria Sorge, et Riccardo Bellé pour les figures (Paris, Vrin, 2009), I, qu. 6, 120: "Nulla humana cognitio videtur omnem gradum erroris excludere; et sequitur consequenter quod valde est suspectum aliquam posse humanitatem habere demonstrationem de rebus naturalibus, si per demonstrationem intelligis processum excludentem omnem gradum errori".

27 According to Blasius, science is a quality capable of degrees: see QP, 1a red., I qu. 4, 232[ra]: "quelibet cognitio distincta est certe intensionis et certi gradus"; a distinct knowledge has some gradual amplitude, making it more or less distinct: "latitudo huius cognitionis distincte remissibilis est usque ad non gradum et per consequens continet totam latitudinem usque ad non gradum" (ibid.). See also QP 2a red., I, qu. 2, 64[va]. Cf. Graziella Federici Vescovini, *Astrologia e scienza. La crisi dell'aristotelismo sul cadere del Trecento e Biagio Pelacani da Parma* (Firenze, Nuovedizioni Enricho Vallecchi, 1979), 191–199.

28 QP, II, qu. 6, 95[ra]: "licet verisimile sit quod *a* ignis nunc producatur ex hiis straminibus tamen nec mihi nec tibi, supposita diligenti advertentia, est evidens quod aliquis ignis producatur."

29 Ibid., 95[ra]: "Et etiam patet quod sepe fiunt delusiones et apparentie quibus nihil correspondet in facto. Patet ex alio quia sepe putas tantam habere cognitionem et notitiam hora qua dormis de combustione tue domus, ut quod tua domus tunc comburatur et pro tunc non habes aliquam notitiam, nisi forte in habitu."

30 Ibid., 95[ra].

All this reasoning, from the first errors of the senses to illusions and dreams, leads to a radical doubt about the possibility of demonstrating any causal relation, a doubt that one can specify as "hyperbolic," because, all things considered, it is generally held that some knowledge is possible, but all means are engaged to question this conviction. As with any hyperbolic doubt, it might be overthrown thanks to some particular point of support. However, at this stage, the argumentation is a little disappointing. The 9th conclusion indeed asserts that it must be asserted firmly (*est firmiter asserendum*) that a cause is and has been productive of a new effect.[31] But the argument is based only on the Bible, with a quotation from Genesis.[32] From this initial acceptance of causality, all effects can be reintroduced.[33] However, we may question the *philosophical* force of this reversal.

Blasius of Parma will not limit himself to this initial line of argument. We may, however, offer a partial assessment: his arguments suggest, on the basis of our natural capacities and their limits, a radical doubt about the possibility of demonstrating a causal relation between one thing and another (the production of an effect by a cause), a conclusion that only will be overcome *in extremis* by an appeal to the authority of the Bible. Nevertheless, Blasius of Parma implicitly distances himself from this line of argument, since he presents the second one as "more profound."

This second line of argument is offered as conforming to Aristotle's doctrine, even if it partly departs from it. In fact, this line of argument shows traces of the debates involving Nicholas of Autrecourt (a notorious anti-Aristotelian), even if Blasius' final conclusions widely differ from those of the Lorraine master.

Indeed, in this second way of reasoning (*secundus processus*), the first conclusions start from the evidence of a first principle, and above all consider the possibility of "reducing" any knowledge to this first principle (that is, the principle of non-contradiction). But one preliminary remark frames the discussion by distinguishing different sorts of evidence according to their relation to the primary evidence of the first principle:

> I note that some evidence is the highest evidence, as is the evidence of the first principle; some is evidence formally dependent on the first, as is

31 Ibid., 95ra: "Ultima conclusio: quod aliqua causa sit novi effectus productiva et alias produxit est firmiter asserendum."

32 Ibid., 95ra: "Patet conclusio cum scribitur primo Genesis 'in principio creavit Deus celum et terram'."

33 Ibid., 95ra: "Sic dico de multis aliis, ut quod grave nunc descendat quod prius erat detentum."

> the evidence that formally follows from the first; some is evidence very distant from the first. And these can be distinguished, because some is evidence deducible from an experiential sensible knowledge (*notitia*), as that every fire is hot, which knowledge is deduced from its singulars; some is knowledge that is not so deducible, but arising rather from a plausible conjecture, or from a command of the will.[34]

The appeal to evidence seems to have become more flexible. The evidence of the first principle is erected as an absolute norm to which some other propositions can be "reduced"—this is clearly an echo of the Parisian debate at the time of Autrecourt. The knowledge of this first principle is asserted to be such that no rational person doubts of it, and that the intellect is forced to assent to it from the moment that it understands its terms.[35] Apart from this highest evidence (*evidentia summa*), or deriving from it, we have the sort of evidence that formally follows from this primary evidence. However, such a derivation may itself not be obvious. Certainly a second conclusion actually affirms the possibility of this derivation. Blasius writes, "some knowledge, distinct from the first principle, is possible to humans by evidence reduced to the certitude of the first principle."[36] Yet an objection might be raised: namely how can we know whether a truth, distinct from the first principle, formally follows from it? Blasius shies away from this objection, but in order to reject it he names truths that can seem to be tautological, such as "*si aliquid est, aliquid est*" and "*si chimera est, chimera est*". Here again is probably an echo of Autrecourt's thesis, according to which an identity, at least partially, is required between a subject and a predicate in order to have evidence reducible to a first principle.

The third conclusion, however, retreats from any skeptical interpretation: "by evidence reduced to the certitude of the first principle, we can know that something exists."[37] But this sort of evidence has some limits: firstly, we cannot know whether God exists, because this would suggest a demonstration, which is not possible, since we cannot assign a cause to God's existence. Nor can we

34 Ibid., 95ra: "noto quod quedam est evidentia summa, ut est evidentia primi principii, quedam est evidentia formaliter dependens a prima, ut est evidentia que formaliter sequitur ex prima, quedam est evidentia multum distans a prima. Et ista potest distingui, quia quedam est talis deducibilis ex notitia sensitiva experimentali, ut quod omnis ignis est calidus, que notitia deducta est ex suis singularibus; quedam est notitia non sic deducibilis sed potius ex verisimili coniectura, vel mediante imperio voluntatis."

35 Conclusio 1, ibid.

36 Ibid., 95rb: "evidentia reducta ad certitudinem primi principii est homini aliqua notitia possibilis, primo principio condistincta."

37 Ibid., 95rb: "evidentia reducta ad certitudinem primi principii sciri potest quod aliquid est."

know that some thing differs from another, because considering any individual thing as distinct from others, we can doubt that it exists. Finally, the doubt is solved only by affirming the existence of "something" (that is, an intellect) that doubts.

The conclusions that follow, referring to "experiential knowledge", mark a break. They recognize that we actually have the experience of sensations and the judgments of the senses. So we can, through experiential knowledge (which is not formally reducible to a first principle), know that a plurality of things exist. What, then, about causality?

We know that something is the productive cause of another. But it quickly appears that this thesis remains general: there is causality. In contrast, we cannot know precisely that one specific thing is the cause of another: "nor by the knowledge of the first principle, nor by one reducible to the first principle, nor by experiential knowledge, do you know about some particular agent that it is the cause of that effect."[38] Any causal relationship can be put into doubt: if fire burns my feet, I do not know evidently whether it is the fire or another cause that acts; for example, an interior heat, or an astral influence.[39]

Other texts, however, restore the role of causality, particularly in question 4 of Book I of the first redaction of the *Physics*: "Does perfect cognition of a natural thing presuppose the perfect cognition of its causes?"[40] But even within this framework, limits are set. First, we can know something without knowing all its accidental causes, which are infinite in number, nor all its remote causes. However, to know a substance we must know its essential parts, that is, its intrinsic causes. Primarily, only the *per se* and total cause is able to make known its effect. The illustration provided by Blasius here is the epistemic causality of premises on a conclusion, because, if the principles of natural things can make things known to us, this is possible only through terms and propositions, these propositions being premises of arguments or antecedents of consequences:

> Always the total and *per se* cause, or its cognition, is able to cause or produce the cognition of its effect in the human intellect. This conclusion is evident from its terms, because in the conclusion it is said "total and *per se* cause", and by these expressions it is given to be understood that such

38 Ibid., 95va: "nec notitia primi principii nec reducibili in primum principium nec notitia experimentali scitur de aliquo agente particulari quod illud sit causa istius effectus."

39 Ibid., 95va.

40 "Utrum perfecta cognitio rei naturalis presupponat perfectam cognitionem omnium causarum istius?" (ms. Vat. Chig. O IV 41, 230vb *sqq.*)

a cause is sufficient, due to the fact that it is sufficient to make its effect evidently known.[41]

I shall emphasize two points. Firstly, the causality here considered (and often used as an example) is efficient causality. Secondly, Blasius's concern about the process of production is not so much to identify necessary relations, than to find what is sufficient to make an effect evidently known. Objective necessity is undermined by the indigence of the causal power of particular causes: "no natural thing alone is productive of an effect. This is obvious, because then the first and universal cause would be superfluous, the opposite of which is written in book XII of the *Metaphysics*."[42] As a result, one can never necessarily conclude that a particular natural agent is the cause of any effect, because the universal cause can suspend this power.[43] "If the universal cause would suspend its action, any secondary cause would not act."[44]

3 The Hierarchy of Causal Powers

Before examining more in detail the nature of this universal causality, it would be advisable to specify as much as possible how the network of particular causes is organized. For in spite of the difficulty of identifying any determined relation of causality, it cannot be denied that there exists a network of particular causes. But such a network does not display a purely horizontal causality, with particular causes all on the same level. Particular causes are not only subordinated to an absolutely first cause, but are "essentially subordinated to each other." The pattern is reminiscent of the *Liber de causis*,[45] and at the same

41 QP, 1a red., I, qu. 3, 229vb–230ra: "Semper causa totalis et per se vel eius cognitio nata est causare vel efficere cognitionem sui effectus in intellectu humano; patet ista conclusio ex suis terminis eo quod in conclusione dicitur 'totalis causa et per se' per quas particulas datur intelligi quod talis causa est sufficiens eo modo quo sufficit in notificando evidenter eius effectus."

42 QP, 1a red., I, qu. 3, 230ra: "nulla res naturalis se sola est productiva alicuius effectus. Patet quia tunc prima causa et universalis superflueret, cuius oppositum scribitur XII° *Metaphysice*."

43 QP 2a red., II, qu. 6, 95va: "ubi causa universalis suspenderet eius actionem, quilibet effectus tenderet ad non esse"; II, qu. 9, 100vb: "ubi ad ymaginationem vel quomodolibet aliter prima causa eius actionem subtraheret ad extra, tunc omne quod non est Deus tenderet ad non esse"; cf. QP, 1a red., I, qu. 3, 2a conclusio, 230ra.

44 QP, 1a red., II, qu. 8, 275rb: "in causis essentialiter invicem subordinatis, prima non agente, secundarie et dependentes non agunt."

45 We find in Blasius' work some few references to the *Liber de causis*: see QP, 2a red., II, qu. 9, 100vb: "Et ob hoc scribitur primo *De causis* propositione prima 'causa universalis plus influit in causatum quam causa particularis'"; *Questions on the Meteorology*, qu. 1,

time is a theological expression of omnipotence[46] and a reference to Aristotle's *Metaphysics*.[47] But a trivial example leads to a generalization, namely, it is the soul that moves the stick, which moves a stone, so that the soul is then called the "primary cause."[48]

This fragmentation of the idea of primary cause results from the fact that the world is structured according to degrees of being and perfection.[49] This hierarchical organization is universal. It is essential for our consideration of the relation between qualities and substantial form,[50] but also between simple and mixed bodies,[51] and between inanimate and living beings (some of which have a superior degree of perfection). There is in nature a whole scale of perfections, so that natural things can be situated on this scale according to their degree of complexity (like the mixed, relative to the simple) or according to their nobility (like the intellective soul, relative to the sensitive soul). The natural universe is structured according to degrees and ratios of perfection: "between whatever natural thing and another, be it of the same species or of another, there is some ratio (*certa proportio*) according to essential perfection."[52] And in this context a being of a lesser perfection cannot produce a being of a greater perfection.[53]

---

ms. Vat. lat. 2160, 64ra, where this work is attributed to Aristotle: "Dicit etiam Philosophus in prima propositione *De causis* 'omnis causa primaria plus influit in causato quam causa secundaria et particularis'."

46 QP, 1a red., II, qu. 8, 275rb: "Tertia propositio: in causis essentialiter subordinatis prima agit vel agere potest non agentibus causis secundariis et dependentibus. Istud patet quia cum primaria causa non indigeat causis secundariis, quia ubi indigeret videretur habere dependentiam ab eis, constat quod prima causa potest agere secundariis non agentibus, et per hoc assumitur infrascripta propositio tamquam principium teologicum. Propositio est hec: quicquid potest Deus mediante causa secunda potest ipse se solo."

47 *QP*, 1a red., I, qu. 3, 230ra.

48 QP, 1a red., II, qu. 8, ms. C 275rb: "Secunda propositio: in causis essentialiter invicem subordinatis, prima non agente, secundarie et dependentes non agunt. [...] Et patet per experientiam quia cum baculus moveat lapidem et bracchium moveat baculum et anima moveat bracchium, clarum est quod isti motores dependent ab ipsa anima ..."

49 See Joël Biard, "Nature et degrés de perfection. L'échelle des étants chez Blaise de Parme", in Luca Bianchi, Onorato Grassi e Cecilia Panti (eds), *Edizioni, traduzioni e tradizioni filosofiche (secoli XII–XVI). Studi per Pietro B. Rossi* (Rome, Aracne, 2018), 397–410; Id., *Hypothèse matérialiste et pensée radicale*, ch. 2, 95–99.

50 QP, 1a red., II, qu. 8, 275ra: "numquam qualitas in perfectione essentiali equatur ipsi substantie"; *Questiones super De generatione et corruptione* [QGC], II, qu. 8, ms. Vat. Chig. O IV 41, 52va (old numeration): "... isto supposito, ut quod omnis substantia perfectior sit quocumque accidente"; cf. also QP 2a red., II, qu. 8, ms. V1, 98va.

51 See e.g. QP, 2a red., II, qu. 8, 100ra: "forma ferri est maioris perfectionis quam forma ignis. Est enim quodlibet mixtum omni elemento perfectius".

52 QP, 2a red., II, qu. 3, 90rb: "cuiuslibet rei naturalis ad aliam, sive eiusdem speciei fuerit sive alterius, est secundum perfectionem essentialem certa proportio".

53 QGC, II, qu. 8, 52va: "Nihil agit ultra perfectionem quam habet"; this is repeated several times in the question.

This pattern of organization applies to various possible relations of causality. Because of this difference of perfection, one will have to deny that a mixed body is generated solely by the action of elementary bodies.[54] This organization also applies to the relation between qualities and substantial form.[55] Indeed, the main question here, recurrent in Blasius of Parma's texts, concerns the issue of the causal power that can be attributed to qualities.[56] As often, it is not easy to attribute a clearly declared position to Blasius among the different theories that he presents as theoretically tenable. This issue overlaps with a debate in other texts on natural philosophy, medicine, and astrology, since at least the end of the thirteenth century, on whether qualities can, by themselves, produce a substantial form. Even if Blasius considers it rational to think that qualities can, in certain cases, produce a substantial form, most of the time he upholds the thesis that proportions of qualities only condition the production or the emergence of a form.

From this point of view, a quality, which is an accident, is an instrumental form—even if in truth it is an accident of prime matter and not of any particular substantial form.[57] This position is asserted especially in his commentary on the *De generatione et corruptione*.[58] Elements and primary qualities are also said to be "accidental causes" of generation and corruption, since, by their mutual action and reaction, they are in a determinate proportion that predisposes matter to receive the form of a mixed body.[59] These accidental forms can be specified as "active causes;"[60] however, they are not the principal active causes of any substance.[61]

---

54 See QGC, I, qu. 16, 26rb: "non contingit aliquod mixtum generari in vigore et ex sola actione quattuor elementorum. Patet ista conclusio quia quodlibet mixtum est perfectius quam sint elementa".

55 This principle is applied to the causal relationship, in QP, 1a red., II, qu. 8, 274rb: "si qualitas posset producere formam substantialem, tunc minus perfectum posset perfectius eo producere."

56 See particularly QGC, II, qu. 8; QP, 1a red., II, qu. 4; Qu. Phys, 2a red., II, qu. 8. This issue is crucial for reflexion about the status of matter.

57 See J. Biard, "La matière est-elle le sujet des accidents? Un débat au XIVe siècle", *Acta philosophica*, 31, 2022, 15–31.

58 QGC, II, qu. 8: "Utrum qualitates prime sint cause active generationis et corruptionis."

59 QGC, II, qu. 8, 53ra.

60 Ibid., 53ra

61 Ibid., 52vb: "cause active sunt duplices. Quedam sunt principales, ut que principaliter concurrunt ad effectum producendum, sine quibus nullo modo effectus est producibilis. Quedam sunt cause inessentiales, ut sunt cause que diriguntur a principalibus causis in producendo aliquem effectum, que cause inessentiales nullo modo agerent vel concurrerent ad actionem nisi dirigerentur a principalibus causis."

To deepen our understanding of these causal relationships, we can turn to Blasius' commentary on the *Physics*, which introduces a new distinction among several modes of production of an effect: formally, virtually, and consequently (*formaliter/virtualiter/consequenter*). On this point the two redactions of the *Physics* commentary are very close to each other.[62] One thing formally causes another thing when the effect is of the same species as the cause; for example in generation, a donkey generates a donkey, or a man generates a man. Next, to cause virtually is to cause something specifically different, but including a perfection equal or superior to the effect. So, the sun produces fire, though the sun is not fire; similarly, the sun produces the vital heat of a plant. Here, the principle of hierarchical organization of perfections is decisive: "each natural agent must contain in itself virtually or formally a perfection equal or greater than the essential perfection of its effect."[63] Finally, to cause consequently (*consequenter, consecutive*, or *exsecutive*) is to cause somehow secondarily, when an effect follows from what is formally or virtually produced, like shivering follows from a fever, or a certain movement from the generation of fire.

Thanks to this analysis which organizes in a hierarchy the causal powers of "particular agents", Blasius of Parma returns to the relation between qualities and substantial form. No quality is formally productive of a substance, and neither is a quality virtually productive of a substance, because a substance has a degree of perfection superior to the quality. We can admit by contrast that a primary quality (but not a quality of a different sort, such as whiteness) is productive of an effect similar to it: thus heat produces heat.[64] More interestingly, perhaps, a quality can produce "consecutively" a substantial form. A great heat and a great drought can produce consecutively fire or heat and humidity to prepare the matter of an egg for the generation of a chicken, from wherever this heat comes.[65] More generally, one or several qualities can dispose matter to receive a specific substantial form.

---

62 QP, 1a red, II, qu. 4, 273vb *sqq*.: "utrum qualitas in vigore proprio possit producere formam substantialem, que questio sub alia forma queri potest ut dicatur utrum sola forma substantialis sit causa et principium cuiuslibet motus et mutationis"; 2a red., II, qu. 8, V1, 98rb *sqq*.: "utrum inter agentia particularia, supposita semper generali influentia superiorum, possit qualitas una vel plures formam substantialem producere."

63 QP, 1a red., II, qu. 4, 274rb: "omne agens naturale debet in se continere virtualiter vel formaliter perfectionem equalem vel maiorem quam sit perfectio essentialis effectus ipsius."

64 On that point, the formulation differs from the QGC.

65 QP 1a, II, qu. 4, 274vb: "Et per hoc etiam certa caliditas cum certo gradu humiditatis preparabat materiam ovi ad generationem pulli, sive illa caliditas esset galline vel hominis vel limi."

The causal power of qualities seems therefore limited. In contrast, a substantial form can produce, through its own power and formally, another substantial form.[66]

However, Blasius of Parma also adds some nuances to the question, or at least complicates the picture.

Firstly, given the presentation of this first theory, one cannot completely reject the thesis that one or several qualities could act as principal agents. The examples given above are all examples of physical local movement, that seem to result solely from the qualities of motion, but we could also find the effect in other qualities: "it does not seem that at the moment of the generation of fire, that anything else is required other than the matter sufficiently arranged by heat, lightness, and rarity, and this can be made by one or several qualities."[67] Such an hypothesis, whereby the proportion of qualities would by itself permit the emergence of a substantial form, is also presented (although as a "fantasy") when Blasius questions the number of principles required by any movement, including generation.[68]

Secondly, in the first redaction of the *Physics*, Blasius presents another possible theory, overlapping only partially the first one.[69] The function of this second theory is to emphasize the importance of universal causality, and completely subordinate to it the causality of particular agents. I shall try to clarify, in the following section of this paper, how one has to understand this universal causality, on which the very existence of nature and its regular course depends, but a causality that can also directly interfere with the course of nature.

In this other possible theory, no quality can by itself produce a substantial form, be it formally, virtually, or consecutively. Does this not contradict what seemed to be stated in the former theory? If we collate the indications present in the two versions of the *Physics*, we can hold that no quality can by itself produce a disposition that would *necessitate* the introduction or the emergence of a particular substantial form,[70] nor a disposition that would necessarily prevent the introduction of a form. Does this mean that every generation is based on substantial form? Rather, substantial form itself is now relieved of sufficient

66 QP, 1a, II, qu. 4, 279vb; or 275rb: "forma substantialis est causa et principium motus et status et aliarum mutationum."

67 QP, II, qu. 8, 99va: "non apparet nobis quod hora generationis ignis, requiratur aliud quam quod materia sufficienter disponatur per caliditatem, levitatem et raritatem, et hoc potest fieri per unam qualitatem vel plures."

68 See Biard, *Hypothèse matérialiste et pensée radicale*, ch. 1.

69 If we do not find explicitly this second theory in the second redaction, its conclusions are nevertheless present, like the function attributed to the universal cause.

70 See e.g. 1a red., II, qu. 4, concl. 5, 275ra.

causal power: "no natural agent can on its own produce any effect".[71] And this general assertion can then be applied as much to substance as to quality. The reason for this dispossession is the appeal to a general or universal causality, one that does not only guarantee the common course of nature, as held by Oresme or Buridan, but on one hand must support any particular causation, on the other hand can substitute for it.

4 The Universal Cause

The second determination proposing an answer to the question of whether qualities can produce a substantial form, or substantial form is the only cause of change, is said by the author *magis gratam*; in other words, he prefers it. The theses or "conclusions" are preceded by "propositions", and the first of these affirms the subordination of this mundane world to a superior world.[72] The second proposition, already mentioned, introduces the hierarchical organization of causes, both primary and secondary, and claims that secondary causes never act without the assistance of the primary. The third proposition specifies that a primary cause can, by contrast, act without the secondary causes acting.

These principles imply that no natural agent can act alone, by itself, to produce any effect, because the assistance of a primary cause must be involved. This also holds for the production of a living being by another of the same species, as a donkey by a donkey. The consequence is that the causal powers of any natural being whatsoever seem to be abrogated, and all efficiency is assigned to the universal cause. In fact, during his examination of one of the difficulties, when questioning the validity of the Aristotelian definition of nature as the principle of movement and rest, Blasius qualifies this conclusion. He returns some causal efficacy, at least in this case, to the form. This does not negate the assistance of the primary cause, no more than other agents, but form is the principal agent "here below" among the secondary causes.[73] Granted this framework and these limits, it is then substantial form that is said to be the principle and cause, while a quality is a cause that can dispose something

---

71 Ibid.: "nullum agens mere naturale potest se solo aliquem effectum producere."

72 QP, 1a red., II, qu. 4, 275[rb]: "fatendum est hunc mundum corruptibilium a superiori dependere et successive in esse conservari."

73 QP, 1a red., II, qu. 4, 276[ra]: "[Aristoteles] vellet dicere quod inter agentia particularia hic inferius reperta, forma substantialis est principium principale et causa illius in quo est etc., ita quod licet Deus concurrat et multa agentia concurrant ad augmentationem et ad generationem, tamen principale agens hic inferius est ipsa forma."

(understood: the matter) to the natural production of an effect.[74] Furthermore, we even find the notion of "instrumental cause" applicable to the heat of the water that warms up my hand, but which results from the fire that previously warmed up the water, since the latter is in itself cold.

The universal cause is nevertheless regularly mentioned.[75] What is the universal cause? Formulations vary.[76] Some of them clearly designate God as primary cause: "glorious God, on whom the whole heavens and earth depend."[77] Again there is an allusion of the same sort, although more abstract, in the text when it mentions "an incorporeal agent that has no magnitude, non-distant, and assisting the whole universe and each of its parts. And it cannot be anything other than God himself."[78] But most of the time the allusion is part of other descriptions more properly philosophical. Generally, what is suggested is a continuity of the superior world with the inferior world, and consequently the governance of the latter by the former:

> First proposition: you must say that this world of corruptible things depends on the superior and is continuously maintained in being by it. This proposition comes from Aristotle in the first book of the *Meteorology*, when he says that it is necessary that this inferior world is continuous to the superior, so that by this any power is preserved.[79]

But a contiguity, and even a continuity, moves away from the theological relationship of God to the created world. Blasius' general framework is completed by various references, explicit or implicit: sometimes to the *Liber de causis*, as we have already seen; elsewhere to book VIII of the *Physics* for the idea that God

---

74 QP, 1a red., II, qu. 4, 276ra: "causa qualitative disponens concurrit ad effectum producendum, et forma substantialis meretur dici principium et causa."

75 QGC, II, qu. 8; *Questiones de anima*, III, qu. 9; QP, 1a red., II, qu. 4; QP, 2a red., II, qu. 7.

76 For broader developments, see Biard, *Pensée radicale et hypothèse matérialiste*, 79–90.

77 QP 1a red., II, qu. 4, 275vb: "gloriosus Deus a quo totum celum et natura dependet;" *Questiones methaurorum* [*Qu. Meth.*], I, qu. 1, ms. Vat. lat. 2160, 63vb: "… ad quemcumque effectum substantialem productum vel producendum in hunc mundum inferiorem concurrit immediate ipse gloriosus Deus."

78 QP 1a red., II, qu. 4, 275[va]: "agens incorporeum nullam habens magnitudinem, indistans et assistens toto universo et cuilibet eius parti. Et hoc non potest esse nisi ipse Deus." The "holenmerist" presence, used to specify the type of presence of the soul in the body, is also used to describe the mode of the presence of God in the world.

79 Ibid., 275[rb]: "Prima propositio: fatendum est hunc mundum corruptibilium a superiori dependere et successive in esse conservari. Ista est Aristotelis primo *Methaurorum*, dicentis quod necesse est hunc mundum inferiorem esse superiori continuum ut omnis virtus inde conservetur."

assists the world;[80] and we also find, refered to the *Metaphysics* of Aristotle, the idea that God is an animal:

> And the Philosopher seems to have consented to this conclusion when he says, in book XII of the *Metaphysics*, that God is a sempiternal animal, from whom the whole of the heavens and earth depend, from whom result being and living.[81]

But the most important claim is that the universal cause is celestial or astral causality. In some cases, the heavens and the celestial bodies operate something like the vicar of the creator: this is what we find for example in the second redaction of the *Physics*:

> By universal cause, we understand the whole heavens etc. [that is, the whole heavens with its celestial bodies exerting a certain influence by light and movement], without removing the influence that the first cause regularly exerts, because if it did not have an influence, the heavens would have none either.[82]

Most often, it is suggested that the universal cause is simply the causality of the heavens, understood broadly, i.e. including the celestial Intelligences and all the celestial bodies exerting their influence through light and movement. In the *Questions on the Soul*, the well-known authority from the *Meteorology* about the contiguity between the superior world and the inferior world is bookended by two references to astral causality. The first is a reference to the *Centiloquium* of the pseudo-Ptolemy:

---

80 Ibid., 275vb: "... quod agens, ut dicetur in VIII° huius, erit ipse gloriosus Deus a quo totum celum et natura dependet".

81 QGC, I, qu. 16, 26rb: "Et huic conclusioni videtur Philosophus consensisse, dum dicit Deus est animal sempiternum a quo totum celum et natura dependet, XII° *Metaphysice*, a quo derivatur [*coniect.*] esse et vivere"; QP, 1a red., I, qu. 3, 230 ^{ra}: "tunc prima causa et universalis superflueret, cuius oppositum scribitur XII° *Metaphysice* per infrascripta verba 'Deus est animal sempiternum a quo celum et natura dependet';" QP, 2a red., II, qu. 9, 100vb: "Deus est animal sempiternum a quo totum celum et natura dependet". See *Aristoteles latinus*, 25. 3. 1, *Metaphysica*, XII, 7, 1072b14–29: "ex tali igitur principio dependet celum et natura [...] Dicimus autem Deum esse animal sempiternum."

82 QP, 2a red., II, qu. 7, 97vb: "Per causam universalem intelligitur totum celum etc. [i.e. totum celum cum suis astris taliter lumine et influentia et motu influentibus] non circumscribendo a proposito influentiam que assidue fit a prima causa, quia si illa non influeret nec celum influeret."

> The conclusion is obvious because, as hold the astronomers and particularly Ptolemy in the 9th proposition of his *Centiloquium*, this inferior world is subject to celestial images, and so this inferior world depends upon the superior world, as upon the celestial bodies that everywhere are natural agents which cannot be impeded in their actions.[83]

The second is a more general reference to astrological science:

> Likewise, [without this continuity,] what in astronomy would be the value of the books written by the astronomers about nativities, a science by which the accidents of newborns and the duration of their lives are revealed, and everything about the body and the soul, even after life?[84]

The first question in the *Meteorology* enters into more detail. After the mention of some well-known phenomena, such as the influence of the moon on the tides or the conservation of heat due to the sun, we find relations among specific planets and the effects that they exert on animate bodies.[85] Even medical practice, with its astrological prognostications, testifies to this celestial influence.

From this perspective, natural agents exert an actual causal power, but it is dependent upon astral influence, either at the moment of generation or at the moment of the exercise of this power. This effect concerns substances, including living beings, but also the qualities that are disposed according to some proportions because of this influence.

This theory is not refutable by natural reason. It is only contrary to the Catholic faith because, from the viewpoint of faith, it would be inappropriate to eliminate the efficaciousness of a human being. From a *natural* point of view, this theory could extend even unto the human being itself. The point of debate consequently concerns whether it is possible to generate the substantial form

83 *Questiones de anima*, III, qu. 9, 214rb (new numeration): "Patet conclusio per hoc quia, ut volunt astronomi et specialiter Ptholomeus nona propositione sui *Centiloquii*, iste mundus inferior subiectus est celestibus ymaginibus, et sic iste mundus inferior dependet a superiori ut ab astris qui utique sunt naturalia agentia, que in suis actionibus nequaquam impediri possunt."

84 Ibid.: "Item ad quid valerent libri astronomorum facti de nativitatibus in astronomia, per quam scientiam depre⟨he⟩nduntur accidentia nati et duratio vite sue, et quicquid de corpore et de anima et post eius vitam."

85 *Qu. Meth.*, I, qu. 1, 64ra: "Habent etiam hec astra influentias suas supra corpora animata et dispositiones eorum". For further details, see Biard, *Pensée radicale et hypothèse matérialiste*, 87.

of a human being solely through the actions of qualities, arranged in some specific proportion by astral influence.

Before considering in more detail the case of humans, it is important to specify where exactly Blasius departs from previous commentators. John Buridan is also someone who admits some causal influence of celestial bodies on the things of this world.[86] He even justifies this claim by an appeal to astrology.[87] Nevertheless he insists, contrary to Blasius, on the insufficiency of these celestial causes, once we attempt to remove divine omnipotence:

> Fifthly, we must also say that to save in a natural way the many diversities and arrangements of the changes and natural effects in this world, God, the celestial bodies, and primary matter are not sufficient, if we do not suppose other various particular agents ordering them.[88]

The reason for this insufficiency is the uniformity of celestial influence; even in the case of generation by putrefaction, it is not a sufficient explanatory cause:

> Also in the things that are generated by putrefaction, the particular causes of diversity are, along with the celestial powers, the various dispositions of matter; and this is why frogs are produced from one portion of matter, and houseflies from another, though the heavens are everywhere uniformly influent.[89]

Celestial influence, since it is uniform, does not produce the determinate proportions of qualities sufficient to generate a specific form. Its influence depends upon the dispositions of matter, due to particular causes. Blasius, as we saw, would reverse this order of importance.

---

86 *Quaestiones super octo libros Physicorum Aristotelis (secundum ultimam lecturam)*, II, qu. 13, ed. by Michiel Streijger & Paul J.J.M. Bakker, vol. I: libri I–II (Leiden-Boston, Brill, 2015), 340: "manifestum est quod corpora caelestia sunt causae ordinativae diversitatis plurimarum transmutationum et plurimorum effectuum in hoc mundo provenientium."

87 Ibid.: "et aliter periret omnino scientia astrologiae."

88 Ibid., 341: "Quinto etiam dicendum est quod ad multas diversitates et ordinationes transmutationum et effectuum naturalium in hoc mundo salvandas modo naturali non sufficiunt Deus, corpora caelestia et prima materia, nisi ponantur alia agentia particularia diversa haec ordinantia."

89 Ibid., 341: "etiam in his quae ex putrefactione generantur causae particulares diversitatum sunt cum virtutibus caelestibus diversae dispositiones materiae, propter quod ex una portione materiae fiunt ranae et ex alia muscae, licet caelo hic et illic uniformiter influente."

In the case of humans, we find two possible descriptions, which, however, are not radically opposed. According to one, the universal cause and a particular agent (a man, a donkey, etc.) contribute to the production of a substantial form, because the primary cause could not create a specific effect, without becoming itself a particular cause.[90] So this first description emphasizes the complementarity between the universal cause and particular causes. But according to another theory, attributed to Albert and Avicenna,[91] the universal cause could, in certain cases, operate without a particular cause. Indeed, the celestial bodies cause all "elementary" qualities, those which are permanent as well as those which appear and disappear contingently, due to the influence of the Sun and other celestial bodies (the coolness of Saturnus, the humidity of the Moon, etc.). When these influences are combined, all possible ratios among qualities can be produced, as well as any amplitude (*latitudo*) of these ratios: "by the celestial bodies, here below any ratio of elementary qualities and any amplitude of the ratio of these qualities can be caused".[92] Because of this astral influence, prime matter is arranged in some specific disposition, and in some specific proportion of qualities, so that some specific form can be generated. Therefore, an individual of some species, such as a donkey, can be generated, even if no other individual of its species exists to perform the function of a particular agent.

What could lend some weight to this theory? Perhaps, the phenomena of spontaneous or equivocal generation. Blasius of Parma regularly appeals to these phenomena where there is no transmission or production of a form by an individual of the same species; for him, these provide a sort of experiential evidence. He occasionally mentions spontaneous generation by putrefaction.[93] More spectacularly, he appeals to the appearance of frogs out of the sky; there are no frogs at the source of this phenomenon:

> And it is also obvious by reason, because it is the case that, when though no particular frog exists, it is possible that some frog be generated, sometimes in some porosity of the earth, and sometimes in the middle region of the air. For often from the middle region, along with rain, many frogs fall that were generated there.[94]

90 QP, 2a red., II, qu. 7, 97[ra]: "ad cuiuslibet rei substantialis generationem, celum videtur concurrere et natura. Et per naturam in presenti intelligo agens particulare."

91 Yet Blasius of Parma rejects the theory of the Giver of Forms (*dator formarum*), favoring instead the emergence of the form in matter.

92 QP, 2a red., II, qu. 7, 97[rb]: "ab astris hic inferius omnis proportio elementalium qualitatum et omnis latitudo proportionum earumdem qualitatum est possibilis causari."

93 QGC, I, qu. 16, 27[rb]: "videmus enim multa animalia generari ex putrefactione."

94 QP, 2a red., I, qu. 3, 65[vb]–66[ra]: "Et patet etiam hoc per rationem, quoniam constat quod

> From the middle region often fell frogs, fishes, vipers, and calves, as Avicenna reports in his *Metaphysics*.[95]
>
> In the middle region, often by the force of some celestial body, or several, animals of various species are generated that we see falling, such as frogs, fishes, oxen, etc.[96]

Debates over the possibility of equivocal generation are ancient.[97] But Blasius of Parma seems to be the only one who extends it to the case of human beings. In truth, Blasius is cautious and balanced. In the *Questions on the Soul*, he admits as rationally demonstrable, though somehow relativizing the force of the argument, that the generation of a human being could result from a certain mixture of qualities, dependent on astral influence. Resuming and expanding an argument from Avicenna, he invokes the universal flood as proof of this sort of generation. Indeed, Avicenna accepted the possibility of the spontaneous genesis of a human being: the human species could disappear and be regenerated by the elements.[98] But this implies the action of the *Dator formarum*, who has no place in Blasius' doctrine.[99]

Blasius of Parma mentions the flood in question 8 of Book I of the *Questions on the Soul*. In this long question, dealing with the separability of the human soul, he juxtaposes a discussion according to the point of view of the faithful (*ut fidelis*) or according to faith (*secundum fidem*), and another, more detailedly developed, presented as conforming to the schools of the philosophers (*secundum scholas philosophorum*), a discussion which states that it cannot be proven

---

nulla particulari rana existente contingit rana generari, modo in aliqua porositate terre, et aliquando in media regione aeris. Unde sepe a media regione cum pluvia multe rane ceciderunt que ibidem sunt genite."

95 QP, 2a red., II, qu. 7, 96rb: "a media regione sepe ceciderunt rane, pisces, vipere et vitulus, ut narrat Avicenna in eius *Metaphysica*."

96 *Questiones de anima*, I, qu. 8, 129vb (n.n.): "in media regione, sepe vigore alicuius astri vel plurium, generata sunt animalia diversarum specierum, visa a nobis cadere deorsum, ut sunt rane, pisces, boves, etc."

97 See Maikke van der Lugt, *Le Ver, le Démon et la Vierge. Les théories médiévales de la génération extraordinaire* (Paris, Les Belles Lettres, 2004); esp. 176–181.

98 This is not the case for Albert the Great, whom Blasius associates wrongly with Avicenna on that point; see M. van der Lugt, op. cit., 170–175. See also A. Bertolacci, "Averroes against Avicenna on Human Spontaneous Generation: The Starting-Point of a Lasting Debate", in A. Akasoy and G. Giglioni (eds.), *Renaissance Averroism and its Aftermath* (Springer: Dordrecht 2013), pp. 37–54.

99 Ead., ibid., 165–167. Concerning Avicenna, see *De diluviis*, ed. M. Alonso Alonso in "Homenaje a Avicena en su milenario. Las traducciones de Juan Gonzáles de Burgos y Salomón", *Al-Andalus* 14, 1949, 291–319, text 306–308.

naturally that the soul is separable from the body. It is in this context that he mentions the flood, also considered from a purely natural point of view (consequently without paying attention to the fables of women—"*nec in hoc velis credere paraboles mulierum*").[100] This means that he refuses to believe the story of Noah's Ark. He says that Noah had other preoccupations, and that the animals would not have coexisted on the Ark.[101] If the flood truly happened, then all species would have been destroyed, and there would have been a time at which there were no longer any humans. If, then, some human being appeared anew after the flood, it must be supposed that that person, also, due to a certain astral configuration, was generated from putrefaction.[102] One cannot exclude that this event might reoccur, if it should happen that a certain astral configuration appropriately arranges matter at some place.

What concerns us now is not the materialist anthropology of Blasius of Parma, but his doctrine of causality, one that culminates with a conception of a universal cause that prioritizes celestial action. For the universal cause that supports particular causes is *par excellence* the heavens and the celestial bodies. "The heavens along with nature seem to contribute to the generation of any substantial thing. And by 'nature' I here understand a particular agent."[103] This celestial cause remains in continuity with the inferior world because its influence transpires by the intermediary of heat and light. This is why the Sun is often taken by metonymy as the whole heavens. Thusly Blasius tackles the issue in question 7 of Book II of the *Physics*: "Is the impulse of a universal cause required for the generation of whatever natural thing, besides the particular agent, a universal cause that is said to be the Sun?"[104]

One may object that this scheme does not satisfy the principle of the hierarchical organization of causes, because ultimately it would be qualities that produce form. Blasius of Parma seems to be aware of the objection. In the first question on Book I of the *Meteorology*, he considers the production of fire by

---

100 *Qu. An.*, I, qu. 8, 129vb (n.n.).

101 See ibid.: "In tanto enim et arduo casu putandum est quod Noe habuit aliam curam et sollicitudinem: non enim homo libenter cum leone tripudiat, nec agnus cum lupo."

102 Ibid.: "Undecima conclusio: evidens est ex putrefactione hominem naturaliter posse generari. Patet evidenter ista conclusio quia alias mundus fuit sic reparatus ut quod, post tempus diluvii, natus est homo ex putrefactione."

103 QP, 2a red., II, qu. 7, art. 2, secundus processus, concl. 2, 97ra: "ad cuiuslibet rei substantialis generationem, celum videtur concurrere et natura. Et per naturam in presenti intelligo particulare ⟨agens⟩."

104 QP, 2a red., II, qu. 7, 96ra *sqq.*: "utrum ad cuiuscumque rei naturalis generationem preter particulare agens requiratur influxus cause universalis, que causa universalis dicitur Sol."

heat alone, and rejects it.[105] Immediately after, he reminds us of the principle of the greater perfection of the agent relative to the effect; and further on, that a quality cannot produce a substance. The counterpart is the proper efficiency of universal causality, first identified as God, and later, according to natural evidence, as celestial bodies.[106] But the first question of Book IV provides a further qualification. At the outset, generation is preceded by alteration, therefore by the action of qualities; but these are "instrumental principles" and not "principal principles." But the examination of the second difficulty questions this analysis. In some cases, quality produces a form, like the rays of the sun acting on putrid matter, or also on a concave mirror.[107] At this point Blasius raises the objection: these are only instrumental causes, the main cause being the sun itself. In fact, the progression cannot stop here, for we might imagine that God removes the sun while maintaining its rays; or even, operating naturally, that the rays remain for a time after the disappearance of the sun. The results would be the same. We must therefore admit that a ray of light has in itself the power to heat, and consequently that a quality can, by its own power, produce a form.

What may we conclude? It does not seem at first glance, in accordance with the normal course of nature, that astral causality violates the principle of the hierarchy of causes, insofar as light and heat can be considered as *instrumental* causes, and that the *principal* cause is then a substance (the heavens, the celestial bodies, and Intelligences). In other words, in the cases of spontaneous generation that suggest the possibility of an emergence of a form from a proportion of qualities, the required astral configuration can lead back to substances ontologically superior. But qualities (at least elementary qualities) are by themselves active and productive. Previously, Blasius has presented the ray of light as a quality.[108] And a ray, which seems to be a qualitative disposition of matter, is by itself *calefactivus*.[109] In any case, the intent of Blasius is actually to allow for the possibility of generation from qualities, operating subordinate to the general principle of universal causality.

---

105 *Qu. Meth.*, I, qu. 1, ms. Vat. lat. 2160, 63va.

106 Ibid., 64ra–va^.

107 *Qu. Meth.*, IV, qu. 4, ms. Vat. lat. 2160, 130vb: "Sit hec conclusio: qualitas in vigore proprio potest producere formam substantialem in materia. Et licet dixerim oppositum, probatur conclusio quia radius Solis occurrens materie putride disponit ipsam et producit formam substantialem. Et hoc idem facit occurrens speculo concavo, producit ignem."

108 *Qu. Meth.*, I, qu. 1, 63vb: "radius Solis est una qualitas."

109 In qu. 1 of book IV, Blasius mentions other rays, such as those emitted from a cat's eyes, or some weapons, that are also *calefactivi*, so that the sun is not the cause of the heat.

5 Conclusion

Different from the materialists of the eighteenth century, or even from Avicebron—at least as he is read by Latin Medievals—Blasius of Parma judges that matter, when considered in itself, is purely passive. That is why the great question of his natural philosophy is the second principle of the dynamics of nature. This is what leads to the alternatives, continually reintroduced, between a monist temptation that favors matter, and the requirement of a functional duality that takes different forms. Nature is growing, developing, transforming itself; the subject of physics, ever since Aristotle and even since the Presocratics, is to understand the principle of these changes. Such is the *dunamis* of the *phusis*. Does this mean that Blasius adopts, roughly speaking, the Aristotelian doctrine? Actually, Blasius of Parma completely redefines the principles of change, by transforming the connection of matter and form. If matter is not active and by itself productive, a certain proportion of qualities is transmitted and conditions the appearance of a new form. In this context, which entities have the power to cause the emergence of a new substantial compound?

The natural world, according to Blasius, is not a homogeneous totality, but an ordered arrangement according to degrees of complexity and perfection. This thesis may seem commonplace; in truth, it is less trivial if you remember that this characterization is based on the idea that any form, accidental or even substantial, implies some amplitude (*latitudo*) that allows it on one hand to be placed in the scale of being; on the other hand, at least in some cases, it allows possible variations without specific change. In this arrangement are to be found the potencies or powers that account for the dynamics of nature.

This view led Blasius to first define different modes of causal action, that is, formally, virtually, and consecutively. This distinction allows us to describe the possible causal effect of qualities (accidental forms) on substances—a crucial point in understanding the production and conservation of bodies, simple or compound. But along with this view arose a manner of accounting for natural change that accentuates the primacy of a universal cause by weakening the power of particular causes. This analysis does not make everything depend directly and exclusively upon a universal cause, since in the normal course of nature it does not turn into a particular cause. But we can discern a theory according to which the universal cause (designated by various names, but ultimately situated in the heavens and celestial bodies) conditions the appearance of a form, by creating various proportions of qualities in matter. This equivocal causality (even if it is exceptional, though never designated a miracle) bears witness to the possibility of the generation of a form, of any sort, from a certain proportion of qualities, dependent on an astral configuration.

The description of causality is therefore fragmented, not only into the four traditional types of cause whose relationships are re-defined by a new conception of matter and form, but also into different and usually complementary types of agents. In this entanglement, no causal relation in itself is necessary. Thereby is raised the issue, recurrent since the mid-fourteenth century, of the possibility of knowing causality. Even if there is no doubt that there are causal relations in nature, we still cannot demonstrate with absolute evidence that some thing is the necessary cause of such an effect. All we can do is seek, find, and expose sufficient explicative causes.

CHAPTER 3

Blasius of Parma on Motion in the Sublunary World

Erik Åkerlund, Henrik Lagerlund and Sylvain Roudaut

1 Introduction

It is an old claim, sometimes modified and rejected, but that never has quite disappeared, that seventeenth century physics or mechanics should be traced back to the fourteenth century and the discussions of motion at the time in Oxford and Paris.[1] The discussion of the medieval background to early modern mechanics includes mention of the Oxford Calculators and what was then called the school of John Buridan, which by Pierre Duhem was thought to include Albert of Saxony and Nichole Oresme as well as Themon Judaeus and Marsilius of Inghen. We now know that there was no real School of John Buridan,[2] but there was a group of interconnected thinkers who shared some initial presuppositions, but who also were quite different. Buridan was one of several figures in this group, not unlike John Mair and the group around him in early sixteenth century Paris. Within this extended group a considerable rethinking of the physics of the day took place. To this extended group of like-minded and innovative thinkers, Blasius of Parma needs to be added.

Regarding theories of motion, Blasius certainly qualifies as an innovative thinker, but he is in other respects also quite traditional. The most remarkable and innovative claim he makes is that all motion in the sublunary world is violent motion. This, first of all, rejects the Aristotelian distinction between

1 Pierre Duhem, *Le système du monde. Histoire des doctrines cosmologiques de Platon à Copernic*, 10 vols. (Paris, 1913–1959), VIII: 200: "La mécanique de Galilée, c'est, peut-on dire, la forme adulte d'une science vivante dont la mécanique de Buridan était la larve." For further discussion of this thesis see also John E. Murdoch, "Pierre Duhem and the History of Late Medieval Science and Philosophy in the Latin West," in *Gli studi di filosofia medievale fra otto e novecento*, eds. Ruedi Imbach and Alfonso Maierù (Rome, 1991), 253–302, and Johannes M.M.H. Thijssen, "The Debate over the Nature of Motion: John Buridan, Nicole Oresme and Albert of Saxony. With an Edition of John Buridan's '*Quaestiones super libros Physicorum, secundum ultimam lecturam*', Book III, Q. 7", in *Early Science and Medicine*, Vol. 14, No. 1/3, *Evidence and Interpretation: Studies on Early Science and Medicine in Honor of John E. Murdoch* (2009), 186–210.

2 For a treatment and rejection of the idea of a school of Buridan see Johannes M.M.H. Thijssen, "The Buridan School Reassessed, John Buridan and Albert of Saxony," *Vivarium* 42, 2004, 18–42.

 | DOI:10.1163/9789004734968_005

two kinds of motion, natural and violent; but, secondly, it also suggests that the analysis of falling objects is like that of projectiles. Projectile motion is not unnatural; it is simply a motion that has an explanation similar to that of natural motion, which, for example, suggests that the account of impetus can be applied to falling bodies as well. This view was not completely new at this time—impetus was used to analyze all sorts of motion—but it certainly indicates that we must not draw distinctions between what in nature is natural and not. It is in keeping with a completely mechanistic account of nature.

As a means of achieving this result, Blasius provides an innovative analysis of lightness and heaviness considered as principles of motion and rest. Together with his view that a body is neither heavy nor light when located in its proper place, this leads to the view that a body does not have any principle of motion or rest when located in its natural place.

There are other elements of Blasius' thought that are quite traditional, however. One such aspect is his use of Aristotelian terminology, although occasionally that terminology is given a new meaning. Perhaps the most significant aspect is his analysis of the 'desire' and the 'appetites' of bodies striving for their proper place. Another traditional aspect is his appeal to magnets to explain the attraction of a proper place, a move which rejects the involvement of teleology. A further consequence is his view that the desire for life is a manifestation of a more universal tendency of things to persist in existence and to always remain in the same state. Considered as a whole, Blasius' discussion of motion offers a rich reconceptualization of a notion that at the time was undergoing significant change, involving both traditional and new ideas.

In this essay, we aim to situate Blasius' view of motion in the context of an extensive discussion, of course going back to Aristotle, but which also includes Averroes as well as Ockham and Buridan. It will bring out the novel character of Blasius' discussion of motion.

2 Aristotle and Averroes on Motion

There is a case to be made that change or motion (*kinesis*) is the very defining notion for Aristotle's natural philosophy.[3] Indeed, as he writes:

3 For Aristotle's treatment of motion, see Christopher Shields, *Aristotle*, second edition, Routledge, 2014, ch. 5.1 (232–240); István Bodnár, "Aristotle's Natural Philosophy", *The Stanford Encyclopedia of Philosophy* (Summer 2023 Edition), Edward N. Zalta & Uri Nodelman (eds.), URL = https://plato.stanford.edu/entries/aristotle-natphil/

> Nature is a principle of motion and change, and it is the subject of our inquiry. We must therefore see that we understand what motion is; for if it were unknown, nature too would be unknown.[4]

Understanding what motion or change is amounts to understanding what nature (*physis*) is.

In the background of Aristotle's treatment of change stands Parmenides and his challenge to the very reality of change. How can something both be and not be? How could any kind of knowledge of change be possible?

Aristotle's answer to these questions when it comes to nature is entwined with his metaphysics of act and potency. In a sense, to think in terms of act and potency blurs the line between being, on the one hand, and non-being, on the other, a line from which Parmenides and his follower Zeno drew paradoxical conclusions about change. For over and above having or being something actually, there is also having or being something potentially. To move from having or being something potentially, to having or being this actually, is what change is:

> We have distinguished in respect of each class between what is in fulfillment and what is potentially; thus the fulfillment of what is potentially, as such, is motion[.][5]

When distinguishing among different kinds of change, Aristotle makes use of other parts of his metaphysics, namely his theory of the categories, since change always takes place within a specific category. Aristotle enumerates four such categories: substance, quantity, quality, and place. Since change takes place within one of these categories, nothing can change from, say, being blue to being 5 feet tall; it rather changes from being blue to being green, or from being 2 feet tall to being 5 feet tall. Further, this change also entails the loss of something, over and above the actualization of a potentiality. As Aristotle writes:

> It is always with respect to substance or to quantity or to quality or to place that what changes changes. But it is impossible, as we assert, to find anything common to these which is neither 'this' nor quantity nor quality nor any of the other predicates. Hence neither will motion and change have

4 *Physics* III.1, 200b12–14; Barnes 1984, I: 342.

5 *Physics* III.1, 201a10–12; Barnes 1984, I: 343.

> reference to something over and above the things mentioned; for there is nothing over and above them. Now each of these belongs to all its subjects in either of two ways: namely, substance—the one is its form, the other privation; in quality, white and black; in quantity, complete and incomplete. Similarly, in respect of local motion, upwards and downwards or light and heavy. Hence there are as many types of motion or change as there are of being.[6]

This last comment, that "there are as many types of motion or change as there are of being", opens up the possibility that the notion was meant to be used in a flexible and wide-ranging way with respect to nature, and not just in line with the categories he has explicitly mentioned.

The notions "actual" and "potential" do not divide the world into two neat, separate classes of things. One and the same thing can be potential as well as actual, although not in the same respect at the same time:

> The same thing can be both potential and fulfilled, not indeed at the same time or not in the same respect, but e.g. potentially hot and actually cold. Hence such things will act and be acted on by one another in many ways: each of them will be capable at the same time of acting and of being acted upon.[7]

If we have two bodies, one hot and the other cold, they can interact insofar as the hot body acts on the cold one in actualizing the hotness that is potentially in it, whereas the cold one acts on the hot one in actualizing the coldness that is potentially in it. The act takes place in the body that is acted upon; it is there that a change or motion is caused.

Finally, motion or change is not something over and above the process whereby the potential becomes actualized. Hence, when the process has reached its conclusion, there is no longer motion or change, but only the actuality of the end result itself. Aristotle sometimes speaks, though, about the motion or process itself as an actuality. Aristotle, again:

> [M]otion occurs just when the fulfillment itself occurs, and neither before nor after. For each thing is capable of being at one time actual, at another not. Take for instance the buildable: the actuality of the buildable as buildable is the process of building. For the actuality must be either this or the

6 *Physics* III.1, 200b33–201a9; Barnes 1984, I: 342–343.

7 *Physics* III.1, 201a20–23; Barnes 1984, I: 343.

> house. But when there is a house, the buildable is no longer there. On the other hand, it is the buildable which is being built. Necessarily, then, the actuality is the process of building. But building is a kind of motion[.][8]

When a house actually exists, it is no longer buildable. While the house is being built, it is the actualization, not of the house *per se*, but of the buildable as buildable. If it were the actualization of the house, the house—the end result—would already be there, and there would be no process. Hence, motion is a certain kind of actuality, according to Aristotle. The explanation of motion still hinges on his scheme of act and potency, however. In the end, he does not seem to think of change as anything over and above the "things" themselves that change.[9]

In his long commentary on the *Physics*, Averroes distinguishes between two different ways of understanding motion.[10] Averroes associates the passages in the *Physics* referred to above with Aristotle's treatment of motion in the *Categories*,[11] and pits against each other the two positions on the status of motion. According to the one view, motion is nothing other than the end result of the motion, although in a diminished or yet-not-perfected state. According to the other view, motion must be something distinct from the thing that is the end result of the motion. Motion must, then, be in its own genus, differing in kind from the thing that results from it:[12]

---

8 *Physics* III.1, 201b7–14; Barnes 1984, I: 343.

9 "There is no such thing as motion over and above the things." (*Physics* III.1, 200b33; Barnes 1984, I: 342.)

10 For a succinct presentation, see Cecilia Trifogli, "The Reception of Averroes' View on Motion in the Latin West. The Case of Walter Burley," in Paul J.J.M. Bakker (ed.), *Averroes' Natural Philosophy and Its Reception in the Latin West*, Leuven: Leuven University Press, 2015; 127–139, especially 128–129. For a fuller treatment, see Cecilia Trifogli, *Oxford Physics in the Thirteenth Century (ca. 1250–1270). Motion, Infinity, Time, and Place*, Leiden: Brill, 2000, especially 47–51. In both of these works, Trifogli refers her readers to a fuller treatment in Anneliese Maier, *Zwischen Philosophie und Mechanik*, Rome: Edizione di storia e letteratura, 1958, especially 59–143.

11 Aristotle writes about motion in *Categories* 9, 11^b1–8: "Doing and being-affected admit of contrariety and of a more and a less. For heating is contrary to cooling, and being heated to being cooled, and being pleased to being pained; so they admit of contrariety. And of a more and a less also. For it is possible to heat more and less, and to be heated more and less, and to be pained more and less; hence doing and being-affected admit of a more and a less." (Translation from Jonathan Barnes [ed.], *The Complete Works of Aristotle*, vol. 1, Princeton, NJ: Princeton University Press, 1984, 18.)

12 Translation from Trifogli, *Oxford Physics*, 48, who in her turn takes it from Marilyn McCord Adams, *William Ockham*, vol. 2, Notre Dame, IN: University of Notre Dame Press, 1987; 802–803.

> [I]nsofar as motion does not differ from the perfection to which it attains except according to more or less, it must exist in the genus of that perfection. For motion is nothing but the generation of one part after another of the perfection to which motion is directed until it is perfected and exists in act. Thus, it is necessary that the motion that is in substance be found in the genus of substance, and the motion that is in quantity in the genus of quantity, and likewise for motion in place and quality. But insofar as motion is the way to that perfection, where the way is other than the perfection, it must be a genus in itself. For the way to a thing is different from the thing, and to that extent, it was assumed [to be] a category in itself. And the latter way is the more famous, but the former is the truer. Therefore, Aristotle introduces the more famous way in the *Categories* and the true way in this book.[13]

Averroes himself goes along with the view that motion is nothing other than that which comes into being, although with a reduced status in the process. The dichotomy between this "truer" view, on the one hand, and the "more famous" view, according to which motion is or has a genus all to itself, would form the background of the debates concerning the status of motion in the Latin West.

3 Blasius of Parma's Background: William Ockham and John Buridan

Averroes seems to have set the question that initially came to dominate the fourteenth century, namely the question about the ontological status of motion. Later on, new questions about motion began to be asked and the discussion changed somewhat, but for William Ockham the most important aspect

13 Averroes, *In* Physicam, in *Aristotelis opera cum Averrois commentariis*, vol. 4, *De Physico auditu libri octo. Cum Averrois Cordubensis variis in eosdem commentariis*, Venice: apud Iunctas, 1562; III, 87ra–b. '... motus secundum quod non differt a perfectione ad quam vadit nisi secundum magis et minus, necesse est ut sit de genere illius perfectionis. Motus enim nihil aliud est quam generatio partis post aliam illius perfectionis ad quam intendit motus, donec perficiatur et fit in actu. Unde necesse est ut motus qui est in substantia inveniatur in genere substantiae, et motus qui est in quantitate in genere quantitatis et similiter qui est in ubi et qualitate. Secundum autem quod est via ad perfectionem, quae est alia ab ipsa perfectione, necesse est ut sit genus per se. Via enim ad rem est aliud ab ipsa re, et secundum hoc fuit positum praedicamentum per se. Et iste modus est famosior, ille autem est verior. Et ideo Aristoteles induxit illum modum famosum in *Praedicamentis* et istum modum verum in hoc libro'.

of his discussion of motion is framed by Averroes. Ockham's view is in line with his general philosophical stance and involves a reduction of motion to the moving thing. He presents his arguments in the section on motion of the *Tractatus de successivis*.[14]

Ockham divides all change into immediate change (*mutatio subita*) and successive change (*motus successivus*). In the first way, a form is acquired or lost in an instant as part of a generation or corruption; the second way, he points out, is motion in its proper sense (*motus stricte*) and includes augmentation, diminution, alteration, and local motion. In neither of these ways is motion something distinct from the thing undergoing the motion.

If motion is real, that is, something distinct from the thing moving, it would have to be a successive being, which Ockham thinks is impossible, since such a being would have to have temporal parts. Ockham, for example, writes that "a successive [being] never exists without having one part in the past and another part in the future, and when one part of the successive [being] exists, the successive [being] exists."[15] Such entities cannot exist, Ockham thinks, since they have parts that do not exist (past and future ones). His rejection of motion as something above the moving thing rests on his rejection of successive beings.

The reason previous thinkers thought motion needed to be a successive entity was that it could not be a permanent entity. A substance undergoing a change from black to white will have to have all its parts as a permanent entity while undergoing alteration. For the reason stated above, Ockham thought this was a ridiculous and self-defeating view.

In an argument using his famous razor, he shows that all that is needed is the permanent entity, first in one state and then in another. He writes:

> And proceeding in this way with regard to other [places] it is clearly evident that besides a body and the parts of place and other permanent things, it is not necessary to posit another thing [to explain the locomotion of the body]. It is only necessary to posit that the body is at a particular time in this place ... and at another time it is not.[16]

---

14 *Tractatus de successivis* (ch. iv, n. 33) 32–69. This treatise seems to be an excerpt from his much larger and incomplete work called *Exposition of the Physics*.

15 William Ockham (1984), *In Phys.* III.2.6 (Opera phil. IV: 431).

16 *Tractatus de successivis*, 46. Translation by Blake D. Dutton in "Nicholas of Autrecourt and William of Ockham on Atomism, Nominalism, and the Ontology of Motion." *Medieval Philosophy and Theology* 5, 1996, 63–85 at 73.

It is simply superfluous to posit anything above the substance and the place in which it exists. As is true of many philosophical discussions in the fourteenth century, Ockham's position was the starting point and set the tone for the rest of the century and even centuries to come.

The other philosopher from this time providing a background to Blasius' treatment of motion is John Buridan. His most famous discussion of motion is in Book III, q. 7 of his commentary on Aristotle's *Physics*. He there rejects Ockham's argument presented above. His reasoning is as intriguing as it is famous.[17] He imagines a local motion of the outermost heavenly sphere. This sphere is not in any place, since it lacks a surface of the surrounding body (place is defined by Aristotle as a surrounding surface). God can, however, Buridan thinks, move the entire universe in a straight line and also, consequently, rotate it in its entirety. This would be a motion, but it cannot be accounted for in terms of the thing moving (the universe) and a place, since the universe is not in any place. Motion would, hence, have to be explained in some other way than the one given by Ockham. For Buridan, this is by succession; local motion, therefore, is a successive being, holds Buridan.[18]

Buridan is perhaps most famous for another aspect related to motion and that is his introduction (even though he was not the first one to do this) of an impressed force (impetus) to explain projectile motion. In Book VIII, q. 8 of the same commentary, he asks whether a projectile after leaving the hand of the projector is moved by the air or by something else. It is Aristotle's position that it is moved by the air, but Buridan rejects this and replaces it with his theory of impetus. He holds that only an internal moving force, transmitted from the mover to the moved, could explain the motion of a projectile. The idea has been seen as similar to what we know as momentum in modern physics.

---

17 However, it is not really 'his' argument, since it is already in Scotus (John Duns Scotus, *Quaestiones quodlibetales*, in *Opera omnia*, vol. 25, Paris, Louis Vivès, 1895, q. 11, 441–442), who himself follows the 1277 Parisian condemnation, which states this possibility explicitly. Buridan here uses it against Ockham.

18 John Buridan, *Quaestiones super octo libros Physicorum Aristotelis (secundum ultimam lecturam)*, eds. Michiel Streijger and Paul J.J.M. Bakker (Brill: Leiden, 2016), III, q. 7, 79: "Sexta conclusio est quod motus ultimae sphaerae est res pure successiva, cuius scilicet est pars prior et pars posterior non manentes simul, quia si esset res permanentis naturae, tunc secundum illam ultima sphaera non se haberet aliter et aliter prius et posterius plus quam secundum eius magnitudinem vel figuram vel alia eius accidentia permanentia, quod est falsum." Buridan never actually says in this question that motion is on a par with a quality, but it is often interpreted as that. See, for example, Thijssen, "The Debate over the Nature of Motion", 197. Buridan seems to indicate that motion is like 'white' and that seems to be a quality. See John Buridan, *Quaestiones super octo libros Physicorum Aristotelis (secundum ultimam lecturam)*, 79.

An impetus is still a quality, for Buridan, which makes his view different for example from Galileo's, and impetus is still mainly viewed as an explanation of projectile motion and not motion in general, as suggested later by Blasius.[19] There are, however, many other aspects of impetus that are more in line with later, modern, accounts of motion. In his *Metaphysics*-commentary (Book XII, q. 9), Buridan suggests that an impetus does not lose power on its own, that is, if it were not for the counteraction of the air's resistance, an impetus would continue on infinitely. He also notes, in the *Physics* this time, that the strength of the impetus is dependent on the speed and quantity of the matter of the moving object. This explains for him why we can throw a stone farther than a feather. Buridan sets the discussion of motion on a new track that is picked up by Blasius and formed into something that does indeed look like a proto-Galilean account.

4 Blasius of Parma on Motion in the *Physics*

Blasius of Parma's theory of motion represents a significant departure from earlier theories of motion. Despite employing an Aristotelian terminological framework similar to that of Thomas Aquinas, William of Ockham, or John Buridan, to name but a few of his most eminent medieval predecessors, Blasius reaches conclusions about motion that set him apart from them.

Over the course of his career, Blasius reflected on nearly all the questions raised in his time about motion and its different types (local motion, alteration, quantitative change), ranging from the general definition of motion to precise queries about particular types of motions, including a mathematical analysis of the rules of motion.[20] In what follows, we shall focus on what arguably con-

19 However, it should be mentioned that Buridan, in the *Physics*, applies his impetus hypothesis to the motion of the heavens and to natural motion as well, showing that it allows for an explanation of acceleration in free fall (natural) motion.

20 On Blasius' theory of local motion, see N. Weill-Parot, "Une position originale? Le contact entre le moteur et le mû dans les *Questiones disputatae super octo libros Physicorum* de Blaise Parme," in J. Biard and A. Robert (eds.), *La Philosophie de Blaise de Parme. Physique, psychologie, éthique* (Florence: SISMEL—Edizioni del Galluzzo, 2019), 197–217; S. Rommevaux, "Les règles du mouvement de Blaise de Parme dans les deux versions de ses '*Questions sur le traité des rapports de Thomas Bradwardine*'," in J. Biard and S. Rommevaux (eds.), *Mathématiques et théorie du mouvement, XIVe–XVIe siècles* (Villeneuve d'Ascq: Presses Universitaires du Septentrion, 2008), 31–52. A. Harrison, "Blasius of Parma's Critique of Bradwardine's *Tractatus de proportionibus*," in A. Poppi (ed.), *Scienza e filosofia all'Università di Padova nel Quattrocento* (Padua / Trieste: Edizioni Lint, 1983), 19–69. On qualitative motion, see J. Biard, "Blasius of Parma on the Calculation of the Variation of

stitutes the most radical departure by Blasius from the Aristotelian concept of motion, namely his peculiar way of defining the natural or violent character of local motion. In short, Blasius characterizes the concepts of natural and violent motions so that, strictly speaking, *all* local motions that happen beneath the sphere of the moon (the sublunary world) are considered to be violent, in complete contraposition to the standard way of describing motion as found in Aristotle and the earlier medieval scholastic tradition.

To understand this point, let us follow the analysis of the natural or violent (non-natural) character of motions Blasius offers in his second question on the second book of Aristotle's *Physics*, which enquires whether all natural beings possess an intrinsic principle of motion and rest.[21] Blasius' standpoint on the different types of motion is grounded on his particular interpretation of what a 'principle of motion and rest' is, since the concept of nature (hence, of a 'natural motion') in the Aristotelian tradition is based on the possession of an internal principle of motion and rest.[22] At first glance, Blasius seems to be completely faithful to this way of defining nature. He writes:

> Every natural body whose form is educed from matter's potency has in itself an active principle of motion and rest. [This] conclusion is obvious, because any such natural body belongs to the genus of heavy or light things; hence the conclusion is true. And I say in the conclusion 'whose form is educed from matter's potency', because heaven does not have in itself a principle of motion by which it moves.[23]

In this passage, the exclusion of the heavens from the class of beings possessing an internal principle of motion is worth noting, but it can only be understood from the criterion chosen by Blasius to have such a principle, namely being

Qualities and Aristotelian Physics," in D.A. Di Liscia and E.D. Sylla (eds.), *Quantifying Aristotle. The Impact, Spread, and Decline of the 'Calculatores' Tradition* (Leiden / Boston: Brill, 2022), 232–254; J. Biard, "L'être et la mesure dans l'intension et la rémission des formes (Jean Buridan, Blaise de Parme)," *Medioevo* 27, 2002, 415–448; G. Federici Vescovini, "La '*Quaestio de intensione et remissione formarum*' di Biagio Pelacani da Parma," *Physis* 31, 1994, 433–535.

21 Text 2 (II, q. 2) in the present volume.

22 Aristotle, *Physics* II.1, 192b13–15; 192b21–23.

23 See text 2 (II, q. 2) in the present volume, n. 35. Let us note that this text is included in the 'first determination' that Blasius puts forward in this question. He later in the same question contrasts two alternative ways of addressing the issue of the existence of an internal principle of motion and rest in natural beings. Therefore, it is important to keep in mind that this passage does not necessarily reflect Blasius' ultimate view on the matter.

light or heavy. According to Blasius, only these qualities, which he calls motive qualities (in contrast with altering qualities such as heat and coldness), account for the fact that beings may be said to have a principle of motion and rest. As apparent from his understanding of weights, Blasius thinks that the opposite qualities of lightness and heaviness are the two causes that suffice to explain the motion of sublunary corporeal substances, as far as local motion is concerned.

In itself, this claim does not carry any particular originality, since the characterization of heaviness and lightness as motive qualities was common by Blasius' time. The originality of Blasius lies rather in his belief that 'being light' and 'being heavy' are relative properties such that a body placed in a medium of the same nature as its own is neither heavy nor light, such as when some part of air reaches its natural place, that is, the sphere of air located above the sphere of water and below that of fire.[24]

The identification of lightness and heaviness as principles of motion and rest, combined with the view that bodies are neither heavy nor light when located in their proper place, should logically lead Blasius to the view that a body does not have a principle of motion and rest when it is located in its natural place.

As will emerge shortly, such a conclusion is in some sense accepted by Blasius. But his peculiar view on the status of natural and violent motions is not directly established, in his second question on Aristotle's second book of the *Physics*, by appealing to the relative character of lightness and heaviness, although his analyses of this issue may be viewed as complementing the development of this question. Indeed, in this second question on Book II, Blasius' approach consists in more thoroughly specifying the constellation of terms connected to the notion of 'principle', namely terms such as 'inclination' or 'appetite.'

Blasius explains that "for something to have an intrinsic principle of motion is for it to be naturally inclined to move."[25] A principle of motion is therefore nothing else than an inclination. An inclination refers to the same thing as the 'appetite' a thing has for moving or behaving in a certain way: "a natural appetite is neither different nor anything else than a natural inclination for a thing not present."[26]

---

24 On this point, see the study "Scales and Weights. Introducing Mechanics in Natural Philosophy" in this volume.

25 See text 2 (II, q. 2) in the present volume, n. 58 (3a det.).

26 See text 6 (IV, q. 3) in the present volume, n. 14.

What is an inclination? According to Blasius, an inclination refers to the same thing as having a desire (*desiderare*) or an appetite (*appetere*), two terms borrowed from the sphere of psychological experiences but then employed for describing natural processes in the Aristotelian tradition ever since Aristotle himself.[27]

In the same way as the desire of matter to be united with a form is identical with the matter, since it is the privation of form, the appetite, desire, or inclination of a body to move in a certain direction is identical with the moving quality inhering in it. In his *Questions on the Physics*, John Buridan, one of the main sources of Blasius' own questions on the same text, identified the inclination—or desire—of a body to move toward a certain place as its active principle, which he thought corresponds to either the substantial form of a body (say, the form of a stone) or one of its qualities (e.g., heaviness).[28] Blasius' way of reducing 'desires' or 'appetites' to a movement in a particular direction is close to Buridan's. The main difference between their views is that Blasius unequivocally sees the motive quality (heaviness or lightness) of a body and not its substantial form as the active principle of this body.

However, from here, the resemblances end between Blasius and Buridan or to any of the authors previously mentioned. Indeed, unlike them, Blasius draws conclusions, from this characterization of inclinations as active principles, that are in complete opposition to the standard Aristotelian conception of motion. Blasius thinks that one should carefully distinguish, when talking about an 'active principle of motion and rest,' between the two terms included as the objects of the active principle or inclination. Strictly speaking, rest is the end term of this inclination, whereas motion only constitutes a *means* to attain a state of rest. It follows from this observation that only rest constitutes a natural state for a body. Why is that so? Because, according to Blasius, who relies here on a principle originating in Averroes, all bodies aim at conserving themselves: "every natural being desires (*appetit*) to persist and to be conserved in being."[29] The state of rest constitutes the ideal expression and manifestation of every natural

---

27 Aristotle describes matter as desiring form (*Physics*, I.9, 192a20–23), and the elements as moving toward their proper place in virtue of a desire to reach them (*On the Heavens* IV.3–4).

28 John Buridan, *Quaestiones super libros Physicorum* (*secundum ultimam lecturam*). *Libri I–II*, eds. M. Streijger and P.J.J.M. Bakker (Leiden / Boston: Brill, 2015), II, q. 2, 253, ll. 1–6.

29 See text 2 (II, q. 2) in the present volume, n. 59. This principle, applicable to every being (*omne ens* …), becomes important in the fourteenth century in debates over the tendency of natural beings for self-conservation. On this, see A. Robert, "Désir de persévérer dans l'être et mort volontaire chez Nicole Oresme," *Quaderni di Noctua* 5, 2019, 199–239.

being's inclination to persevere in being. By contrast, motion only represents a way to reach this state:

> The following argument relying on the motion of heavy things is not valid: 'A heavy thing acquires its natural place downward; therefore such a motion is natural.' Nor is this one: 'A heavy body desires to be in a natural place; therefore it desires to move toward its natural place.' Because in the same way it could be proved that 'you willingly take rhubarb or turbith, because you desire health; therefore you desire to take rhubarb' and so on. Who naturally desires an end, indeed, does not naturally desire the requisites or means by which the end is acquired. And it is thus written that "⟨a cat⟩ wants fish, not to fish."[30]

As can be seen from this passage, Blasius' move consists in interpreting the intentional scope of the term 'desire' in a special way. It is true that a particular body desires to be in a precise place. Being proper to the type of thing it is, this desire is natural to it. The place toward which the body moves may also be said to be 'natural,' the term 'natural' being predicated of a place because it has a conservative effect on the bodies that tend toward this place.[31] But it does not follow from the characterization of the body's inclination and the place toward which it is directed as natural that the body naturally desires to move. A desire, indeed, aims at an end and not at the means to this end.

A consequence of this precision is that, strictly speaking, it is not in the nature of a body to move locally. Indeed, what is natural to a body is its intrinsic tendency or appetite to be in a certain place, this 'appetite' being a term denoting the moving quality inhering in this body. This appetite, despite its particular manifestation depending on the type of body (moving downward for heavy bodies; upward for light bodies), reflects a universal tendency to remain in its own state.[32] However, whereas this appetite is in a sense natural, motion understood as a process of acquiring a proper place is not. Blasius deduces from these considerations the conclusion—eminently paradoxical for an Aristotelian—that all sublunary local motions, that is, all motions occurring in bodies possessing an intrinsic principle of motion and rest, are *violent* motions:

---

30 See text 2 (II, q. 2) in the present volume, n. 70.

31 See text 5 (IV, q. 2) in the present volume, n. 24, where Blasius relies on Aristotle, *Physics* IV.3 210b34–211a1.

32 See text 2 (II, q. 2) in the present volume, n. 59; IV, q. 2, n. 27.

> Nevertheless, what has been said may be said of local motion in a similar way, namely that every sublunary natural body naturally desires to rest. This is proved: Every natural body desires to be in its natural place in order to be preserved and to rest in it. And it follows consequently that no sublunary natural body desires to move locally. This is clear, because every natural appetite is determined in such a way that it is not toward the opposite. Indeed, a natural appetite is not indifferent toward contraries [*idem contrariorum*], like a rational power is with respect to opposites. Now, it is at least clear that by natural appetite every natural being desires to rest, therefore not to move locally. And it follows consequently that every sublunary local motion is violent. This is obvious since it is against the inclination of the natural body.[33]

The affirmation that all sublunary local motions are violent is definitely radical. Reaction should be toned down, however, by the following three remarks.

First, it is important to note that Blasius does not claim that every sublunary motion is violent. It is clear that his reasoning applies here only to local motion. No similar conclusion is deduced for other types of motion, in particular for qualitative motions occurring due to the action of heat or coldness.

Second, it is important to note that this conclusion is located in one of the three possible answers to the general question raised by Blasius on the principle of motion and rest present in natural beings (II, q. 2), which Blasius calls three possible 'determinations' of the question. This conclusion is placed after the third and last determination. This being said, the conclusion does not appear to depend on the content of this particular determination and could have been placed elsewhere. Indeed, it essentially follows Blasius' peculiar analysis of the terms 'inclination' and 'appetite' that is common to all three of the possible determinations.

Third, it is important to note that Blasius does not always stick with his conclusion about the violent character of local motion, but often qualifies instances of sublunary local motions as natural.[34] There is a tension in Blasius' account of motion. He acknowledges the Aristotelian criterion of something having an internal principle of motion for being a natural motion, which explains that he often calls sublunary local motions 'natural.'[35] But at the same time, he clearly does not take for granted that such motions are natural.[36]

---

33 See text 2 (II, q. 2) in the present volume, n. 69.

34 See text 2 (II, q. 2) in the present volume, nn. 33–34; II, q. 2, n. 79. Cf. text 6 (IV, q. 3), n. 17.

35 See, e.g., text 6 (IV, q. 3) in the present volume, n. 15.

36 For instance, presenting an argument on the motion of bodies toward their natural place,

In the second question of Book IV of the *Physics*, he reiterates the analysis delivered in Book II (q. 2):

> Second conclusion and response: Every natural being existing out of its natural place desires to move toward its natural place. [...]
>
> Third conclusion: Every natural being existing beneath the heavens naturally desires to rest. This is obvious because every such ⟨being⟩ desires to be in its natural place and not to be removed from it; therefore, it desires to rest.
>
> Fourth conclusion: No natural body existing beneath the heavens desires to move absolutely [*simpliciter*] ⟨speaking⟩. This is obvious because every such ⟨being⟩ naturally desires to rest; therefore, etc. The argument holds because moving and resting are opposites.[37]

Should we then conclude that Blasius is inconsistent whenever he calls sublunary local motions 'natural'? Possibly. At any rate, we must keep in mind that Blasius' peculiar way of interpreting the non-naturalness of certain motions is at odds with his way of defining natural and violent motions, which remains standardly Aristotelian. As recalled above, his anti-Aristotelian stance regarding the status of lightness and heaviness may also help understand his particular conception of local motion; but we also should keep in mind that Blasius was well aware of the critical reception of his ideas by his peers, which already earned him a condemnation by the time he was writing the second version of his *Questions on the Physics*. It cannot be excluded that, while proposing obviously innovative ideas, Blasius also took care to conform to more orthodox theses or, at least, to formulate these ideas in an acceptable way. This may explain the apparent contradictions we find in Blasius' description of sublunary local motions as sometimes natural and sometimes violent.

In the second question on Book II, Blasius tries to justify further the description of sublunary local motion as violent. As we have explained, the appetite of a body for a certain place is its inclination toward it. Ontologically speaking, this inclination is nothing other than the moving quality that causes the body's motion, that is, heaviness or lightness. But what is it that ultimately accounts for the tendency to move to a particular place (for a heavy body, for instance,

---

he notes: "And this follows from the fact that you think that natural beings move naturally toward their natural place" (IV, q. 2, n. 7).

37 See text 5 (IV, q. 2) in the present volume, nn. 28–30.

to move downward)? What type of causation, if any, is exercised by the natural place itself? Blasius explains this fact in the following way:

> If you ask: By what then is a heavy body moved when it is out of its natural place and is not restrained ⟨there⟩? I say that it is moved in an inchoative way [*inchoative*] by the place. The natural place exerts indeed an attracting power on the located thing, like the magnet on iron. ⟨It is moved⟩ successively, however, by the place and the impulse [*impetus*] acquired through motion with the assistance of the higher Intelligences, about which the Commentator wrote in the twelfth ⟨book⟩ of the *Metaphysics*, "each natural being is directed toward a natural place by an Intelligence in which there exists no error."[38]

Blasius in this passage compares the attraction exercised by a place on a body to a magnet's action. In a sense, a place attracts a body in virtue of the quality (heaviness or lightness) this body possesses. One may wonder why this analogy appears in this particular context. The answer is given by Blasius himself a few paragraphs later, where it appears clear that, rather than reinjecting some type of teleology into his account of motion, the analogy of a magnet is precisely meant to offer a non-teleological explanation of motion toward a place, by emphasizing the *violent* character of attraction:

> Second difficulty: Is iron moved by a magnet in a natural way? I answer negatively. This is proved, because if iron were moved by a magnet in a natural way, the iron, no matter how large, would be moved toward the magnet, which is false. For it appears that a small piece of iron is moved toward the magnet. And when a large piece of iron is presented, it does not move, whereas it should move faster. And this is so only because the motion of the iron toward the magnet is nothing but a motion of traction, in such a way that the magnet attracting some small piece of iron cannot draw to itself anything larger.[39]

Blasius adds that a magnet is not "exhausted" by exerting an attracting force on iron, just like the internal organs of an animal are not exhausted when they attract nutrition, as when a person feeds herself.[40] The model of the magnet for

---

38 See text 2 (II, q. 2) in the present volume, n. 70.
39 See text 2 (II, q. 2) in the present volume, nn. 75–76.
40 See text 2 (II, q. 2) in the present volume, n. 77.

illustrating motion is wrongly called 'natural' and this, hence, justifies the violent character that Blasius attributes to motions in the sublunary world. At the same time, it also justifies the constancy of such motions, which do not vary across time.

As a matter of fact, this passage is far from being the only place where Blasius appeals to magnets to analyze natural phenomena.[41] This is the case in particular in his *Questiones super Tractatum de ponderibus* where Blasius defends the view that the motion of iron toward a magnet is *natural*. There, he describes this motion by saying that it should not be viewed as a motion of traction, since a motion of traction is not natural, but violent.[42] By contrast, in the second version of his questions on the *Physics*, Blasius envisions a special theory of magnets—a topic about which he thus seems to have oscillated—that serves to ground a novel account of the naturalness and non-naturalness of sublunary motions.

Blasius' heterodox views on motions derive from a particular understanding of a definition of natural motion coming from Aristotle that he likes to quote, namely that a natural motion, by contrast with a violent one, is due to an intrinsic principle conferring force to the patient.

Blasius' views on motion tend to blur the distinction between natural and violent motions.[43] As Federici Viscovini has shown, Blasius seems to read this definition as implying that whenever a motion comes from an intrinsic force moving an object, it is natural. On this basis, she argued that Blasius was led to conclusions completely opposite both from the classical Aristotelian understanding of motion and from the new theories of motion inherited from the fourteenth-century Parisian tradition.[44] Indeed, it follows from this under-

41 Cf. also text 2 (II, q. 2), n. 7; text 5 (IV, q. 2), n. 13; and VII, q. 9, Vat. lat. 2159, f. 207rb–vb^; cf. *Questiones circa Tractatum proportionum magistri Thome Braduardini*, eds. J. Biard and S. Rommevaux (Paris: Vrin, 2005), q. 11, 171–172. Cf. *Questiones super Tractatum de Ponderibus*, ed. J.E. Brown in ead., *The* Scientia de ponderibus *in the Later Middle Ages*, Ph.D. Dissertation (Madison: University of Wisconsin, 1968), 353–354, 365–370. On this topic, see S. Rommevaux, "Magnetism and Bradwardine's Rule of Motion in Fourteenth- and Fifteenth-Century Treatises," *Early Science and Medicine* 15, 2010, 618–647.

42 Cf. *Questiones super Tractatum de ponderibus*, 353–354, 365–370. Cf. text 2 (II, q. 2) in the present volume, n. 7.

43 Aristotle, *Nicomachean Ethics* III.2, 1110b15–6. See text 6 (IV, q. 3) in the present volume, n. 15. Blasius already employed the same definition in the first redaction of his questions on the *Physics*. Blasius of Parma, *Quaestiones super VIII libros Physicorum* (*redactio prima*), II, q. 2, ms. Vat. Lat., Chig. O IV 41, 269vb: Motus localis est duplex: quidam est motus naturalis et est ille motus qui fit a principio intrinseco passo vim conferente, et quidam est motus violentus et iste est a principio extrinseco passo vim non conferente.

44 Federici Vescovini, *Astrologia e scienza*, 281.

standing that a motion resulting from an impetus, defined by Parisian authors as a violent motion, is a natural one. On the contrary, all motions due to some external principle—i.e., the paradigmatic case of natural motion acknowledged in the 'Aristotelian' model according to which everything that moves is moved by something else—are violent motions.

In the second redaction of his *Questions on the Physics*, Blasius does not reject the Parisian theory of impetus. He merely opposes two different ways of explaining the motion of projectiles, one corresponding to the Aristotelian model, the other corresponding to the impetus theory.[45] But it is true that his views on local motion lead him to reassess the impetus theory as a way merely to explain violent motions in contrast with natural ones, since these two types of motions do not oppose each other as in the orthodox interpretations of Aristotle's theory of motion. In a motion where a force is transmitted from an external agent to a mobile one, the motion may be described as violent at the initial stage when the external agent transmits a force or impetus to the object, but once the object is set in motion, this motion is not violent any longer, given that the force operates as an intrinsic principle of motion. One and the same motion may be described, indifferently, as natural or violent. Thus, in his questions on the *De caelo*, Blasius asserts that "every violent motion is natural."[46]

Having explained Blasius' motivation for his unorthodox theory of local motion, what sense can we make of his radical propositions? Blasius' innovative ideas about local motion can only be understood in the context of his more general reflection about self-conservation in nature. The appetite proper to natural bodies to remain in a state of rest is treated as a part of the more general problem of understanding how bodies tend to remain in the same state. In the context of this more general problem, the case of local motion is just a particular aspect of the issue at stake, since the problem also concerns the qualitative dispositions of bodies and their tendency toward change or stability. In other words, Blasius also seeks in this context to understand whether bodies insofar as their internal qualities are concerned are subject to intrinsic change, or if qualitative changes only come about due to external factors. A brief look at what he says about this problem will help us achieve a more complete understanding of his general account of motion.

---

45 Blasius of Parma, *Quaestiones super VIII libros Physicorum (redactio secunda)*, VIII, q. 10 ('Utrum per rationes naturales demonstrari possit a quo proiecta moveantur contra inclinationes naturales cum ab impellente recesserunt'), ms. Vat. Lat. 2159, 222ra–223rb.

46 Blasius of Parma, *Quaestiones de caelo*, I, q. 4, ms. Milan, Biblioteca Ambrosiana, P 120 sup., 6rb–7va: "omnis motus violentus est naturalis", cited in Federici Vescovini, *Astrologia e scienza*, 290. Let us note, however, that Blasius' perspective in this text is different. There, he seems rather to envision a reduction of different types of motion to natural motion, since motion is never distinct from the natural things that move. In both contexts, however,

Unlike the case of local motion, where Blasius provides an unequivocal answer regarding the appetite to remain in a state of rest proper to bodies, he appears quite hesitant about how to analyze the qualitative structure of bodies and their intrinsic equilibrium. A clear sign of this hesitation can be seen in the fact that he gives not only two possible 'determinations' (that is, possible answers) for this problem, as is often the case when he decides to leave open a particular question, but three of them.

In the first determination of the question regarding the tendency of natural bodies to remain in a certain state, Blasius presents several conclusions in favor of the view that simple bodies (that is, the elements) would always remain in the same place if they were not affected by external actions or if they were placed in a void.[47] According to this first way of answering the problem, however, mixtures always tend by themselves to non-being, because they are composed of different elements whose intrinsic tendency is to oppose each other.[48] Still, in all mixtures, a substantial form exists whose aim is to maintain its own being through the cohesion of the body's parts. According to this first determination, in other words, a natural body constitutes a dynamic equilibrium among the different elements which, taken in themselves, are perpetually stable, but whose interaction generates a disaggregative effect on the body's structure. The persistence of the mixture is made possible thanks to the unifying effect of the substantial form which, however, can only ensure the persistence of the body for a certain period of time.

Blasius' second determination of the problem offers another ontological analysis of the structure of natural bodies, one which sees all bodies, be they simple or mixed, as tending intrinsically toward non-being. The key thesis of this alternative way of answering the question of the persistence of bodies lies in the claim that all bodies (even simple elements) always contain some heat and wetness, without which they could not exist as bodies. However, insofar as the effect of fire is to 'consume' wetness, all bodies are, so to speak, doomed to self-corruption, since their internal wetness is destined to be destroyed by their internal heat. In this view, it is impossible for any body to last eternally.[49]

Instead of explicitly deciding between these two competing perspectives on natural bodies, Blasius puts forward a third option, which appears to be closer to the first determination, but goes even further. It consists in affirming that, naturally, no body—be it simple or mixed—possess any intrinsic tendency to deviate from its own nature, and therefore, no tendency toward non-being:

Blasius tends to minimize the distinction between natural and violent motions (at least with regard to local motion).

47 See text 2 (II, q. 2) in the present volume, n. 29.

48 See text 2 (II, q. 2) in the present volume, n. 42.

49 See text 2 (II, q. 2) in the present volume, nn. 53–55.

> No natural being, be it simple or mixed, has in itself the active principle of its alteration toward a disposition contrary to its nature [*sibi disconvenientem*]. [...]
>
> Second conclusion: No natural body, be it simple or mixed, has originally in itself the active principle of its corruption.[50]

Unlike the first determination, this third position claims that elementary bodies tend to remain in their own state, so that they would never be corrupted if all external influences were removed, but it extends this claim to mixtures as well.

An important consequence of this view is that, if it is true of all mixtures, then it is impossible for a mixture to tend by itself toward its own corruption. In the case of animated mixtures, this point amounts to denying the possibility of suicide, that is, of a thing willing its own death (with the notable exception of human beings, whose free will can override this natural tendency).[51] This also leads Blasius to the conclusion that, considered from the point of view of a living body, death always comes from the outside, being necessarily due to external circumstances and never to any intrinsic tendency of living things:

> Here it is replied that there exists no natural death, where by 'natural death' you exclude all extrinsic things actively concurring to the alteration contrary to a thing's nature [*sibi disconvenientem*]. First, indeed, with respect to place, which is never such that it does not alter it [i.e., the thing], and with respect to constellations and time. The Philosopher says indeed about time in the fourth [book of the *Physics*] that it is 'the cause of corruption in itself and of all evil.' The stars have been discussed by astronomers who want to reduce such effects to universal causes on account of the stars.[52]

This insistence on the extrinsic character of death, and more generally corruption, confirms the priority given by Blasius to rest rather than motion in the definition of nature. Just as motion in the sublunary world is extrinsic to

50 See text 2 (II, q. 2) in the present volume, nn. 61–65.

51 On some aspects of this problem, see A. Robert, "L'éthique de Blaise de Parme dans ses *Questions sur la Physique*," in J. Biard and A. Robert (eds.), *La Philosophie de Blaise de Parme. Physique, psychologie, éthique* (Florence: SISMEL—Edizioni del Galluzzo, 2019), 33–58.

52 See text 2 (II, q. 2) in the present volume, n. 68.

a thing's nature, which only desires rest, death does not follow from a thing's nature. From this point of view, corresponding to Blasius' third determination, the intrinsic tendencies of bodies to remain in the same state are in line with Blasius' view that rest constitutes the only natural state of bodies, strictly speaking.

5 Concluding Remarks

Ultimately, the main conclusions reached by Blasius in this question (II, q. 2) derive from the principle according to which all things seek to persist and remain in the same state. The violent character of local motion and the external corruption of bodies represent two consequences of this same law, which manifests itself in the case of animated beings with a perpetual desire to live.[53]

Admittedly, Blasius was not the first to see the desire for life as a manifestation of a universal tendency of things to persist in existence and to always remain in the same state. By contrast, he was certainly original in drawing the conclusion that all (local) motions taking place in the natural world may be violent motions, marking a radical departure from the traditional conception of motion, paving the way for a deep reevaluation of the conceptual framework employed for understanding nature.

53 Whether some of the ideas of Blasius should be viewed as an anticipation of inertia has been discussed by V.P. Zoubov, "Blaise de Parme et le principe d'inertie," *Physis* 5, 1963, 39–42.

CHAPTER 4

Rarefaction and Condensation in Blasius' *Physics*

Erik Åkerlund

1 Introduction

The case of rarefaction and condensation, and the question of how these changes are possible, presents an interesting test case for notions such as matter, place, quantity, and extension.

When a void or a vacuum is allowed in one's natural philosophy, condensation and rarefaction seem quite straightforward phenomena to explain. Condensation occurs when more matter occupies a given space, i.e., there is less void in that space; rarefaction is the reverse process. However, given a plenum theory of matter, with vacuum considered as contrary to nature, the phenomena of condensation and rarefaction are not as straightforward to explain. As Dennis Des Chene describes the problem for those who deny the possibility of a vacuum:

> Since a vacuum was thought to be impossible, rarefaction could not consist in the dispersion of particles into a larger, otherwise empty space, or in the enlargement of vacuous pores in a spongelike connected mass.[1]

Solutions differed for different thinkers.

Below, we shall first briefly examine some accounts of condensation and rarefaction given before Blasius, in Buridan, Ockham, and Crathorn. Then Blasius's account will be presented, based mainly on his treatment in his seventh question on the fourth book of Aristotle's *Physics*.[2] As will be shown, Blasius's treatment focuses on the case of simple bodies or elements, where the extension of the sublunary world sets some limits to the possibility of condensation and rarefaction. Generally, Blasius synthesizes from different thinkers ways of dealing with the problem, although he seems to come closest to Ockham's answer that ultimately density has to do with the closeness of the parts of something "all the

1 D. Des Chene, *Physiologia. Natural Philosophy in Late Aristotelian and Cartesian Thought* (Ithaca, NY: Cornell University Press, 1996), 107. Des Chene treats the question of condensation and rarefaction on 107–109.

2 Text 7 in the present volume.

 | DOI:10.1163/9789004734968_006

way down." In his treatment, he also refers to some breathtaking thought experiments, especially in countering a proposed solution to the problem found in, among others, William Crathorn.

2 The Background to Blasius' Account of Rarefaction and Condensation

By the time of Blasius, in the late fourteenth century, there were many different attempts to account for how condensation and rarefaction are possible.[3] The two main lines are represented by Buridan and Ockham.[4]

Buridan held the common view that rarefaction means that a given quantity of matter acquires a larger quantity attached to it, and the opposite for condensation. As he writes:

> Dimension [≈ quantity] yields something extended just as heat yields something hot, or light something luminous. So it seems reasonable that if more heat or light yields a subject that is more hot or luminous, and less heat or light yields one that is less hot or luminous, so too more dimension yields something more extended and less yields less. So just as more quality yields a more intensely qualified thing, so more dimension yields a more extended thing.[5]

One feature of Buridan's theory is that he viewed quantity as "additive," similar to the way that luminosity was usually understood. Hence, when something acquires a larger quantity, this is not a case of one quantity being replaced by another; rather, the original quantity is increased by more quantity in a case of rarefaction, and vice versa for a case of condensation.[6]

---

3 For a good overview, see R. Pasnau, *Metaphysical Themes 1274–1671* (Oxford, UK: Oxford University Press, 2011), ch. 15.1 ("Condensation and Rarefaction: Buridan," 301–308). The period is also treated in X. Wang, "Rarefaction and Condensation," in M. Sgarbi (ed.), *Encyclopedia of Renaissance Philosophy* (Springer eBook, 2022), 2764–2770.

4 Buridan's and Ockham's respective understandings of rarefaction and condensation, in relation to their understanding of quantity, have recently been extensively treated in F. Bruno, *The Quantity Debate in Late Medieval and Early Modern Europe*, PhD diss. (Cornell University, 2019). (Oresme is also treated in this work.) See also M. Roques, "Blaise de Parme et la Quantité," in J. Biard and A. Robert (eds.), *La Philosophie de Blaise de Parme: Physique, psychologie, éthique* (Firenze: Sismel—Edizioni del Galluzzo, 2019), 175–196.

5 *In Phys.* IV.11, f. 78rb. Translation from Pasnau, *Metaphysical Themes*, 306.

6 For Buridan's position, see Pasnau, *Metaphysical Themes*, 304–308. Buridan's position is

For Ockham, the explanation of condensation and rarefaction was closely connected to his rejection of quantity as real. Instead of explicating condensation and rarefaction as a certain quantity of matter changing to a different quantity, Ockham explained that in condensation the parts of the body come closer together. As he writes:

> [M]atter is made to have a greater or lesser quantity not through its receiving any absolute accident, but through condensation and rarefaction alone. This is nothing other than for the parts of matter to come more or less close to each other, which can happen through local motion alone with respect to those parts—that is, through the parts of matter being dilated and contracted.[7]

It is important to remember, though, that Ockham denied the existence of vacuum, so there is no free space into which these parts move. Instead, in order for two parts to come closer together, all the parts in between also have to come closer together, "all the way down,"[8] that is, throughout.

Such were the two main positions before Blasius (although there were of course many variants within these different lines of explanation). One further position that may be interesting to mention, though, is the "eccentric" position of William Crathorn.[9] According to Crathorn, condensation can be accounted for by the "co-location" of matter. In other words, when something condenses, some parts of the matter of this thing become co-located, that is occupy the same space. In this way, the same amount of matter can take up less space. As will be seen, Blasius also refers to this position, or some variant of it, although only *en passant*, rejecting it offhand, albeit with a quite surprising counter-example.

---

treated, in relation to Ockham's, in J.M.M.H. Thijssen, "The Buridan School Reassessed. John Buridan and Albert of Saxony," *Vivarium* 42(1), 2004, 18–42, especially 30–38.

7 *Summula* I.13, *Opera phil.* VI: 194. Translation from Pasnau, *Metaphysical Themes*, 303.

8 For Ockham's position, see Pasnau, *Metaphysical Themes*, 303–304. For Ockham's view on condensation and rarefaction, especially from the point of view of theological debates and in relation to Duns Scotus, see also M. McCord Adams, "Independent Accidents: Scotus and Ockham," In M. McCord Adams (ed.), *Some Later Medieval Theories of the Eucharist: Thomas Aquinas, Giles of Rome, Duns Scotus, and William Ockham* (Oxford: Oxford University Press, 2010), 197–226, here 224–226.

9 The label "eccentric" is due to Pasnau, though referring to Crathorn's views in general. On Crathorn, see Pasnau, *Metaphysical Themes*, 303.

3 Blasius' View of Rarefaction and Condensation

At the outset of the presentation of his own view,[10] Blasius refers to the views of exactly Buridan and Ockham, though not by name. As he writes:

> In the second article, moving to the determination, noting that I do not presently consider whether some new quantity is acquired by the act of rarefaction, or whether a new quantity is not acquired but the parts just become more distant [from each other] than before.[11]

He seems to set aside these opinions, then, at the outset. The two positions appear to be precisely those of Buridan and of Ockham, respectively. It is not clear why he chooses to not deal with these views more at length. That it is the views of Buridan and Ockham that are presented is not surprising, however.[12]

Continuing with positions he will not accept, Blasius also refers to thinkers who draw a distinction between *simple* bodies—or elements—which do not have the ability to condense and rarefy, and *mixed* bodies which *can* condense and rarefy. The idea here is that mixed bodies, consisting of different simple bodies or elements, can contain simple bodies or elements that are more rarefied. If these are expelled, the body would become condensed. A stone, for example, might have holes or "pores" of air within it. If this air is pushed out, and the pores closed by earth (of which stone is otherwise basically constituted), this mixed body—the stone—would become condensed. One way to describe it is to say that the parts of the stone have come closer together. This explanation is only possible for mixed bodies, though, as it presupposes that the given body consists of different elements, each of which could not itself condense or rarefy.[13]

---

10 Over and above text 7 in the present volume (IV, q. 7), devoted to the matter at issue, Blasius also treats condensation and rarefaction in text 5 (IV, q. 2, especially n. 44) and in text 8 (V, q. 1, especially nn. 4, 6, 36, and 38) in the present volume.

11 See text 7 (IV, q. 7) in the present volume, n. 13.

12 It is not unusual for Blasius, not only to relate to Ockham and Buridan, but also to contrast their relative positions when they differ. On Blasius's commentary on Peter of Spain's *Logic*, and more specifically on the treatise on the *Categories* contained in it, it has been written: "Un autre trait distinctif du traité de Blaise de Parme sur les *Catégories* est qu'il s'y révèle sensible aux divergences entre les partisans les plus célèbres de la position réductionniste, en l'occurrence Ockham et Buridan, qu'il connaît manifestement très bien." (Roques, "Blaise de Parme et la Quantité," 177–178).

13 See text 7 (IV, q. 7) in the present volume, n. 13. This would later be a model used by Descartes to explain, or account for, condensation and rarefaction. See his *Principles of Philosophy*, pt. II, § 6: "But with regard to rarefaction and condensation, anyone who

In his own positive treatment, Blasius starts out by focusing on simple bodies or elements. The most fundamental case of condensation and rarefaction for him seems to be that of different elements changing into one another. Regarding this transformation, he writes:

> And this is the first conclusion: it is possible that matter informed by the form of earth or water is after this informed by the form of fire.[14]

Further, this transmutation of one element into another is indeed the basic case of condensation and rarefaction for simple bodies:

> The second and responsive conclusion: rarefaction is possible, as well as condensation. The first part of the conclusion holds, because matter that is now informed by the form of earth can become informed by the form of fire.[15]

So, elements changing one into another involves rarefaction and condensation. The underlying assumption here is that different elements are differently rarefied. Or, rather: matter is rarefied to different degrees depending on—at least among other things—what kind of substantial form it is presently informed by, that is, what kind of element it is presently a constitutive part of.

But can one and the same element rarefy and condense in itself, without this transformation to another element taking place? In other words: can different bodies of the same kind be differently rarefied? Blasius does not really answer this question directly, though he seems at least implicitly to think that this is possible. In treating the case of transmutation, for example, Blasius points out that fire cannot exist with the same density as earth.[16] There are, then, some limits as to how dense fire, for example, can become. Further, according to Blasius, it is possible for something to condense all the way into non-quantity, occupying less and less place with more and more matter.[17] Also, according

attends to his own thoughts, and is willing to admit only what he clearly perceives, will not suppose that anything happens in these processes beyond a change of shape. Rarefied bodies, that is to say, are those which have many gaps between their parts—gaps which are occupied by other bodies; and they become denser simply in virtue of the parts coming together and reducing or completely closing the gaps." (CSM I, p. 225; AT VIIIA:43).

14 See text 7 (IV, q. 7) in the present volume, n. 15.

15 See text 7 (IV, q. 7) in the present volume, n. 16.

16 See text 7 (IV, q. 7) in the present volume, n. 16.

17 See below.

to Blasius, it is possible to have uniform rarefaction.[18] Blasius' treatment as a whole suggests, then, that rarefaction and condensation are possible on a continuous spectrum, but that there are natural limits as to the density of different elements.

So far we have considered simple bodies and elements. As regards bodies in general, the possibility of condensation and rarefaction depends on the circumstances. Blasius claims that it is not naturally possible for just any body to rarefy or condense. The limits are here set either by the impossibility of a vacuum or by the size of the sublunary world. If everything in the world would together become more dense and at the same time, the whole would occupy less space, resulting in a vacuum, which is impossible. A similar argument applies to rarefaction, but in such a case the heavens would have to expand outwards (*dilatentur*).[19]

We have, then, an implicit assumption here that there is a constant sublunary space, setting limits to the possibility of condensation and rarefaction of bodies. It is of course possible otherwise for natural bodies, or elements, to rarefy or condense;[20] this possibility, however, is contingent upon other bodies also rarefying or condensing.[21] The sublunary world as a whole might be thought of as a mixed body with fixed dimensions, in which the simple bodies or elements can rarefy or condense, and also change into one another, but there is an interconnectedness among them that governs that possibility, an interconnectedness determined by the fixed dimensions of the universe.

We have now, with Blasius, established that condensation and rarefaction are possible, although within certain limits. But what *is* rarefaction and condensation? Before delivering his own answer, Blasius first rejects other answers to this question, ones that amount to stating that a given body in rarefaction changes so as to occupy more place (*locum*) than before.[22] One counterexample he gives is that of an animal growing; the animal does not become more rarefied just because of this process.[23] Also, Blasius rejects the view that something rarefies just because its outermost parts become more removed from the center.[24] Instead, something rarefies when—without change of fig-

18 See text 7 (IV, q. 7) in the present volume, nn. 34–37. Note, here, that Blasius uses terminology from the Oxford Calculators. See, for example, Treatise 5 (*De raritate et densitate*) and 6 (*De augmentatione*) of Richard Swineshead's *Liber calculationum* (Venice, 1520).

19 See text 7 (IV, q. 7) in the present volume, n. 24.

20 See text 7 (IV, q. 7) in the present volume, n. 24.

21 See text 7 (IV, q. 7) in the present volume, n. 25.

22 See text 7 (IV, q. 7) in the present volume, nn. 18–21.

23 See text 7 (IV, q. 7) in the present volume, n. 18.

24 See text 7 (IV, q. 7) in the present volume, n. 22.

ure—*all* parts of a thing become more distant from each other, and this without any matter being added or removed:

> Seventh conclusion: from this [it can be said that] something is said to rarefy because any two parts become pointwise more distant than before, and this in accordance with all diameters, its matter by no means neither added to nor subtracted from.[25]

This answer seems to mark Blasius' opinion as similar to Ockham's, in terms of the basic explanation of these phenomena.[26] However, Blasius does not explicitly contrast this explanation *against* one involving quantity. When treating the question of whether something can condense to non-quantity, Blasius affirms that this is possible. He does not seem to have any problem here with treating condensation and rarefaction—and along with these, density—in terms of the relation between matter and quantity:

> To the difficulty it is to be said, then, that something can condense toward non-quantity by the sole bringing together of the parts and the removal of quantity. And in this case it is to be conceded that this non-quantity is infinitely dense, because in such parts brought together there is continually more and more matter in less place, and finally in a non-quantity place.[27]

It is unclear whether Blasius is here using the word "quantity" in a similar way as does Buridan in the context of the question of rarefaction and condensation.[28] It is clear, though, that on a surface level, Blasius is comfortable with treating condensation as the "removal of quantity," alongside with the more Ockhamist way of regarding it as a "bringing together of parts."

Thus it may be seen that Blasius combines expressions linked to both Buridan and Ockham when treating condensation and rarefaction; Blasius offers

---

25 See text 7 (IV, q. 7) in the present volume, n. 23.

26 Blasius holds that matter is infinitely divisible. See J. Biard, "Blasius of Parma Facing Atomist Assumptions," in C. Grellard and A. Robert (eds.), *Atomism in Late Medieval Philosophy and Theology* (Leiden: Brill, 2009), 221–234, especially 234.

27 See text 7 (IV, n. 7) in the present volume, n. 45.

28 In his commentary on the *Categories*, Blasius criticizes Ockham's as well as Buridan's distinction between continuous and discrete quantity as "ill-founded." Also, Blasius there makes a distinction among three senses of "quantity"—continuous, discrete, and understood as a multitude (uncountables)—that are irreducible to one another. Cf. Roques, "Blaise de Parme et la Quantité," 188.

also a thought experiment fascinating in its own right: the hypothetical case where matter is brought together closer and closer until it occupies only a place that has no quantity.[29]

4 Conclusions

In summary, then, Blasius's account of rarefaction and condensation exhibits similarities to Ockham's solution to the problem, though including expressions that also make him sound more Buridanean.[30] The case of natural elements transmuting into one another is central to his account, as is the condensation and rarefaction of these elements themselves. Taken as a whole, the possibility of rarefaction or condensation of any particular body in our natural world is dependent on the simultaneous condensation or rarefaction of other bodies; the space of the sublunary world is fixed, and so, for anything to rarefy something else must condense, and vice versa.

Finally, it is intriguing to note that Blasius also mentions—and rejects—the account of rarefaction and condensation found in Crathorn.[31] Interestingly, Blasius refers to the supralunary world and a rather surprising example to support his own position:

> Fifth difficulty: whether penetration of bodies is condensation. It would seem that this is so. Because there is more matter than before in the same place. The opposite holds [though], because when the bodies of the blessed transverse the spheres all the way to the eighth heaven, this happens without condensation; wherefore, etc. It is answered that the penetration of bodies is not condensation, as the argument proves.[32]

---

29 This could presumably be described also as place without extension, or a point-like place, but to do so would require a more thorough presentation of Blasius' understanding of the relations among place, space, extension, and quantity. For the relation between matter and accidents in Blasius, see J. Biard, "La matière est-elle le sujet des accidents? Un débat au XIVe siècle," *Acta Philosophica. Rivista internazionale di filosofia* 31(1), 2022, 15–31.

30 For a similar conclusion, involving Blasius' understanding of quantity in his *Categories* commentary, see Roques, "Blaise de Parme et la Quantité," 178.

31 For a fascinating presentation of Crathorn's understanding of matter generally, see A. Robert, "William Crathorn's Mereotopological Atomism," in C. Grellard and A. Robert (eds.), *Atomism in Late Medieval Philosophy and Theology* (Leiden: Brill, 2009), 127–162. The case of rarefaction and condensation is treated on 154–160. Robert calls Crathorn's answer "quite radical" (157).

32 See text 7 (IV, q. 7) in the present volume, nn. 46–49.

Although able to incorporate different lines of reasoning in his own answer to how condensation and rarefaction are possible, Crathorn presents a solution that is a bridge too far for Blasius. He needed, though, to ascend all the way to the heavens to find a counterexample to it.

CHAPTER 5

Blasius of Parma's Account of Artifacts in the *Physics*

Henrik Lagerlund

1 Introduction

A changing view of artifacts was part and parcel of the new view of nature in the seventeenth century. This can perhaps be seen most clearly in the views of Francis Bacon (1561–1626) and René Descartes (1596–1650) about artifacts as natural things. In their opinion, there is no ontological distinction between a clock and a dog; that is, they are not significantly different ontologically, since they are both mechanisms and explainable by the same science.[1]

In *De dignitate et augmentis scientiarum* (1623), Bacon writes these famous words:[2]

> I find nature in three different states… She is either free, and follows her ordinary course of development as in the heavens, in the animal and vegetable creation, … or she is driven out of her ordinary course by the perverseness … of matter … as in the case of monsters; or lastly, she is put in constraint, molded and made as it were new by art and the hand of man; as in things artificial.

Important to note here, of course, is that he is speaking of nature in three ways, one of which is the artificial. In the third sense, nature can be molded by humans and turned into an artifact. He comments further on this a little later on the same page:

1 For more on this issue see Deborah J. Brown and Calvin Normore, *Descartes and the Ontology of Everyday Life* (Oxford: Oxford University Press 2019); and Deborah J. Brown, "Nature, Artifice, and Discovery in Descartes' Mechanical Philosophy", *Philosophies* 8, no. 5 (2023): 85. https://www.mdpi.com/2409-9287/8/5/85

2 See Francis Bacon, *The Works of Francis Bacon*, ed. James Spedding, Robert Leslie Ellis, and Douglas Denon Heath (London: Longmans, 1857–1874); reprinted (New York: Garett, 1968), showing volume and page number, with original cited in square brackets, here 4.294 [1.496].

 | DOI:10.1163/9789004734968_007

> And I am the more induced to set down the history of the arts as a species of natural history, because an opinion has long been prevalent, that art is something different from nature, and things artificial different from things natural; whence this evil has arisen,—that most writers of Natural History think they have done enough when they have given an account of animals or plants or minerals, omitting all mention of the experiments of mechanical arts. But there is likewise another and more subtle error that has crept into the human mind, namely, that of considering art as merely an assistant to nature, having the power indeed to finish what nature has begun, to correct her when lapsing into error, or to set her free when in bondage, but by no means to change, transmute or fundamentally alter nature.

The fascinating aspect of this passage is that he so clearly places the mechanical arts on a par with animals, plants, and minerals, also suggesting that nature can be altered as a result of human activity. He also on the same page suggests that there is a difference between nature and the artificial since "the artificial does not differ from the natural in form or essence, but only in the efficient." The difference is the efficient cause. He is clearly suggesting that animals, plants, and minerals have their efficient cause in God while the artificial have their cause in humans. The difference is not ontological, but causal or in the agent—a feature that we also find in the fourteenth century.

Descartes on the other hand is less concerned than Bacon about pointing out the errors of the old (or Aristotelian) view. He instead emphasizes that art or mechanical art can imitate nature to such an extent that we are not able to tell the difference between something created by art or by nature. He writes in *The Search for Truth by Means of the Natural Light* that:[3]

> Then I shall lay before your eyes the works of men involving corporeal things. After causing you to wonder at the most powerful machines, the most unusual automatons, the most impressive illusions and the most subtle tricks that human ingenuity can devise, I shall reveal to you the secrets behind them, which are so simple and straightforward that you will no longer have reason to wonder at anything made by the hands of men. I shall then pass to the works of nature, and after showing you the cause of all her changes, the variety of her qualities, and how the souls of

3 See René Descartes, *The Philosophical Writings of Descartes*, translated by John Cottingham, Robert Stoothoff, and Dugald Murdoch (Cambridge: Cambridge University Press, 1984 (vol. II) and 1985 (vol. I)), shortened as CMS, II.405.

> plants and of animals differ from ours, I shall present for your consideration the entire edifice of the things that are perceivable by the senses.

He argues that the principles are the same for the natural and artificial. We cannot tell them apart because they are in fact the same. A similar view is presented in the *Discourse*:[4]

> I made special efforts [in the *Traité de l'Homme*] to show that if any such machines had the organs and outward shape of a monkey or of some other animal that lacks reason, we should have no means of knowing that they did not possess entirely the same nature as these animals.

Here he focuses on our inability to know the difference between art and nature, and that art can become as sophisticated as nature. There were artifacts and mechanical automata in Descartes' time that to many seemed indistinguishable from what could be produced by nature alone, lending credit to his assertions.[5] His main claim, of course, is that artifacts are ontologically indistinguishable from natural things such as animals or plants. There, however, might be other differences distinguishing them.

The fourteenth century also saw a changing view of artifacts, not unlike in the seventeenth century. A more mechanical view of nature combined with a nominalist metaphysics brought on a rejection of Aristotle's sharp distinction between nature (*physis*) and art (*technê*). A rejection of the distinction, at least ontologically, was essential for the development of science, since there is no science of art or artificial things, according to Aristotle. Making nature and art ontologically equal allows for a science of artifacts, mechanisms, and tools. It opens up the possibility of a new science of nature where mechanics plays a crucial role and where mechanical models can be used to investigate as well as understand nature.

Blasius of Parma is no exception in this regard. He deals in Book II, question 1, of his commentary on Aristotle's *Physics*[6] with artifacts, what they are, and whether they can be distinguished from natural things. He there asks the question whether the house-builder in building a house creates something distinct from natural things. It might seem an odd question, but it is exactly the kind of question one would expect given Blasius' overall view of the physical

4 See CMS I.139

5 See Brown, "Nature, Artifice, and Discovery in Descartes' Mechanical Philosophy", for more on this issue in Descartes.

6 See text 1 in the present volume.

world. He is attempting to investigate what distinguishes artifacts from natural things, if there is anything.

In this chapter, I shall first have a look at how thinkers like William Ockham and John Buridan dealt with artifacts. Their views are similar to Blasius', but not quite the same. It is clear, however, that they serve as the philosophical background to and inspiration for Blasius. They are all nominalists, and reject any ontological criteria for distinguishing between nature and art.

2 The Background to Blasius: William Ockham and John Buridan on Artifacts

The sharp distinction between nature (*physis*) and art (*technê*) presented by Aristotle in *Physics*, II.1 came under attack in the fourteenth century, and it became more and more difficult for thinkers such as William Ockham, John Buridan (and, as we shall see, Blasius) to, at least ontologically, draw a distinction between them. They all seem to have thought that there is a distinction to be drawn, but it needed to be based on something other than self-motion, change, or some other ontological criteria like the one Aristotle seems to have had in mind. Aristotle attributed the distinction primarily to an internal principle of change or motion not present in artificial things. He writes, famously:[7]

> All the things mentioned plainly differ from things which are not constituted by nature. For each of them has within itself a principle of motion and of stationariness (in respect of place, or of growth and decrease, or by way of alteration). On the other hand, a bed and a coat and anything else of that sort, qua receiving these designations—i.e., in so far as they are products of art—have no innate impulse to change. But in so far as they happen to be composed of stone or of earth or of a mixture of the two, they do have such an impulse, and just to that extent—which seems to indicate that nature is a principle or cause of being moved and of being at rest in that to which it belongs primarily, in virtue of itself and not accidentally.

Natural things such as plants and animals have such a principle, but beds do not, he seems to think. A bed has the natural tendency to move towards the

7 Aristotle, *Physics*, II.1, 192b9–23; p. 329. The translation is by R.P. Hardie and R.K. Gaye and modified by Jonathan Barnes in his edition *The Complete Works of Aristotle* vol. 1 (New Jersey: Princeton University Press, 1984).

center of the earth only because it is made up of natural things, which means that it only has this motion accidentally. Similarly, artifacts are not substances for Aristotle, and have only accidental being. They are made of substances, as the bed is made of wood, and borrow their motion, but as a thing in itself it only has accidental being. Now, there is a lot to say about this suggestion of the difference between nature and art and a lot has been said,[8] but what is clear is that even though most thinkers held there is some difference between natural and artificial things, it became less and less clear that Aristotle's way of drawing the distinction was the correct one.

A brilliant example of this is Ockham's discussion of artifacts.[9] In maintaining his reduced metaphysics and the principle of parsimony, Ockham rejects Thomas Aquinas' account of artifacts, which introduces artificial forms,[10] or John Duns Scotus' account, which introduced "forms of the whole". In Scotus' view, a "form of the whole" unifies the matter and the form into one thing. In a natural thing, it is the essence of the substance.[11] *Summula philosophiae naturalis*, chapter 20 (*De forma rei artificialis*), is Ockham's most sustained treatment of artifacts.[12] He is there looking for an account of artifacts that does not commit him to postulating new entities.

---

8 See, for example, Phil Corkum, "Aristotle on Artifactual Substances", *Metaphysics* 6 (1) 2023, 24–36. For an overview of the medieval discussion see Kamil Majcherek, *Medieval Metaphysics of Artifacts 1250–1500* (PhD dissertation, University of Toronto, 2022). It also contains a study of Blasius' view of artifacts.

9 See J. Zupko, "'Nothing in Nature Is Naturally a Statue:' William of Ockham on Artifacts," *Metaphysics*, 1(1) 2018, 88–96. My discussion here is very similar to Zupko's. Ockham is also discussed in a similar fashion in Kamil Majcherek, "Paul of Venice's Metaphysics of Artefacts", *British Journal for the History of Philosophy*, 28(1) 2020, 29–48.

10 See the discussion in M. Rota, "Substance and Artifact in Thomas Aquinas", *History of Philosophy Quarterly*, 21(3) 2004, 241–259.

11 See Richard Cross, "Duns Scotus's Anti-Reductionistic Account of Material Substance", *Vivarium* 33(2) 1995, 137–170, for Scotus' view of the form of the whole.

12 See Zupko, "'Nothing in Nature Is Naturally a Statue': William of Ockham on Artifacts", 89 note 7, for a list of other places where Ockham deals with artifacts. In *Brevis Summa Libri Physicorum*, Ockham writes: "in the generation of artificial things, no new thing is produced that is really distinct from a natural thing; but due to local motion, there is a new designation, which is called 'generation', broadly speaking. And the reason for this is that a plurality of things should not be posited unless it is necessary to do so [*in generatio artificialium non est productio alicuius novae rei distinctae realiter a re naturali; sed propter motum localem est ibi nova denominatio quae vocatur generatio, largissime loquendo. Et huius causa est quia plura non debent poni sine necessitate*]" (*Brevis Summa* II, cap. 1 (*Opera Philosophica* VI, p. 25, ll. 21–25)). The quote is from Zupko, "'Nothing in Nature Is Naturally a Statue': William of Ockham on Artifacts", 89 note 6. It is interesting that he here uses local motion as a way of distinguishing an artificial thing from a natural one.

Ockham's ontology includes individual substantial forms, individual accidental forms, and individual matter. A composite substance is composed of its essential parts, namely its matter and form. Besides its essential parts, a substance also has integral parts, like flesh, bones, hands, and so forth. Anything with integral parts is extended and material. All things with essential parts are composed of matter and, hence, also have integral parts and are extended, according to Ockham. Everything extended is a quantity, and every quantity is divisible into quanta; hence there is no indivisible matter or substance.

Ockham insists that a substance is nothing but the parts that constitute it. This is in contrast to Aquinas and Scotus, who held that, although substances have integral parts, these parts depend ontologically on the whole of which they are parts. All parts of a substance are actual and not dependent on anything that makes them actual, Ockham argues, again contrary to Aquinas and Scotus.[13]

On this view of substance, the question is how artifacts fit in. An artifact will also be constituted by the parts that constitute it. In this passage, he suggests that its parts are unified in a distinct way. He writes:

> … it is different as regards artifacts, because one part can remain distinct from another such that the parts can exist simultaneously in the natural order even though they are not united; thus, that whole is not always and by necessity those parts that exist, but it is those existing parts only when the parts are united in the right way, and properly and spatially situated. But when they are separated or not spatially arranged in the right way, then the whole—a house, say—is not those parts. And since it is possible that those remaining parts are sometimes properly arranged and situated and sometimes divided from each other, 'the house is its parts' is therefore sometimes true and sometimes false; nevertheless, whenever those parts are arranged in place in the right way, then the house is those parts.[14]

---

13 See Calvin Normore, "Ockham's Metaphysics of Parts", *Journal of Philosophy* 103(12) 2006, 737–754, for more on Ockham's account of substance. See also H. Lagerlund, "Material Substance", in J. Marenbon (ed.) *The Oxford Handbook of Medieval Philosophy* (Oxford: Oxford University Press, 2012), 468–485, for further discussion of this view.

14 *Summula* I, cap. 19 (*OP* VI, pp. 206–207): "… de artificialibus est aliud, quia una pars potest manere distincta ab alia, ita quod partes potest simul esse in rerum natura quamvis non sint unitae, ideo illud totum non semper et necessario est illae partes existentes, sed tunc solum est illae partes exsistentes quando partes sunt debito modo unitae et localiter situatae et convenienter. Quando autem sunt separatae vel inconvenienter ordinatae secundum locum, tunc totum, puta domus, non est illae partes; et quia hoc est possibile quod illae partes manentes sint quandoque debite ordinatae et situatae et quandoque a

This is an interesting passage. Obviously, an artifact cannot be its essential parts since it does not have any. The parts of an artifact might themselves have essential parts, but the artifact itself does not, which means that the artifact is not simply the parts that make it up, since those parts can exist separately or be united in some other way. The parts that make up the house might not be arranged in a proper way to make a house, but when they are, they constitute a house. The house is its parts, but only if those parts are correctly unified to make a house. Furthermore, if we destroy the house, no thing has been lost, Ockham thinks. The only event that has occurred is that the parts are not unified in the right way anymore.[15]

The focus in Ockham's account is on the artisan and how the house builder reconfigures, arranges, or unifies natural things to produce an artifact. Artifacts are not new things, since they are simply rearrangements of natural things. Even though there is nothing new there, in the sense of a new thing, there remains a difference as to the unity of natural and artificial things, since a natural thing has a *per se* unity while an artificial thing has only a *per accidens* unity.[16] Ockham here differs from his fellow nominalist John Buridan, who rejects this sort of distinction, insisting it is only verbal.

Buridan's treatment of artifacts can be found in his *Physics* commentary, II, question 1: "Whether artificial things are distinct from natural things".[17] Buridan does not focus on arrangements of the parts but on the notions of shape (*figura*) and motion. He elaborates both a view where shape and motion are real things, beside the thing shaped or moved, and a view where shape and

---

se diversae, ideo haec est vera quandoque 'domus est illae partes' et quandoque est falsa; tamen quandocumque illae partes sunt debito modo ordinatae secundum locum, tunc domus est illae partes." The translation quoted here is from Zupko, " 'Nothing in Nature Is Naturally a Statue:' William of Ockham on Artifacts".

15 See Zupko, " 'Nothing in Nature is Naturally a Statue': William of Ockham on Artifacts", 92. See also J. Pelletier, "Getting Real: Ockham on the Human Contribution to the Nature and Production of Artifacts", *Philosophies* 7 (90) 2022. https://doi.org/10.3390/philosophies7050090.

16 One might ask, if a unity or arrangement, *per se* or accidental, is an ontological criterion. It seems to depend upon the way the thing exists, one way being natural and the other artificial or accidental. 'A way' something exists is not part of Ockham's ontology and he does not see it, I suspect, as an ontological criterion, but he clearly attributes a difference between nature and art that is stronger than in later thinkers. Both Buridan and Blasius will spell out more clearly what they mean with the 'way' something exists.

17 See John Buridan, *Quaestiones super octo libros Physicorum Aristotelis (secundum ultimam lecturam)*. Libri I–II, ed. M. Streijger and P.J.J.M. Bakker (Leiden: Brill 2015). See also P.J.J.M. Bakker, "John Buridan on the Ontological Status of Artifacts: Interpreting His Commentaries on Aristotle's *Physics*", in S. Johnston and H. Lagerlund (eds.) *Interpreting Buridan: Critical Essays* (Cambridge: Cambridge University Press, 2024), 101–134.

motion are nothing apart from the thing shaped or moved. The first, 'realist' view, regards an artifact as an actual thing. There is a real shape of an artificial thing and a real motion that went into making the artificial thing. But on the other, the 'nominalist' view, there is nothing that is *the* artifact, only the parts of real, natural things. Buridan seems to be defending the 'nominalist' view,[18] and on that view the only reason we can speak about the artifact, that is, the house, although it is not a thing, is that our vocabulary extends or is amplified to include houses among things, and we then speak as though they were real things distinct from natural things. The distinction is merely verbal.

On both Ockham's and Buridan's views the role of the artist is important. In Ockham's view, the artist gives the artifact the order it has, which is what makes it the artifact it is. Buridan also discusses the role of the artist. He writes in question 1, Book II, that:[19]

> everyone calls the artifact that is made by our intellects and will by a common name and the artifact such as a house or an ax would never have existed without our voluntary action. Or if the artifact is not 'simply' made by us, it is still said to come from us in the way it is (*taliter se habens*) by our intellect and will. However, it would never have existed without our voluntary action, just like the iron, when the ax was made of it, would not have been the way it is (*taliter qualiter se habet*) without our voluntary action.

Even though there is nothing over and above the parts of an artifact (hence, in some sense there is no artifact), we call it 'a house' or 'an ax' by name. As such it would never have come to exist without us, humans, making or building it. It is the way it is because of us making it, that is, the parts are shaped

---

18 I say 'seems' but it is not clear which view Buridan in the end defends. He seems to hold what can perhaps be called a realist view of local motion, since he rejects Ockham's reduction of motion to the thing moved (see in this volume Essay 3 on motion) and he defends a view of a shape as really distinct from the substance, but how this in the end plays into his view on artifacts is not clear. It seems to me that in the question on artifacts he inclines towards a nominalist view, but does not want to clearly commit himself.

19 John Buridan, *Quaestiones super octo libros Physicorum Aristotelis* (*secundum ultimam lecturam*): Libri I–II, eds. M. Streijger and P. Bakker (Leiden: Brill, 2015), II, q. 1, 245–246, "Sed artificiale communi nomine dicitur omne illud quod fit a nobis per intellectum et voluntatem et numquam fieret sine actione nostra voluntaria, ut domus vel securis. Vel si simpliciter sic non fiat a nobis, tamen artificiale dicitur illud quod a nobis per intellectum et voluntatem fit taliter se habens, qualiter numquam se haberet sine actione nostra voluntaria, ut ferrum numquam sine actione nostra voluntaria se haberet taliter qualiter se habet, quando ex eo facta est securis."

and formed the way they are by us. At the end of the quote he notes that not all artifacts are 'simply' made by us; for example, the iron of the ax is a natural thing that we have shaped into an ax; hence, the iron would not have been the way it is or howsoever it is (*taliter qualiter se habet*) had it not been for us. We add an arrangement in which the natural parts are, to use Buridan's vocabulary. As we shall see below this formulation of Buridan is very close to what Blasius ultimately says.

3 Blasius on Artifacts

In the beginning of question 1, Book II, of the *Physics* commentary (text 1 in the present volume), Blasius asks "whether a house-builder when he is building a house makes something distinct from a natural thing?" He quickly divides the question into two parts, namely, (1) 'whether building a house is making anything at all?' and (2) 'whether artificial things are distinct from natural things?' He notes already at the outset that he will answer 'yes' to (1) and 'no' to (2). He also introduces these questions in a different way by saying that if a house and other artificial things are distinct from natural things, then the house-builder does not make stone, wood, or any other natural thing; but on the other hand if artificial things *are* natural things then a house is nothing but the stone or the wood it is made of.

A difficulty that seems to bother Blasius is that, if one adheres to the principle that 'a whole is nothing but its parts', which was a standardly-held metaphysical principle in the nominalist tradition in which Blasius operates, as we have seen,[20] then a house should be nothing but the parts that make it up. A house then is reduced to the stones or the wood of which it is built, or at least it is nothing above those parts. On the other hand, clearly a house is something on its own and not just any random collection of stones or wood; but how does one maintain the principle, and still distinguish a house from a heap of stones, without adding to one's metaphysics entities that are not natural things?[21] Blasius in this regard is clearly influenced by Ockham.

20 See Lagerlund "Material Substance". See also A. Arlig, "Medieval Mereology." *The Stanford Encyclopedia of Philosophy*, 2015, E. Zalta (ed.). URL: https://plato.stanford.edu/entries/mereology-medieval/

21 This issue is all the more emphasized by Blasius' materialism, which is quite thoroughgoing. See J. Biard, *Hypothèse matérialiste et pensée radicale: La philosophie de la nature de Blaise de Parme* (Paris: Vrin, 2022).

There is, however, another principle that is important for Blasius' treatment of artifacts, and also is equally important for the whole nominalist tradition stemming from Ockham.[22] Blasius formulates the principle in the following way:[23]

> (Separability:) If *a* does not belong to the essence of *b*, and *b* does not belong to the essence of *a*, then God can separate them and conserve them independently of each other.

Things that are independent and can exist independently, God can also separate and conserve.

Blasius uses this principle to prove several auxiliary principles, such as:[24]

> (Motion:) Local motion is nothing but the motion of the moving thing.

The proof of this is that, if motion were something other than the moving thing, then it would not belong to the essence of the moving thing but be something added to it; thus (by separability) God could separate it and conserve it. There would, hence, exist a motion which is not a motion of anything; this Blasius holds to be absurd. This is exactly the same argument we find in Ockham (as we saw above).

Similarly, he proves the following principle:[25]

> (Shape:) Shape is nothing but the thing shaped.

If shape were something other than the thing shaped, then shape (by separability) could be separated and preserved by God by itself, since it does not belong to the thing's essence. What would a shape be without the thing shaped? Nothing, answers Blasius. A realist could argue here, as some did, that such a shape is a form (Aquinas); perhaps an accidental form.[26] But according to Blasius' principles it would have to be able to exist by itself, and that is not possible, he thinks. Buridan argues for the same position in his presentation of the 'nominalist' position, as seen above.

Using these principles, Blasius goes on to draw several important conclusions pertaining to artifacts. The first one is that a house is nothing but the

22 See John Boler, "Ockham's Cleaver," *Franciscan Studies* 45, 1985, 119–144,
23 See text 1 (II, q. 1) in the present volume, n. 17.
24 See text 1 (II, q. 1) in the present volume, n. 18.
25 See text 1 (II, q. 1) in the present volume, n. 19.
26 For more on this see M. Rota, "Substance and Artifact in Thomas Aquinas".

stones and the wood that constitute it. He proves this point by assuming that it is false, asking: what would a house be if it were not the parts that constitute it? It would then be something superadded to those parts, he notes; but this superadded thing must either be a shape or the local motion of adding the stones and wood in building the house. However, as the principles proved above, (motion) and (shape), state, these are nothing but the thing shaped and moved.[27]

From this conclusion the next one follows, namely, that all artifacts are only the natural things that constitute them. Hence, the artificial is reduced to the natural. From this conclusion three others follow. The first is that building a house is not really making anything. Furthermore, Blasius notes that houses, cities, and castles are really constructed by nature operating naturally.[28] This may sound like a strange claim, but it is meant, I think, to highlight the fact that artifacts are not outside of nature. They are a part of nature and made by natural things, that is, by humans.

It is important to remember that all these conclusions pertain to the ontological question, that is, what artifacts are, metaphysically speaking. From that perspective artifacts are nothing apart from the natural things (parts) that constitute them. Clearly, this does not exhaust the question of artifacts, according to Blasius. It is also for obvious reasons insufficient as an answer to his initial questions; but, ontologically speaking, we shall not advance any further, he seems to think. In order to more completely answer the questions, he presents another set of conclusions that move us beyond ontology.

The first conclusion of this second set states something that might seem obvious. He notes the fact that a house cannot be built without stones or wood or whatever material is used in it. The second conclusion is more interesting and points towards an ultimate answer. He writes that "the second conclusion is that this or that house is not stones or wood disposed in just any sort of possible way."[29] Here he includes the idea that the material needs to be ordered in a certain way, and cannot be just any heap of stones or wood.

The next conclusion explains this view further. He writes:[30]

> The third conclusion is that for it to be the case that house B, or another house, exists, it is required that the stones, the wood, or whatever is equivalent to them have a certain mark of the artist (*meriti artificis*) that house B did not have before.

---

27 See text 1 (II, q. 1) in the present volume, n. 20.
28 See text 1 (II, q. 1) in the present volume, n. 21.
29 See text 1 (II, q. 1) in the present volume, n. 32.
30 See text 1 (II, q. 1) in the present volume, n. 33.

Since a house is an artifact, it is made by an artist (a builder) and, hence, it has what Blasius calls a 'mark of an artist' (*meriti artificis*). What does this exactly mean? In conclusion four, he develops the idea further:[31]

> The fourth conclusion is that in order to make a house, besides the material parts, which are stones or wood, it is required that they be thus and so (*taliter se habere vel taliter*), which cannot be well expressed in language, namely so that the thing is thus and so as it is.

The 'mark of the artist' is a 'thus and so' of the stones and the wood.[32] To explain this further, Blasius notes that it is the wrong question to ask, "What is a house?" in the same way as it is the wrong question to ask, "What is Socrates running?". The right question to ask instead is, "How is Socrates running?", which has an answer like 'fast' or 'slow'; in the same way we should ask how the stones and the wood are, that is, "How are they?" The answer is, "Like a house". He then concludes by asking: "Are we to say that artificial things are natural things? No, since artificial things are nothing, only thus and so."[33]

Blasius says that natural things can be seen as artificial things, since artifacts are made of natural things; but in another sense he says that they are not, since stones and wood are not a house, and a house is not stones and wood. It takes something more, namely, an artist, to make artifacts, and nature cannot in this way make artifacts. Put in another way, the only difference between a natural and artificial thing is the different 'artists'. Humans and human-like agents make artifacts, while God makes natural things.

---

31 See text 1 (II, q. 1) in the present volume, n. 34.

32 Blasius uses the same Latin phrase, '*taliter se habere*', as Buridan to express the way an artist makes an artifact into what it is. Perhaps it is best expressed as an adverbial theory of artifacts. A house is thus wood and stone existing 'housely'. By expressing it this way, both Buridan and Blasius clarify that this is not an ontological criterion, a point that was left unclear by Ockham. His use of this phrase is also reminiscent of the use of '*aliqualiter*' in the theory of signification. According to Blasius (and before him Peter of Ailly), a syncategorematic term signifies '*aliqualiter*', which does not mean that it signifies anything in the external world or in the mind, but signifies a *way* or a *how* of the categorematic term; that is, a syncategorematic term affects the way a categorematic term signifies. See J. Biard, "Blaise de Parme et la théorie de la signification: 'Doctor parisinus'?" in *Ratio et superstitio: Essays in Honor of Graziella Federici Vescovini*, eds. Giancarlo Marchetti, Orsola Rignani, and Valeria Sorge (Turnhout: Brepols Publishers, 2023), 221–242. The similarity between this description and how an artifact is described is not an accident. The adverbial formulation does not pick out anything superadded, but some extrinsic relation.

33 See text 1 (II, q. 1) in the present volume, nn. 36–37.

4 Final Reflections

According to an influential account of artifacts among contemporary thinkers, "artifacts are objects made intentionally, in order to accomplish some purpose."[34] The definition has three parts, in that artifacts (i) are intentional, (ii) are modifications of some materials, and (iii) must have a purpose (reason or end). Whatever is made intentionally is some modification of some materials with some purpose. A hammer and a house are, hence, by this definition artifacts.

Blasius does not give an ontological criterion for artifacts, but focuses on the intentional aspect and perhaps the final cause of an artifact—to put it in a medieval way. This is also somewhat reminiscent of Bacon's view, which, however, focused on the cause, supposedly the efficient cause of the artifact. A major problem, from a medieval perspective, about this view is that it does not actually provide us with the means to distinguish an artifact from a natural thing, since, as a medieval thinker would have it, natural things are also made with intentions and for a purpose. Perhaps criterion (ii) is on its own sufficient to distinguish the two, since God creates from nothing and does not merely modify any material already existing.[35]

In the end, Blasius' account is quite similar to Buridan's nominalist account of artifacts. There is further focus on the artist in Blasius' account, or what he calls the mark of the artist. The mark results in the way (or a thus and so) that the things (materials) make up the artifact, making those materials exist in a certain way. They exist as 'hammerly' or 'housely'.

This is quite an original view, but one of its important consequences is that one cannot draw an ontological distinction between nature and art. They are the same material things and, therefore, are also subject to the same scientific laws, just as Bacon and Descartes would have it.

---

34 Risto Hilpinen, "Artifacts and Works of Art", *Theoria*, 58(1), 1992, 58–82.

35 A good introduction to contemporary discussion can be found in Beth Preston, "Artifact", *The Stanford Encyclopedia of Philosophy* (Winter 2022 Edition), Edward N. Zalta & Uri Nodelman (eds.), URL = https://plato.stanford.edu/entries/artifact/

CHAPTER 6

Building a Unified Theory of Nature: The Functions of the 'New Analytical Languages' in Blasius of Parma's *Questions on the Physics*

Sylvain Roudaut

1 Introduction

In a paper dedicated to the evolution of philosophy in the fourteenth century, John Murdoch characterized the new developments of this period by labeling as "new analytical languages" the emerging philosophical techniques that were beginning to be employed in various fields of philosophical reflection.[1] This expression, since then, has been adopted by many scholars to refer to the different 'languages' analyzed by Murdoch. According to him, these new languages characterizing fourteenth century philosophy are threefold, or can be classified according to a three-tiered hierarchy:

1) the theory of supposition, which constitutes the logical basis for the more analytical languages based on it;
2) the theory of infinity and continuity;
3) the theory of proportions, primarily related to the 'intension of forms' (*intensio formarum*) and the topic of limits.

According to Murdoch, the role of these languages was to provide a set of metalinguistic rules or algorithms for solving problems raised in different disciplines such as physics which, although distinct from logic, relied on it.[2]

In the fourteenth century, none of these theories were entirely 'new' in the sense of being unprecedented in the history of philosophy and science. The

1 J.E. Murdoch, "From Social into Intellectual Factors: An Aspect of the Unitary Character of Late Medieval Learning," in J.E. Murdoch and E.D. Sylla (eds.), *The Cultural Context of Medieval Learning. Proceedings of the First International Colloquium on Philosophy, Science, and Theology in the Middle Ages* (Dordrecht / Boston: D. Reidel Publishing Company, 1975), 271–339, here 286–287.

2 Besides the study quoted in the previous footnote, this view is also developed in J.E. Murdoch, "Propositional Analysis in Fourteenth-Century Natural Philosophy: A Case Study," *Synthese* 40, 1979, 117–146; J.E. Murdoch, "*Scientia mediantibus vocibus*: Metalinguistic Analysis in Late Medieval Natural Philosophy," in W. Kluxen (ed.), *Sprache und Erkenntnis im Mittelalter* (Berlin: De Gruyter, 1981), 73–106.

 | DOI:10.1163/9789004734968_008

theories of continuity and infinity were largely influenced by Aristotle's analysis in the *Physics*, which supported the idea of the continuous nature of bodies while rejecting the existence of actual infinities (especially in *Physics* III and VI). Similarly, the theory of proportions drew heavily from Euclid's *Elements* (particularly Books V and VIII), which had become newly available in Latin. Additionally, the theory of the intension of forms had a longstanding history dating back at least to Plato and Aristotle.[3] The theory of supposition, which underwent significant developments from the 13th century onward, was also inherited from an earlier period.[4]

Still, the degree of sophistication to which these new languages or theories were developed in the fourteenth century was certainly unknown in their earlier stages. Through a spectacular evolution, the appeal to these theories as systematic lenses through which most philosophical problems were now analyzed was such that it is no exaggeration to see the fourteenth century undergoing what can be called a 'linguistic turn' similar to the one Western philosophy underwent in the early twentieth century. It is in this regard that Murdoch could speak of 'new' analytical languages.

Murdoch's way of classifying these new analytical languages into three main groups is probably debatable, as may be debatable his particular way of listing these languages according to what was according to him an order of logical dependence. One could also dispute the notion of 'language' and think of a more encompassing term such as tools (which Murdoch himself occasionally uses), instruments, techniques, methods, or theories (which will often be employed in what follows). But it is hard to deny that these topics were among the most extensively discussed in the logical and linguistic approach to philosophical issues that characterizes the fourteenth century.

Some studies have already examined how these instruments were applied to physics, leading to new '*sophismata physicalia*', especially in the context of Italian physics.[5] According to Murdoch, the main function of these languages

3 The best studies on the long-term history of this problem remain those of Solère; cf. J.-L. Solère, "Tension et Intention. Esquisse de l'histoire d'une notion," in L. Couloubaritsis and A. Mazzù (eds.), *Questions sur l'intentionnalité* (Bruxelles: Ousia, 2007), 59–124; J.-L. Solère, "Plus ou moins: le vocabulaire de la latitude des formes," in J. Hamesse and C. Steel (eds.), *L'élaboration du vocabulaire philosophique au Moyen Âge* (Turnhout: Brepols, 2000), 437–488.

4 See, e.g., S. Ebbesen, "Early Supposition Theory (12th–13th centuries)," in S. Ebbesen, *Topics in Latin Philosophy from the 12th–14th Centuries*, vol. 2 (Aldershot: Ashgate, 2009), 1–14, and the studies gathered in E.P. Bos (ed.), *Medieval Supposition Theory Revisited: Studies in Memory of L.M. de Rijk* (Leiden: Brill, 2013).

5 A. de Libera, "Le développement des nouveaux instruments conceptuels et leur utilisation dans la philosophie de la nature au XIIIe siècle," in M. Asztalos, J.E. Murdoch, and I. Niiniluoto,

was not to equip only physics with finer theoretical notions but, more broadly, to propose a set of analytical tools that could be applied to virtually any field of knowledge, ranging from physics to theology, including ethics, psychology, medicine, or logic. The true aim of these languages was to achieve a 'unity of science' by establishing a unified system of technical terms and associated rules that could be used to analyze various subjects within a shared conceptual framework.[6]

As a philosopher trained in the fourteenth century and profoundly influenced by the nominalist tradition, in which the logical approach to physical problems was prominent, Blasius of Parma was well familiar with the developments of what Murdoch called the 'new analytical languages.' However, acknowledging that Blasius employed these languages does little to elucidate his unique method of utilization. Indeed, it would be misleading to assume uniformity in the application of the new analytical languages among different authors, or to expect similar outcomes from their use.

The aim of this study is to show that, as far as Blasius is concerned, Murdoch's hypothesis regarding the main function of these languages—i.e., enabling the unity of science by establishing a common conceptual framework across diverse disciplines—is indeed validated. Specifically, it will be shown that Blasius' extension of the new analytical languages into various scientific domains is instrumental in underpinning his conception of physics. A central tenet of Blasius' physics is the integration and harmonization of disparate disciplines into a cohesive whole. Blasius' physics represents a somewhat eclectic system, drawing from various sources including the Parisian nominalist tradition, the works of the so-called Oxford Calculators, as well as insights from medicine and astrology. Despite this eclectic foundation, Blasius' work exhibits a high degree of systematicity and coherence. This achievement is attributed to his methodical use of analytical languages that are crossfunctional in different scientific fields.

This study will therefore present specific examples of how Blasius applies these languages in his physics, aiming to elucidate his overall methodology. Certain fundamental aspects of Blasius's interpretation of fourteenth century analytical languages have already been explored. His stance on the problem of the

Knowledge and the Sciences in Medieval Philosophy. Proceedings of the Eighth International Congress of Medieval Philosophy (S.I.E.P.M.), Helsinki 24–29 August 1987, vol. 1 (Helsinki: Yliopistopaino, 1990), 159–196. J.E. Murdoch, "Philosophy and the Enterprise of Science in the Late Middle Ages," in Y. Elkana and S.l Sambursky (eds.) *The Interaction between Science and Philosophy* (Humanities Press: Atlantic Highlands, 1974), 51–74.

6 Murdoch, "From Social into Intellectual Factors," 280.

intension of forms, for instance, to which Blasius dedicated several works, has been extensively examined, along with his views on infinity.[7] For this reason, this study will more narrowly focus on the *application* of these languages in the context of his questions on the *Physics*. What follows will be divided according to the three layers identified by Murdoch in the new languages of the fourteenth century, namely the theory of supposition or signification (Section 2), that of continuity and infinity (Section 3), and finally the theory of the intension of forms, which plays a central role in Blasius' approach to physical problems (Section 4). This overview will enable us to draw conclusions on the overall methodology adopted by Blasius in physics (section 5).

2 Supposition Theory

Having taught logic over the course of his career, and authored logical treatises and commentaries, Blasius knew well the most advanced innovations of terminist logic.[8] Despite having written on the predicables, on the categories, on consequences, and on the 'topics,' however, he did not dedicate any special treatise to the theory of supposition.[9] His use of the notion of supposition shows that he seems to acknowledge, as did John Buridan, only two main types

7 On Blasius' theory of the intension of forms, see J. Biard, "Blasius of Parma on the Calculation of the Variation of Qualities and Aristotelian Physics," in D.A. Di Liscia and E.D. Sylla (eds.), *Quantifying Aristotle. The Impact, Spread and Decline of the Calculatores Tradition* (Leiden / Boston: Brill, 2022), 232–254; D.A. Di Liscia, "Biagio Pelacani da Parma's Geometrisation of Latitudes and the Problems of the Mean Degree Theorem," in J. Biard and A. Robert (eds.), *La philosophie de Blaise de Parme. Physique, psychologie, éthique* (Florence: SISMEL—Edizioni del Galluzzo, 2019), 129–152; G. Federici Vescovini, *Astrologia e scienza. La crisi dell'aristotelismo sul cadere del Trecento e Biagio Pelacani da Parma* (Florence: Enrico Vallecchi, 1979), 205–238; G. Federici Vescovini, "La '*Quaestio de intensione et remissione formarum*' di Biagio Pelacani da Parma," *Physis* 31, 1994, 433–535, intro., 433–466. On Blasius' treatment of infinity, see J. Biard, "Blasius of Parma Facing Atomist Assumptions," in C. Grellard and A. Robert (eds.), *Atomism in Late Medieval Philosophy and Theology* (Leiden / Boston: Brill, 2009), 221–234; Federici Vescovini, *Astrologia e scienza*, 337–343.

8 For an overview of Blasius' theory of signification, see J. Biard, "Blaise de Parme et la théorie de la signification: '*Doctor Parisinus*'?," in G. Marchetti, O. Rignani, and V. Sorge (eds), *Ratio et superstitio. Essays in Honor of Graziella Federici Vescovini* (Turnhout: Brepols, 2003), 221–242; J. Biard, "Le traité sur les *Catégories* de Blaise de Parme," in J. Biard and I. Rosier-Catach (eds.), *La tradition médiévale des catégories, XIIe–XVe siècles. Actes du XIIIe Symposium européen de logique et de sémantique médiévales (Avignon, 6–10 juin 2000)* (Louvain-la Neuve, Leuven / Paris: Éditions de l'Institut supérieur de philosophie / Éditions Peeters, 2003), 365–378.

9 See J. Biard and G. Federici Vescovini, "Introduction," in Blasius of Parma, *Questiones super Tractatus logice magistri Petri Hispani*, eds. J. Biard and G. Federici Vescovini (Paris: Vrin, 2001), 7–36, esp. 15, 25.

of supposition, namely personal supposition (corresponding to the way a term normally refers to what it signifies) and material supposition (corresponding to cases where a term refers to its own occurrence, as in a proposition like 'Socrates is a word').[10]

The problems dealt with by Blasius in his questions on Aristotle's *Physics* vary in their scope and generality. As evidenced by the texts included in the present volume, even within these broader topics Blasius integrates more specific problems. These may appear as supporting arguments for certain positions or as standalone 'problems' (*problemata*) constituting the first articles of his inquiries.

The theory of supposition and, more generally, the theory of signification is put into use by Blasius both in the context of questions about the general definition of terms and about more particular questions about specific phenomena. To better understand the function of the theory of supposition in the context of Blasius' physics, let us see some examples of both cases.

First, Blasius relies on the theory of supposition to explain the meaning of basic terms that constitute the core conceptual framework of physics. Blasius draws on this theory to ground a nominalist explanation of these terms, i.e., to avoid positing entities over and above what is strictly necessary, thus favoring a reductionist ontology in which only substances and qualities truly exist outside the mind. Although utilized by thinkers beyond the nominalist tradition, the theory of supposition provided a powerful tool for explaining how terms seeming to denote abstract objects or universal properties ultimately refer only to particular entities. What is usually called the 'theory of supposition' encompassed not only the notion of supposition, but also two other foundational notions with which it was connected: the notions of signification and connotation (to which other concepts such as 'appellation' were often added).

In this framework, the primitive notion was that of signification. In his questions on Peter of Spain's *Summulae logicales*, Blasius defines signification by saying that a sound (*vox*) signifies when it is able to represent in the intellect a thing, different things, or a way of being (*aliquid, vel aliqua, vel aliqualiter esse*).[11] In line with his metaphysical convictions, Blasius believes that what is signified by a term is always something particular, or a set of particular entities.

10 That Blasius considered this twofold classification of types of supposition to be exhaustive is argued for in Biard, "Blaise de Parme et la théorie de la signification," 236–237.

11 Blasius of Parma, *Questiones super Tractatus logice magistri Petri Hispani*, I, 7, a. 3, 71: Vox significative est illa que nata est intellectui representare aliquid, vel aliqua, vel aliqualiter esse. For the problems raised by the nature of this *aliqualiter esse*, and its probable origin in Peter of Ailly, see Biard, "Blaise de Parme et la théorie de la signification," 231. On

The notions of supposition and connotation can be defined on this basis. According to the standard account of supposition theory, a term supposits for a thing if it signifies it primarily and can stand for it in a proposition. For instance, in the proposition 'Socrates is walking', the term 'Socrates' is said to *supposit* for Socrates, because it signifies him primarily. A term is said to be connotative if it signifies secondarily something on top of what it signifies primarily. For example, in the proposition 'Socrates is a father', the term 'father' signifies Socrates primarily but it also signifies something else secondarily, namely Socrates' child or children. The term 'father' in such a context is said to *connote* Socrates' children.

This general distinction between signification, supposition, and connotation, did not constitute a rigid framework leading all philosophers to adopt the same semantics. This framework was very flexible and could be adjusted to different philosophical stances. Depending on one's philosophical convictions, it could serve to explain abstracts terms such as 'humanity' or 'animality' from a realist perspective or, on the contrary, from a nominalist one. In such diverging perspectives, what an abstract term like 'humanity' supposits for was different, being a common or general property for a realist, and a set of particular individuals for a nominalist.

Blasius employs a nominalist approach to the theory of supposition when it comes to defining physical terms. This is evident in the definition he provides for the term 'generation,' which refers to one of the different types of change acknowledged by Aristotle:

> [...] These terms 'generation' and 'thing generated' are not entirely convertible terms, because this term 'generation' signifies this or that while connoting its actual production by a generator. But this expression 'generated thing' signifies this or that while connoting the past or present actual existence [*actualem existentiam esse vel fuisse*].[12]

Blasius' interpretation of the term 'generation' in this passage is clear: Blasius seeks to avoid positing some entity (in this case, a special type of motion) that

its implications for Blasius' theory of artifacts, see Henrik Lagerlund's contribution in the present volume.

12 Blasius of Parma, *Quaestiones super VIII libros Physicorum* (secunda redactio), I, q. 8, ms. Vat. lat. 2159, 77vb: Pro quo nota quod isti termini 'generatio' et 'res genita' non sunt termini omnino convertibiles, quia iste terminus 'generatio' significat hoc vel illud connotando actualem productionem ipsius a generante. Sed hic terminus 'res genita' significat hoc vel illud connotando actualem existentiam esse vel fuisse.

would be really distinct from the thing that is generated, in line with his general rejection of motion as a thing added to what is moved.[13] The generation and the thing that is generated are strictly identical, with respect to the things those terms primarily signify. The only difference between them is a semantic one, which Blasius explains using the tools offered by the theory of supposition. In this case, the difference between the two terms or expressions lies in the things they connote in addition to their primary signification, namely the production of the thing on the one hand ('generation') and its actual existence on the other hand ('generated thing').

Blasius' approach to this point is very close to the reductionist definitions of terms related to generation—and, more generally, change—offered a few decades earlier in the Parisian milieu by philosophers such as John Buridan and Albert of Saxony. John Buridan, in particular, argued in his last redaction of his *Questions on the Physics* (the so-called *ultima lectura*) that the expressions 'generated thing' (*res genita*), 'to be generated' (*generari*), and 'generation' (*generatio*) all supposit for the same thing, which Buridan took to be the substantial form that comes into being, and not the hylomorphic compound, as in the standard account of substantial change.[14] This view was taken up by Blasius in the first redaction of his *Questions on the Physics*.[15] In the second redaction, Blasius appears to be more hesitant and, while some passages seem to indicate that he still holds the same view,[16] he asserts that the term 'generation' can signify either the composite or its substantial form.[17] This view, more open to

13 On this point, see the study on Blasius' theory of motion as well as the one on artifacts in this volume.

14 John Buridan, *Quaestiones super octo libros Physicorum Aristotelis (secundum ultimam lecturam). Libri I–II*, eds. M. Streijger, and P.J.J.M. Bakker (Leiden/Boston: Brill, 2015), I, q. 17, 171, ll. 23–28. On this see S. Caroti, "*Generatio/Generare*: Ontological Problems in John Buridan's Natural Philosophy," *Medioevo* 27, 2002, 373–413; S. Roudaut, "The Substrate of Change in Buridan's Theory of Nature: A Semantic Reinterpretation of Hylomorphism," in N. Polloni (ed.), *A Philosophical History of Prime Matter. Volume 2: Scholastic and Early Modern Conceptions* (Cambridge: Cambridge University Press, forthcoming).

15 See J. Biard, *Hypothèse matérialiste et pensée radicale. La philosophie de la nature de Blaise de Parme*, Paris, Vrin, 2022, 33, nn. 1–2.

16 See for instance Blasius of Parma, *Quaestiones super VIII libros Physicorum* (secunda redactio), ms. Vat. Lat. 2159, I, q. 8, 78ra: Quarta conclusio. Generare non est forma quae generatur. Patet quia idem esset generare et generari. Et patet consequentia, quia generari attribuitur formae hora qua generatur sic quod forma quae generatur dicitur generari.

17 Blasius of Parma, *Quaestiones super VIII libros Physicorum* (secunda redactio), ms. Vat. Lat. 2159, I, q. 8, 78[rb]: Omne quod generatur est compositum vel eius forma substantialis.

different possible conventions regarding the meaning of this term central to physical theory, was the one favored by Albert of Saxony.[18]

In fact, despite sharing an overall similar 'nominalist' stance on philosophical problems, Blasius devotes less attention than John Buridan, Albert of Saxony, and their immediate followers to the semantic analysis of physical terms according to supposition and signification. One possible reason for this difference in methodological approach could be Blasius' interest in the varying degrees of probability associated with different possible answers to physical problems. This interest may have led him sometimes to refrain from offering one single semantic interpretation of terms, as these terms could have different meanings depending on various ontological hypotheses.

At any rate, Blasius follows his Parisian predecessors in his will to reduce to concrete particular substances terms referring to processes and abstractions, as is clear in the case of terms denoting change:

> Is generation identical with the generated thing? It appears undoubtedly that it is, because otherwise it would be a thing superadded to the generated thing. But this should not be said, since in the generated thing there is nothing more than the matter and form.[19]

Similar appeals to the tools of supposition theory can be detected whenever Blasius seeks to avoid positing, for a given concept, a corresponding entity that he deems unnecessary. This is the case, for instance, for comparative adverbs expressing speed when applied to qualities or forms that are being acquired. Such terms primarily signify forms and not properties really added to them, although they connote the way forms are acquired with respect to one another:

> And from this it follows that this argument is not sound: 'A and B are altered by the same velocity [*aequali velocitate*], therefore A and B are altered equally fast [*aeque velociter*],' because these names 'velocity' and 'equal velocity,' when taken as nouns in this way, supposit for the forms acquired or being acquired. But when they are changed into an adverb,

18 Albert of Saxony, *Quaestiones super libros Physicorum*, ed. B. Patar, 3 vols. (Louvain-la-Neuve / Paris: Peeters / Éditions de l'Institut Supérieur de Philosophie, 1999), vol. 2, I, q. 19, 254–255.

19 Blasius of Parma, *Quaestiones super VIII libros Physicorum* (secunda redactio), ms. Vat. Lat. 2159, I, q. 8, 77vb: An generatio sit idem quod res genita. Et statim videtur quod sic quia alioquin esset res superaddita rei genitae sed hoc non est dicendum cum in re genita non sit aliud praeter materiam et formam.

> like when it is said 'equally fast,' they signify these forms while connoting their mode of acquisition.[20]

Blasius often employs the theory of signification when it comes to the general problems of physics, since these problems typically involve elucidating the meaning of the core terms forming the basis of physical analyses. For instance, in the course of investigating whether something could be made *ex nihilo* ('from nothing'), an idea seeming to contradict the existence of a cause for every natural phenomenon, Blasius draws on supposition theory to analyze what the term 'nothing' can mean.[21]

Indeed, from the perspective of a nominalist usage of supposition theory and its toolkit, where this theory mainly serves to reduce physical discourse to actual particular substances and their qualities, and to avoid entities such as universal properties or *possibilia*, references to nothingness or other privative terms may be turn out to be problematic. This is the case, in particular, of the void, another instance where Blasius appeals to supposition theory:

> [...] Given this proposition, 'the void is the void,' I ask whether the subject and the predicate supposit for the same thing or things. And if this is so, the truth of this proposition is clear, and in this way this ⟨proposition⟩ 'the void is the void' will be true, from which this proposition 'the void is' formally follows. If in fact the subject and the predicate supposit for different things, then the subject supposits for some thing or things for which the predicate does not supposit. Thus, in the affirmative proposition, the subject can be verified for this thing or these things, and in this way this proposition 'the void is' is true.[22]

---

20 Blasius of Parma, *Quaestiones super VIII libros Physicorum* (secunda redactio), VII, q. 6, 196vb [transcr. Rommevaux, provisional edition, slightly modified]: Et ex hoc patet quod hoc argumentum non valet: '*a* et *b* aequali velocitate alterantur, ergo *a* et *b* eque velociter alterantur', quia hoc nomen 'velocitas' et 'equalis velocitas' sic nominaliter sumptum supponunt pro formis acquisitis vel acquirendis. Sed cum transmutantur in adverbia, ut cum dicitur 'aequevelociter', dicunt illas formas connotando modum acquirendi illas.

21 Blasius of Parma, *Quaestiones super VIII libros Physicorum* (secunda redactio), I, q. 9 ('Utrum per potentiam finitam vel infinitam possit aliquid fieri ex nichilo'), ms. Vat. Lat. 2159, 79vb–82vb.

22 Blasius of Parma, *Quaestiones super VIII libros Physicorum* (secunda redactio), IV, q. 4, in J. Biard and S. Rommevaux, "Questions IV, 4 et 5 des *Questions sur la Physique* de Blaise de Parme," in J. Biard and S. Rommevaux (eds.), *La nature et le vide dans la physique médiévale. Études dédiées à Edward Grant* (Turnhout: Brepols, 2012), 317–350, here at 312: Undecimo, data ista propositione 'vacuum est vacuum', quero an subiectum et predicatum supponant pro eodem vel pro eisdem. Et si sic, patet veritas huius, et sic ista erit vera 'vacuum

As can be seen, the argument presented here relies on a supposition-based definition of truth: an affirmative proposition is true when its subject and its predicate suppposit for the same thing or things.[23] Given the formal validity of the argument, the proposition 'the void is' should be true. Blasius will eventually reject the soundness of this argument, not because it is formally invalid but because it includes a wrong premise, namely that in the proposition 'the void is the void' the subject and the predicate supposit for something (they have no supposition, according to Blasius).[24]

Even though the void does not actually exist within nature, according to him, Blasius will argue in the fifth question on book IV of the *Physics* that it must be admitted that it is possible that it exist and, furthermore, that motion in the void could also be possible. In this case, the term 'void' would supposit for something, namely either for a concave body or for a separate space devoid of matter.

> Third difficulty: From the fact that a void thing is, does it follow that a void *A* is, using these words 'a void *A*' for a separate dimension in the imagination that is not a thing or things in truth? It is replied negatively, the cause of which is that, if it follows that a void *A* exists, from the fact that a void thing exists, then in a sound consequence the false would follow from the true, something that learned ones deny. And this argument is clear because this proposition 'a void *A* is' is false, since its subject does not supposit for anything. But whether it follows from the fact that a void *A* is, that a void thing is, it is replied positively, as is clear to anyone intelligent.[25]

---

est vacuum' ad quam formaliter sequitur ista propositio 'vacuum est'. Si vero subiectum et predicatum supponant pro diversis igitur subiectum supponit pro aliquo vel pro aliquibus pro quo vel pro quibus non supponit predicatum, igitur subiectum potest pro illo vel illis in propositione affirmativa verificari, et sic adhuc ista est vera 'vacuum est.'

23 Cf. Blasius of Parma, *Questiones super Tractatus logice magistri Petri Hispani*, I, q. 6, a. 2, 58; I, q. 8, a. 1, 63.

24 Blasius of Parma, *Quaestiones super VIII libros Physicorum* (secunda redactio), IV, q. 4, 335.

25 Blasius of Parma, *Quaestiones super VIII libros Physicorum* (secunda redactio), IV, q. 4, 328: Tertia difficultas: Utrum ad rem vacuam esse sequatur *a* vacuum esse, utendo istis terminis '*a* vacuum' pro dimensione separata apud ymaginationem que secundum verum non est aliquid nec aliqua? Respondetur quod non, cuius causa est quia, ubi ad rem vacuam esse sequitur *a* vacuum esse, tunc in consequentia bona ex vero sequeretur falsum, quod negatum est a sapientibus. Et patet argumentum quia hec propositio est falsa '*a* vacuum est' cum eius subiectum pro nullo supponat. Sed an ad *a* vacuum esse sequatur aliquam rem esse vacuam, respondetur quod sic, ut patet intelligenti.

This passage illustrates the centrality of the theory of supposition when it comes to assessing the soundness of a proposition about physics. In the case of an affirmative proposition in physics, the term must supposit for some thing or things. Propositions following formally from other propositions containing the term 'void' can only be accepted as true if, in the latter, the term void supposits for a thing.

From the perspective of Blasius, the status of the void carries special importance.[26] Indeed, Blasius grants a key role to thought-experiments involving situations or, as he often puts it, 'cases' that are impossible in the normal course of nature or given the actual structure of the world. This method or technique of conducting physical investigations through thought-experiments was extensively developed in the tradition of the Oxford Calculators, where it was often presented as a reasoning according to the imagination (*secundum imaginationem*), although similar thought-experiments were also present—to a lesser extent—in Paris.[27] Many of these thought-experiments involve entities that do not exist in the real world but can be posited by the imagination, such as points, infinite velocities, or the void.

In Blasius' thought-experiments, even when a counterfactual situation violates core principles of the Aristotelian cosmos, such as the impossibility of the void, the theory of supposition serves as a safeguard to prevent from falling into wrong inferences, e.g., whenever (even in a given imaginary 'case') a subject or a predicate is lacking any supposition.

Besides basic definitional questions, Blasius also appeals to conceptual tools pertaining to the theory of signification when it comes to more particular problems. Interestingly, he relies not only on the theory of supposition but also on

---

26 On Blasius' theory of the void, see S. Rommevaux, "Le vide *secundum naturam* dans les *Questions sur la Physique* de Blaise de Parme," in J. Biard and S. Rommevaux (eds.), *La nature et le vide dans la physique médiévale*, 293–316.

27 On fourteenth-century imaginary physics, see L. Queiroz, "On the Relation between Natural Philosophy, Mathematics and Logic in the Investigation on Atomism in the Middle Ages," *Enrahonar*, Supplement Issue, 2018, 135–149; E. Jung, "Mathematics and the *Secundum Imaginationem* Procedure in Richard Kilvington," *Przegląd Tomistyczny* 22, 2016, 109–120; H. Hugonnard-Roche, "L'hypothétique et la nature dans la physique parisienne du XIVe siècle," in Stefano Caroti and Pierre Souffrin (eds.), *La nouvelle physique du XIVe siècle* (Florence: Olschki, 1997), 161–177; H. Hugonnard-Roche, "Modalités et argumentation chez Nicole Oresme," in Pierre Souffrin and Alain-Philippe Segonds (eds.), *Nicole Oresme. Tradition et innovation chez un intellectuel du XIVe siècle* (Paris / Padua: Les Belles Lettres / Programma e 1+1, 1988), 145–163; H. Hugonnard-Roche, "Le possible et l'imaginaire dans la physique d'Albert de Saxe," in Joël Biard (ed.), *Itinéraires d'Albert de Saxe. Paris-Vienne au XIVe siècle* (Paris: Vrin 1991), 161–173; Jürgen Sarnowsky, "God's Absolute Power, Thought Experiments and the Concept of Nature in the 'New Physics'," in Stefano Caroti and Pierre Souffrin (eds.), *La nouvelle physique du XIVe siècle* (Florence: Olschki, 1997), 179–201.

other aspects of the logic of his time. For instance, when trying to explain in what sense a heavy body has an appetite for falling downwards, Blasius uses the distinction between the compound and divided senses of an expression to clarify what can be asserted about such a body:

> The first evidence is the following: this verb 'I desire [*appeto*]' placed in the title of the question, when it is used in some sentence, makes the term following it appellate the form and reason. Because it does not follow 'John's horse I desire, therefore I desire John's horse;' 'the wine of Grado I desire, thus I desire the wine of Grado.' And in this way it does not follow 'the King of Scotland's horse I promise you, therefore I promise you the King of Scotland's horse.' It is true that the common logic writes in the opposite way, which I am not concerned with at present.[28]

Blasius relies in this passage on a piece of Buridanian logic, namely the theory of appellation. The notion of appellation was not invented by Buridan himself, as it was already in use in thirteenth century logic, although not all logicians employed it (Ockham did not, for instance). Buridan employed this notion to characterize the intensional properties of terms following cognitive verbs like 'to know', 'to believe,' and so on.[29]

In a proposition such as '*x* desires *y*', the particular verb 'desires' does not make it possible to differentiate cases where all properties associated with *y* are epistemically accessible to the agent *x* and cases where they are not. I might desire to acquire a particular horse without knowing that this horse is in fact the King of Scotland's horse, to quote Blasius' example. Blasius rightly underscores that in such cases involving referential opacity, inferences between propositions become problematic.

---

28 Text 6 (IV, q. 3) in the present volume, n. 13.

29 Let us note that Blasius conflates here the two concepts of appellation of reason and appellation of form, which refer to different things in Buridan's doctrine. Let us note also that in Buridan the theory of appellation does not only concern cognitive verbs but has other functions involving modalities. On this theory, see C. Panaccio, "Ockham and Buridan on Epistemic Sentences: Appellation of the Form and Appellation of Reason," *Vivarium* 50(2), 2012, 139–160; Peter King, "Introduction," in John Buridan, *Jean Buridan's Logic: The Treatise on Supposition, The Treatise on Consequences*, trans. P. King (Dordrecht / Boston: D. Reidel, 1985), 3–82, esp. 17–24; G. Klima, "The Essentialist Nominalism of John Buridan," *Review of Metaphysics* 58(4), 2005, 739–754; G. Klima, "'Debeo tibi equum': A Reconstruction of the Theoretical Framework of Buridan's Treatment of the Sophisma," in S. Read (ed.), *Sophisms in Medieval Logic and Grammar. Acts of the Ninth European Symposium for Medieval Logic and Semantics, Held at St Andrews, June 1990* (Dordrecht / Boston: Springer, 1993), 333–347.

While mentioning the Buridanian concept of 'appellation' for this particular type of fallacies, Blasius relies in fact on a technique that was systematized by William Heytesbury (fl. 1330s) in his treatise on the compound and divided senses.[30] This technique consists in placing the nominal group included in the scope of an epistemic verb before or after the verb so as to differentiate its divided sense from its compound one.

Remarkably, Blasius applies this technique for answering a special problem related to the downward motion of heavy bodies, namely the problem of understanding *how* heavy bodies desire to fall downwards:

> Second conclusion: A heavy body existing out of its ⟨natural⟩ place by a curve desires to move downward. This clear from what precedes, because after being in equilibrium the heavy body moves downward naturally, and moving downward it follows a curved line; therefore, it desires to follow it and to move downward along it. It is clear indeed that a heavy body in equilibrium moves downward following a curve, because a heavy body fixed to an arm moves following the motion of the arm. Now in such a motion the arm describes a quarter of a circle whose circumference is an arc.
>
> Third conclusion: A heavy body existing out of its natural place does not desire to move by a straight ⟨line⟩. And this is the conclusion replying ⟨to the problem⟩. The conclusion is proved, because if it desired to move by a straight line, thus its natural inclination would be ⟨to move⟩ in a straight ⟨line⟩. Now, 'what is natural is not otherwise accustomed,' as is written in the second part of the *Ethics*. Thus, if the desire were to move along a straight line, it would never move along a curve—which is false and contrary to the previous conclusion. Hence, that 'a heavy body desires to move downward along a straight line' is the same as a heavy body that desires to move along a line according to the nature in virtue of which this line is straight. And thus such an appetite appears to be limited to this. Because of this I said in the first evidence that this verb 'desire' makes the next term appellate its reason. And, therefore, a heavy ⟨body when placed⟩ upward desires to move neither by a straight line nor by a curve, although by a straight line it desires to move. Similarly, by a curve it desires to be move, from which it follows that a heavy ⟨body⟩ does not desire to move along a straight line by which it desires to move. This is clear when taking the first evidence into consideration.[31]

30 William Heytesbury, *Tractatus de sensu composito et diviso*, Venice, Bonetus Locatellus for Octavianus Scotus, 1494, 2ra–4rb.

31 Text 6 (IV, q. 3) in the present volume, nn. 18–19.

A heavy body does not desire to move downward following a straight line if it is meant that the straight line is part of the body's inclination. A heavy body only desires (i.e., is inclined) to move downward.[32] The inclination of a heavy body is determined by a specific place and, consequently, a definite direction, but it does not necessarily entail a particular trajectory.

Yet, it is still legitimate to say that 'by a curve a heavy body desires to move' since, in this formulation, the anteposition of the adverbial phrase 'by a curve' with respect to the verb indicates that the proposition should be understood meaning here in the divided sense. By contrast, in the compound sense, the adverbial phrase would be included in the intensional scope of the verb. It is however incorrect to say that 'a heavy body desires to move by a curve' since in this case the adverbial phrase would be part of the verb's intensional scope.

An interesting aspect of this whole analysis is the hybrid character of Blasius' semantic treatment of a particular physical problem. On the one hand, this problem is not a general definitional problem but a particular puzzle regarding natural motion, namely the natural trajectory of bodies moving in space. This specific problem arises because, paradoxically, in certain contexts like those involving human artifacts (like a balance on which weights are suspended), a moving body does not follow the trajectory that might seem natural to it.

On the other hand, Blasius' account of such a paradox is mostly semantical and draws on two main logical devices: first, Buridan's theory of appellation which describes how cognitive verbs modify the normal way in which terms signify when placed after these terms; second, a method for disambiguating divided and compound senses that was systematically developed in fourteenth century by logicians such as William Heytesbury. The fine differentiation of the different meanings that a term can have depending on its place in a proposition was in late medieval logic important for the distinction between categorematic and syncategorematic uses of the same term—a distinction directly connected to the theory of continuity and infinity, corresponding to the second layer of Murdoch's hierarchy of analytical languages.[33]

---

32 It seems that the translation of '*moveri*' in this text and other similar passages as 'to move' (active form) makes more sense than 'to be moved'. Referring to his questions on Peter of Spain's logic, Federici Vescovini notes that Blasius treats *moveri* as a deponent verb for reasons systematically connected to his account of motion (see *Astrologia e scienza*, 281; cf. *Questiones super Tractatus logice magistri Petri Hispani*, I, q. 19, a. 1, 117–118).

33 For a study on Heytesbury's theory of divided and compound senses, see G. Sinkler, "William Heytesbury's Word-Order Theory of Propositional Sense," *Journal of the History of Philosophy* 27 (3), 1989, 365–377.

3 The Language of Infinity and Continuity

Let us open this overview of how Blasius employs the analytical language of infinity and continuity by citing a passage where the 'Diabolic Doctor' demonstrates both his familiarity with the paradoxes linked to these notions and some sense of self-referential humor:

> Third difficulty: Since a continuum is infinitely divisible, as was touched upon in the determination of the question, it is thus asked whether, given a continuum no matter how small, all the liberal arts and writers with all the sophisms of logicians and the vagaries of jurists could be written on it. The answer is positive, which is said because, given one foot in the first proportional part of it, you can write, for example, one proposition, and in the second proportional part you can write another one, and if they are not enough you can take a third proportional part next to this one, and so, because there is an infinity of such parts, you will never reach the last one.[34]

According to this argument, all the sophisms proposed by all the logicians who have ever existed could be written on any part of a body, no matter how small, since a body, being continuous, is divisible into an infinite number of parts. Let us note that, in this passage, Blasius does not consider a body as composed of an infinite number of parts of the same size but an infinite number of *proportional* parts, that is to say, such that every next part has the relation 1/2 to the previous one (one sophism being written in one half of the body, another on the quarter of the remaining half, another on the eighth of the remaining quarter, and so on *ad infinitum*).[35]

---

34 Blasius of Parma, *Quaestiones super VIII libros Physicorum* (secunda redactio), VII, q. 2, ms. Vat. Lat. 2159, 190[ra] [transcr. Rommevaux, provisional edition, slightly modified]: Tertia difficultas: quia continuum est in infinitum divisibile, ut tactum est in determinatione questionis, ideo queritur utrum, dato uno continuo quantumcumque parvo, possint in eo figurari omnes liberales artes et scribi cum omnibus sophismatibus loycorum et ambagibus legistarum. Respondetur quod sic, quod declaratur quia, dato uno pedali in prima parte proportionali eius, tu potes scribere gratia exempli unam propositionem et in secunda parte proportionali tu potes scribere unam aliam et si non fuerint sufficiens tu potes capere tertiam partem proportionalem iuxta istam et sic ultra, quia infinitae sunt tales, numquam devenies ad ultimam.

35 A similar sophism—this time involving science as a quality—can be found in Blasius of Parma, *Quaestiones super VIII libros Physicorum* (secunda redactio), I, q. 1, ms. Vat. Lat. 2159, 61[ra] [transcr. Biard, provisional edition]: Tertium argumentum: si notitia scientifica esset nobis possibilis, tunc ista esset subiective in anima. Sed probatur quod non, quoniam

3.1 *Imaginary Physics: The Epistemological Virtues of Thought-Experiments*

The paradoxes involving infinity and continuity constitute a crucial aspect in the application of mathematical reasoning to physics.[36] Blasius believes that mathematics does not represent the noblest science from the point of view of the object of science (where theology or astrology might be considered the noblest sciences) but he does see mathematics as the noblest science from the point of view of certitude and the beauty of demonstration.[37]

Despite its amusing character, the argument just quoted hardly adds anything to a well-known property commonly ascribed in an Aristotelian framework to continuous magnitudes, namely being potentially divisible an infinite number of times. Following what was the most dominant opinion on the matter at that time, Blasius indeed believes in the infinite divisibility of material bodies.[38] He does not dedicate any special question to the divisibility of matter, even though this theme appears quite often in his *Questions on the Physics*, in particular in a question devoted to the dimension of matter required for the reception of a substantial form. In the seventh question of the first book of the *Physics* (second redaction), Blasius defends the view that natural forms can in theory exist in any part of a material body, no matter how small.[39] In

tunc quilibet homo per quamcumque modicam cognitionem esset infinite sciens. Et probatur consequentia quia, ubi notitia scientifica informaret animam canis, ista anima esset aliqualiter sciens, et ubi informaret medietatem, ista medietas esset in duplo intensior quam primo erat, et sic in infinitum. Ergo, ubi illa notitia informaret animam indivisibilem, illa notitia esset infinite intensa, et ista anima esset infinite sciens. Cum igitur anima intellectiva sit indivisibilis, sequitur propositum.

36 Among the different questions on the *Physics* (in Blasius' second redaction), the ninth question on Book I ('Utrum per potentiam finitam vel infinitam possit aliquid fieri ex nihilo') contains interesting discussions of infinity, as well as does the seventh one ('Utrum in materia quantumcumque parva forma substantialis hora generationis producatur'). As well, the questions on Book VI, devoted to motion, contain many arguments involving infinity and continuity.

37 See J. Biard, G. Federici Vescovini, "Introduction," in Blasius of Parma, *Questiones super tractatus logice magistri Petri Hispani*, 18.

38 On this, see Biard, "Blasius of Parma Facing Atomist Assumptions;" Federici Vescovini, *Astrologia e scienza*, 337–340.

39 Blasius offers three ways of answering the problem of whether a form can exist in any part of matter, no matter how small. But his overall position appears to be the following: the only obstacle to the existence of a form in a very small piece of matter is the corruptive effect of a medium. Without it, a form could exist in any part of matter (no matter how small). Cf. *Quaestiones super VIII libros Physicorum* (secunda redactio), I, q. 7, 76^{ra-rb}. The first question on Book VI ('Utrum per bonas rationes concludi possit continuum esse ex indivisibilibus compositum') is also directly to this problem, although the problem of the

other words, Blasius refuses any absolute minima either in the structure of matter itself, which can always be divided into smaller parts, or in the capacity or potency proper to matter to receive forms.

When it comes to the fourteenth century, the logic of infinity did not only concern the property of infinite divisibility entailed by the continuity of matter. It encompassed the many sophisms one could form by conflating two distinct uses of the term 'infinite', namely the categorematic and the syncategorematic senses.[40] The term 'infinite' in the categorematic sense behaves like any other categorematic term, i.e., as an independent term that does not require another term to be meaningful. In the categorematic sense, the term 'infinite' refers to something positively infinite. Examples of 'infinite' in the categorematic term include infinite magnitudes or quantities.

In the syncategorematic sense of the term, 'infinite' behaves like an adverb that qualifies another term. When used in this sense, the term 'infinite' does not imply anything positively infinite, i.e., such that nothing greater can be conceived or posited. In the syncategorematic sense, 'infinite' is used to indicate that there can always be more and more of something. A typical medieval example is the series of natural numbers, which are infinite in the sense that, for any given number, there can always be a greater number, although no 'infinite number' exists as such (according to what was the standard view in the Middle Ages).

Blasius is familiar with this distinction. Most of the time, the paradoxes he puts forward in his physics have to do with the infinite divisibility of continuous magnitudes like motion, time, or bodies, and the infinite quantities resulting from their interaction. In fact, one of Blasius' typical ways of arguing for a particular position is to show that it leads to a contradiction when infinite quantities are to be posited. This is the case, for instance, in the following passage where Blasius considers the effect of rarefaction, which is conceived as a continuous natural change, on the color of a body undergoing this rarefaction, which is also a continuous magnitude (since the degrees or hues of a color are continuous):

> Third argument: if the question were valid, then it could be possible that, for an hour, this ⟨body⟩ would become whiter and whiter, and during the

continuum is dealt with in this question from a more general point of view; cf. ms. Vat. Lat. 2159, 161va–164va.

40 While being common in the fourteenth century, the nature of this distinction is far from obvious. On this, see S.L. Uckelman, "The Logic of Categorematic and Syncategorematic Infinity," *Synthese* 192(8), 2015, 2361–2377.

> same hour, it would continuously have less and less whiteness. I deduce this argument. For let *A* be a thing of one foot, half of which is white and the other black. And it is evident that in this case, *A* is neither white nor black. Now, more rarefaction is possible; let the white part be rarefied. And with this the whiteness is continuously attenuated in the white part. And then the proposition is clear. Because at the end of the hour, it is true to say 'this is white', since more than the half is white. And it is continuously true to say 'this is whiter than before,' since the denomination comes from the part. And because a larger and larger part will be white continuously, in the same way the denomination of whiteness will proportionally become greater and greater in a continuous way. The other ⟨part of the argument⟩ is clear, namely, that it will have less and less whiteness continuously.[41]

This argument is typical of the overlap between logical and physical considerations that were proper to the methods employed by the Oxford Calculators.[42] Here, the situation described involves a paradox: during a process of rarefaction, a body is said to become whiter while it simultaneously becomes less white. The paradox relies on a semantic ambiguity of the expression 'to be whiter,' which can be understood either as referring to the total extension of the colored body (which increases due to rarefaction) or to the degree of intensity of whiteness (which decreases in this scenario). The notion of 'denomination' appearing in this passage was a crucial concept that connected physical problems with logical considerations. In the works of the Oxford Calculators, the problem was often to understand under what conditions a predicate 'denominates' its subject, i.e., can be validly applied to it.

As evidenced by this argument, Blasius, much like his Oxfordian predecessors, has a keen interest in paradoxes lying at the intersection of physics and logic. Many of these paradoxes depend on the properties of continuity and infinity, which scholars have identified as an instrumental part of the new physics of the fourteenth century.[43] Mathematical developments on continuity and

---

41 Text 7 (IV, q. 7) in the present volume, n. 4.

42 For up-to-date information on the movement of the so-called Oxford Calculators, see the recent D.A. Di Liscia and E.D. Sylla (eds.). *Quantifying Aristotle: The Impact, Spread and Decline of the* Calculatores *Tradition* (Leiden / Boston: Brill, 2022) and the bibliography therein. See also E.D. Sylla, *The Oxford Calculators and the Mathematics of Motion, 1320–1350. Physics and Measurement by Latitudes* (New York / London: Garland, 1991) for a good overview of some of its main figures.

43 See N. Kretzmann (ed.), *Infinity and Continuity in Ancient and Medieval Thought* (Ithaca / London: Cornell University Press, 1982); P. Geach, "Infinity in Scholastic Philosophy," in

infinity appear to be intimately connected to imaginary physics, as they often arise from imaginary scenarios where bodies acquire infinite speeds or infinite sizes due to conditions that are impossible in natural circumstances. Let us give an example of such an imaginary scenario, which Blasius borrows from Richard Swineshead (fl. 1340s), nicknamed the "Calculator" in the late Middle Ages:

> Sixth ⟨argument⟩. Thus, it would follow that space could only be travelled through in an infinite time. The consequence is proved. And I presuppose two ⟨things⟩. The first thing is that the Earth is not properly located unless its center is the center of the world. I suppose, secondly, that 'velocity follows the proportion of proportions.' Given these premises, the intended ⟨conclusion⟩ is clear, for the deduction of which I imagine the sphere of Earth perforated through its center from both sides. Then let a columnar body made of pure earth be in our hemisphere, which is *AC*, whose middle is *B*. Now let this columnar body be allowed to move according to its nature through the hole. And I want to prove that it will move over an infinite time, because it will move until it is properly located, and consequently until point *B* is at the center of the world. This is clear from what has been supposed first. But before it is properly located, it will move according to a latitude of proportions composed of an infinity of proportions. For the second is half the first, the third is half the second, and so on infinitely. But in virtue of what has been supposed secondly, 'the proportion of velocities follows the proportion of proportions' and so on. Therefore, when this earth begins to move according to a double propor-

---

I. Lakatos (ed.), *Problems in the Philosophy of Mathematics* (Amsterdam: North Holland, 1967), 41–42. On the evolution of these themes in the fourteenth century, see E.D. Sylla, "Infinity and Continuity: Thomas Bradwardine and his Contemporaries," in S. Shapiro and G. Hellman, *The History of Continua. Philosophical and Mathematical Perspectives* (Oxford: Oxford University Press, 2021), 49–81; R. Cross, "Infinity, Continuity, and Composition: The Contribution of Gregory of Rimini," *Medieval Philosophy and Theology* 7, 1998, 89–110; J.E. Murdoch, "Infinity and Continuity," in N. Kretzmann, A. Kenny, and J. Pinborg, *The Cambridge History of Later Medieval Philosophy* (Cambridge: Cambridge University Press, 1982), 564–591; J.E. Murdoch, "Thomas Bradwardine: Mathematics and Continuity in the Fourteenth Century," in E. Grant and J.E. Murdoch (eds.), *Mathematics and Its Applications to Science and Natural Philosophy in the Middle Ages: Essays in Honour of Marshall Clagett* (Cambridge: Cambridge University Press, 1987), 103–137. On the thirteenth-century background, see C. Trifogli, "Matter and Form in Thirteenth-Century Discussions of Infinity and Continuity," in C. Leijenhorst, C.H. Lüthy, and J.M.M.H. Thijssen, *The Dynamics of Aristotelian Natural Philosophy from Antiquity to the Seventeenth Century* (Leiden / Boston / Cologne: Brill, 2002), 169–187.

> tion, there is an infinity of ⟨proportions of⟩ greater inequalities between the double proportion and the proportion of equality, according to which ⟨the earth⟩ moves downward following all of them. It follows that now it moves somewhat slowly, then twice more slowly; then four times more slowly, and so on infinitely.[44]

Here, the scenario portrays a heavy body moving toward the center of the Earth (and so the center of the universe). In this scenario, for the sake of the argument the Earth is imagined as being perforated. The body, so the argument goes, will not reach its final destination, which is the center of the Earth, because the parts of the body having already reached this destination will act as resistance on the other parts of the body. At each instant, while a new part of the body reaches the center, the resistance offered by all the parts of the body having reached the center will increase. It follows from this that a body will reach the center of the Earth infinitely slowly, which means that it will *never* reach it. This example illustrates the role played by infinity in the analysis of motion. The fact that a continuum is infinitely divisible and, thus, that the motion of a body toward the Earth's center may be infinitely slowed by the resistance, generates a paradoxical situation proving a physical impossibility.

3.2 *Mathematical Principles*

We can note that in the passage just quoted, Blasius speaks of a 'latitude of proportions.' This expression, which seems to have originated with John Dumbleton (active in the 1330s or 1340s), one of the so-called Oxford Calculators, is interesting because it amounts to applying the notion of latitude to a set of proportions conceived as a continuous series. The notion of 'latitude' was a technical concept of fourteenth-century physics, belonging to the vocabulary of the intension of forms, which primarily referred to a continuous range of values or degrees that a magnitude (motion, quality …) can take.[45]

Instead of merely considering a discrete series of proportions like 1:2, 1:4, 1:16, 1:32, etc., the notion of 'latitude of proportions' allows for the inclusion of all possible proportions between any two of these proportions, treating thus the set of these possible functions as a continuous (or dense, in modern terms) magnitude. The term 'latitude' thus represents the application of the vocabu-

44 Text 5 (IV, q. 2) in the present volume, n. 7.

45 On this notion and its history, see E.D. Sylla, "Medieval Concepts of the Latitude of Forms: The Oxford Calculators," *Archives d'histoire doctrinale et littéraire du Moyen Âge* 40, 1973, 223–283.

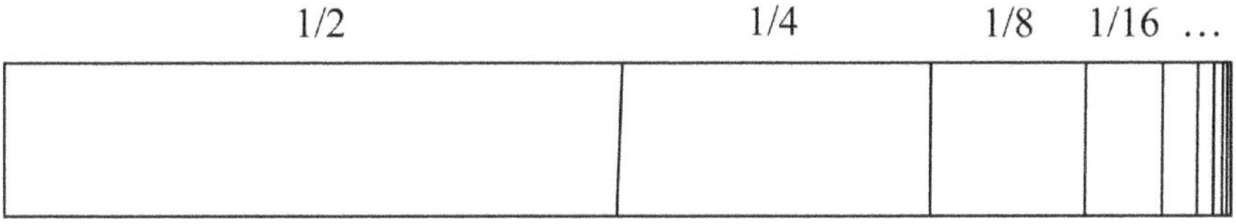

FIGURE 6.1 A body divided into an infinity of proportional parts

lary of the intension of forms to a mathematical function.[46] Here, we are dealing not (?) with a mathematical concept applied to a physical phenomenon, but with rather the reverse procedure: one concept coming from natural philosophy is applied to a mathematical object (a series of proportions) to treat it like a continuous magnitude. The presence of this expression in Blasius is not surprising, as it can be found in the works of other authors dealing with the Calculators tradition, developing similar paradoxes to their greatest extent, and with which Blasius was acquainted.[47]

From this perspective, Blasius' analytical treatment of these notions presents little originality compared to what had already been developed by his Oxfordian predecessors. We can, however, point out two notable features of his use of paradoxes based on continuity and infinity, both of which have to do with accepting actual infinities: (1) Blasius accepts the possibility of dividing bodies into an infinite number of proportional parts, from which he deduces paradoxical conclusions about the infinite power of bodies; (2) he accepts the equivalence between an infinite value and some indivisible state.

Let us examine these two points starting with the first. The initial notable feature of Blasius' treatment of infinity and continuity concerns his acceptance of the idea that a magnitude is composed of an actual infinity of proportional parts (see Fig. 6.1).

By the time of Blasius, accepting that there is an infinity of actual parts in a continuous body was no longer exceptional, as different thinkers of the fourteenth century, even some holding to the 'Aristotelian' view that a continuum is not composed of indivisibles, accepted this view.[48] That an infinity of actual parts like points exist in some continuum did not imply, of course, that these points actually *compose* this continuum.

46 Sylla, "Medieval Concepts of the Latitude of Forms," 264–270; E.D. Sylla, "The Oxford Calculators' Middle Degree Theorem in Context," *Early Science and Medicine* 15, 2010, 338–370, here on 354–357.

47 Cf., e.g., Richard Swineshead, *Liber calculationum* (Venice: Octavianus Scotus, 1520), tr. 1, 2[va].

48 This is the case of Gregory of Rimini (ca. 1300–1358), among others. See Cross, "Infinity, Continuity, and Composition."

At any rate, Blasius subscribes to a stronger thesis, namely that a continuous body may be viewed as composed of an actual infinity of proportional parts. This thesis makes it possible to compose different sophisms such as the one opening this section, where all sophisms that ever existed can be written on a finite continuous body divided into proportional parts. Here, Blasius not only acknowledges that one can divide a body potentially infinitely but he also assumes that the body is actually composed of an infinite number of proportional parts.

In the fourteenth century, both the Parisian and the Oxfordian tradition were interested in the logical analysis of infinity and continuity. However, this particular way of viewing the structure of bodies as a composition of (actually) infinite parts was typical of the way the Oxford Calculators created sophisms at the intersection of mathematics and physics. Many sophisms examined by Richard Swineshead in his *Calculationes*, for instance, involved bodies divided into an infinity of proportional parts. Here, Blasius seems to be primarily indebted to this Oxfordian tradition.

This acceptance of actual infinities insofar as the partitional structure of bodies is in line with Blasius' general stance toward infinite magnitudes or infinite quantities in the categorematic sense. This can be the case for extensive magnitudes like bodies or space, or even for the collection of points contained within a body, as illustrated in the following argument raised against the possibility of rarefaction:

> Fourth argument: if rarefaction were possible, then something could be rarefied. Therefore, it rarefies either uniformly or difformly. But it is proven that it does not, because if something were to rarefy uniformly, then it would acquire an infinite quantity through its rarefaction. And the argument holds, because if something were to rarefy uniformly, then all pairs of points ⟨located⟩ at an equal distance before the rarefaction would be more distant due to the act of rarefaction than before. And since there are infinite points in the body that would be augmented, it follows that infinite non-communicating quantities would be acquired, as I wish to say that between any two points equally distant, there will be equal non-communicating quantities. And thus, if it rarefies uniformly, any quantity of this ⟨body⟩ located between two points rarefies and acquires a new quantity, from which, however small it may be, I deduce that infinite quantities, which are non-communicating and equal to each other, are acquired. Now, such ⟨quantities⟩ appear to make an infinite quantity. If, however, the rarefaction is difform, the same incongruity would be deduced. And I would designate the quantity existing between

> two points very close to the limit that rarefies less intensely. And so the argument proceeds.[49]

The reader will note that when speaking about 'points,' Blasius does not commit himself to the extra-mental existence of these entities as indivisible realities that actually compose the body. In the fourteenth century, a nominalist could well talk about indivisible entities like points without treating these entities as real.[50] More generally, Blasius is happy to invoke in his *Questions on the Physics* all sorts of indivisible entities like limits, points, and instants while demonstrating a general allegiance to nominalism.

In this argument presumed to refute by a *reductio ad absurdum* the possibility of rarefaction, the quantity acquired through rarefaction is infinite because the points contained in a body are infinite and, so, rarefaction would entail infinite distances between these points. The argument is obviously a sophism, one which Blasius rejects. This sophism relies on counting all the points included between two points and, therefore, on counting all proper subparts of a part of a body. This way of counting creates a proliferation of points stemming from the existence of an infinity of points between any two of them. But while the argument draws on this conflation of parts and subparts of a body, Blasius does not see any difficulty with the idea of an infinity of points within a body.

This feature does not only apply to extensive magnitudes but also intensive ones. In the following passage, for instance, it is plain that Blasius does not see any intrinsic contradiction in postulating an infinitely intense quality:

> And for this, consider the case where *A* acquires some whiteness of degree 4, and *B* some infinitely intense whiteness, by way of imagination. And then the proposition becomes clear. Therefore, just as an infinite whiteness is not comparable to a finite whiteness, in the same way the acquisition of an infinite whiteness is not comparable to the acquisition of a finite whiteness, since the acquisition of an infinite whiteness is no different from the infinite whiteness that is acquired or can be acquired.[51]

---

49 See text 7 (IV, q. 7) in the present volume, n. 5.

50 On this, cf. J. Celeyrette "La problématique du point chez Jean Buridan," *Vivarium* 42, 2004, 86–108.

51 Blasius of Parma, *Quaestiones super VIII libros Physicorum* (secunda redactio), VII, q. 5, ms. Vat. lat. 2159, 194va [transcr. Rommevaux, provisional edition]: Et pro isto pone casum quod *a* in hora acquirat albedinem quattuor graduum, *b* vero acquirat albedinem infinitam intensive quoad ymaginationem. Et tunc patet propositum. Unde, sicut albedo infinita

We may note that Blasius in this passage follows his conviction that motion (here in the case of alteration) is nothing other than what is acquired by a body (a quality).[52] The infinite speed through which a quality is acquired does not change anything of this ontological principle. But, most importantly, such an appeal to infinity takes place in what Blasius expressly describes as an imaginary physics (or physics *quoad imaginationem*). For Blasius, imagining a scenario where a situation, though not physically achievable, is nevertheless possible in the sense of lacking formal contradiction, serves as a method of inquiry for studying properties related to qualities, bodies, and speed.

The second point mentioned above as typical of Blasius' physics is the equivalence he admits between an infinite value and some indivisible state. This point is best illustrated through an example that once again involves the dimensional aspect of bodies and their ability to be condensed:

> Then to the problem it must be said that something can be condensed to non-extension (*non quantum*) only by the parts becoming closer and the removal of quantity. And in this case, it must be admitted that the non-extended thing is infinitely dense, because with the parts becoming closer there is continuously more and more matter in a smaller place and eventually in a non-extended place.[53]

We can see that Blasius postulates here an equivalence between an infinite series of iterations and the actual infinite term of this series. In this example, a body undergoing an infinite process of condensation reaches the size of a point, being equivalent to a non-extended body and being therefore infinitely dense. Similar cases involve the equivalence of a point-like body and a body undergoing an infinite process of physical division, as well as the equivalence between an infinite speed and instantaneous motion.[54]

---

non est comparabilis albedini finite, ita acquisitio albedinis infinite non est comparabilis acquisitioni albedinis finite, cum acquisitio albedinis infinite non sit aliud quam albedo infinita successive acquisita vel acquisibilis.

52 On this point, see the essay on Blasius' theory of motion in this volume.

53 See text 7 (IV, q. 7) in the present volume, n. 45: Tunc ad difficultatem dicendum quod per solam approximationem partium et remotionem quantitatis potest aliquid condensari ad non-quantum. Et in hoc casu concedendum est illud non-quantum esse infinite densum, quia in tali partium approximatione fit continue plus et plus de materia in minori loco et tandem in non-quanto loco.

54 See, among similar examples, text 3 (II, q. 7), in the present volume, n. 59, where a thought-experiment involves the sun moving around the Earth with an infinite speed, in such a way that a revolution happens at every instant. Let us note that, in the fifth question on

The important point is that Blasius is not only interested in the mathematical properties (size, volume, or kinetic properties) of a point-like body, but that he analyzes the physical behavior of such a body, as evidenced by the way he mixes these thought-experiments involving indivisible bodies with considerations about the causal powers of bodies. Following the hypothesis that a body's power could be intensified by a reduction of its size or, in other words, by condensation, Blasius shows that this power could become infinite:

> Fourth argument: if a one-foot ⟨body⟩ *A* maximally hot were made indivisible like a point through condensation, then in this case, if the question were true for the affirmative part, *A* could not act on a patient B, because then it would not act on the patient according to any part, because it would have no part. However, the consequent is false, because it would act infinitely faster than when it was earlier extended. And it would induce a greater latitude (*sc.* of heat), which is obvious from this, because in this case it would be as if it were a one-foot uniformly hot ⟨body⟩, and the heat of the whole would be placed in its half. Then this half would be twice as hot as the first. And if it were placed in the fourth ⟨proportional⟩ part, it would be four times hotter than the first. Therefore, if the heat informing uniformly the one-foot ⟨body⟩ were placed in a point, this point would have an infinitely intense heat, and consequently it could induce an infinitely intense heat.[55]

In the *Questions on the Physics*, Blasius does not dwell on the assumptions underlying this type of thought-experiment, nor does he justify his conception of infinite series. However, we know from his questions on Thomas Bradwardine's *Treatise on Proportions* that he was open to admitting the existence of an end term to an infinite series, and to considering infinite iterations as a valid way to transition from a divisible object to something indivisible.

To understand this position, a few words on Blasius' conception of infinity in his treatment of proportions are necessary. In question 7 of his work, Blasius asks whether there exists a ratio between a ratio of greater inequality, i.e., a ratio such as m: n with $m > n$, and a ratio of equality, i.e., a ratio such as m: n with $m = n$.[56] Bradwardine himself denied that any ratio exists between a ratio

---

Book v, Blasius raises a similar doubt involving infinitely fast revolutions of the sun around the Earth (cf. ms. Vat. Lat. 2159, 144vb).

55 See text 8 (v, q. 1) in the present volume, n. 6.

56 Blasius of Parma, *Questiones circa tractatum proportionum magistri Thome Braduardini*, eds. J. Biard and S. Rommevaux (Paris: Vrin, 2005), q. 7, 112–124.

of greater (or lesser) inequality and a ratio of equality.[57] His justification of this point lies in his particular conception of what it means for a ratio to be greater or lesser than another. Accepting a principle classically employed for defining ratios between numbers or magnitudes,[58] Bradwardine posited that two things can be compared by a ratio if and only if the lesser can exceed the greater by a being multiplied a certain number of times, i.e., for any two numbers m and n such as $m > n$, if and only if there is a positive integer a such that $m < na$.

However, in his questions on Thomas Bradwardine's *Treatise on Proportions*, Blasius contrasts this view with another one according to which ratios can be compared when their 'denomination,' i.e., the name used to designate the relation between their two terms, can be compared. According to this alternative approach, ratios of greater inequality may be compared to ratios of equality and even to ratios of lesser inequality, since their denominations can be compared. For instance, the denomination of ratio 3:1 (called in medieval terms the 'triple ratio') is greater than the denomination of the ratio 1:2 (called the 'subduple ratio').[59]

Accordingly, Blasius considers two opposite views that ratios of equality may be compared to ratios of inequality both as probable and having good arguments supporting each perspective. From one point of view, it is perfectly legitimate to say that a ratio of greater inequality is infinitely greater than the ratio of equality:

> First conclusion: every ratio [*proportio*] of greater inequality, whatever it may be, contains infinite ratios, each of which is a ratio of greater inequality. [...] Second conclusion: every ratio of greater inequality contains infinite ratios, each of which is a ratio of equality. [...] Third conclusion: every ratio of greater inequality is infinitely greater than a ratio of equality.[60]

57 Thomas Bradwardine, *Tractatus de proportionibus velocitatum in motibus*, ed. H. Lamar Crosby, Jr., in *Thomas of Bradwardine. His* Tractatus de proportionibus. *Its Significance for the Development of Mathematical Physics* (Madison: The University of Wisconsin Press, 1961), 80, ll. 350–351: Septima conclusio: Proportione aequalitatis nulla proportio est maior vel minor.

58 Euclid, *Elements*, v, def. 4; Archimedes, *On the Sphere and Cylinder*, I, def. 5.

59 On this, see the illuminating explanations of Biard and Rommevaux, "Introduction," in *Questiones circa tractatum proportionum magistri Thome Braduardini*, pp. 11–48, here on 26–27. On Blasius' relation to Bradwardine, see also Federici Vescovini, *Astrologia e scienza*, 354–363.

60 Blasius of Parma, *Questiones circa tractatum proportionum magistri Thome Braduardini*, q. 7, 118: Prima conclusio: omnis proportio maioris inequalitatis, quecumque sit illa,

That a proportion of greater inequality contains an infinity of proportions of equality may seem surprising. Blasius illustrates this position with a physical case—although this analogy may not appear easily transferable to mathematical proportions—where a magnitude *A* having twice the size of another magnitude *B* contains an infinity of parts having the same size as *B*. As can be seen from this illustration, Blasius is clearly interested in the implications of this view for physical problems related to physical quantities and velocities.[61]

The result of this approach to proportions is that a ratio of equality (which is not in itself divisible into component ratios) can be seen as the end term of an infinite series of ratios of inequality. This view, explicitly opposed to Bradwardine's own axiomatization of comparisons between ratios (but found in other theorists of proportions like Oresme[62]) is taken up in the *Questions on the Physics*, where it clearly underlies thought-experiments involving infinite series leading to some actually infinite magnitudes, instantaneous motions, or indivisible states:

> But the proportion of two to one is a proportion of greater inequality, which is infinitely greater than a proportion of equality.[63]

These observations allow us to conclude that Blasius accepts as a general principle the possibility of actual infinite quantities, which he introduces in physical problems involving different aspects of the natural world. His particular way of manipulating the language of continuity and infinity is central to his exploration of imaginary physics, which exhibits a more empirical orientation

infinitas continet proportiones quarum quelibet est proportio maioris inequalitatis. […] Secunda conclusio: omnis proportio maioris inequalitatis infinitas continet proportiones quarum quelibet est proportio equalitatis. […] Tertia conclusio: omnis proportio maioris inequalitatis est proportione equalitatis in infinitum maior.

61 Blasius of Parma, *Questiones circa tractatum proportionum magistri Thome Braduardini*, q. 7, 119, 124.

62 Nicole Oresme, *Questiones super Physicam*, I, q. 11, 80, ll. 62–68: […] Proportio est augmentabilis in infinitum et etiam diminuibilis, quia sequitur inequalitatem; immo est ipsamet inequalitas. ⟨Maior enim inequalitas⟩ et equalitas est in infinitum maior proportio quam minor inequalitas. Patet statim, quia ⟨sicut⟩ proportio dupla, ita ⟨maior⟩ inequalitas minueretur in infinitum antequam veniret ad equalitatem, ut patet in exemplo adducto de inequalitate. Cf. *De proportionibus proportionum*, in *Nicole Oresme: "De proportionibus proportionum" and "Ad pauca respicientes"*, ed. E. Grant (Madison / Milwaukee/ London: The University of Wisconsin Press, 1966), 148, ll. 129–132.

63 Blasius of Parma, *Quaestiones super VIII libros Physicorum* (secunda redactio), VII, q. 9, 207[ra]: Sed proportio duorum ad unum est proportio maioris inaequalitatis, quae in infinitum maior est proportione aequalitatis.

in investigations compared to the more logically focused investigations of the Oxford Calculators. With his liberal attitude toward infinite quantities and his readiness to equate them with indivisible entities, Blasius creates a discursive space where a great number of scenarios can be considered to test more effectively hypotheses about the structure and laws of nature.

4 The Theory of Intensities in Blasius' Physics

The analytical languages most directly connected to the concepts of continuity and infinity were that of the theory of proportions, on the one hand, and that of the intension of forms or, in more contemporary terms, the theory of intensive properties, on the other. These two theories constitute the third and highest level in the hierarchy of analytical languages theorized by Murdoch. In a recent work on this topic, Biard notes that the theory of proportion is primarily applied by Blasius to the analysis of local motion, and only secondarily to the intension of forms.[64] Nonetheless, it is notable that Blasius frequently relies on the theory of proportions to address the issues of the intensity of forms, and that the case of local motion was directly linked, if not encompassed, within the theory of intensive properties, for reasons which will shortly appear. Given that some elements of Blasius' theory of proportions have already been discussed in the preceding section, and that scholars have extensively examined this topic,[65] this section will focus on the utilization of the theory of intensive properties in his *Questions on the Physics*.

The theory or 'analytical language' of intensive properties holds a significant position within Blasius' philosophical and scientific framework, as it recurs in the majority of his writings on natural philosophy. Important results proper to the language of intensive properties, such as the famous Merton theorem of the mean degree, stating that a 'uniformly difform latitude' (a set of degrees chan-

64 Biard, "Blasius of Parma on the Calculation," 253. Cf. p. 242 where Biard notes that Blasius is reluctant to use the two main types of proportions (arithmetic and geometric) to measure intensities.

65 See Biard and Rommevaux, "Introduction," in *Questiones circa tractatum proportionum* S. Rommevaux, "Les règles du mouvement de Blaise de Parme dans les deux versions de ses '*Questions sur le traité des rapports de Thomas Bradwardine*'," in J. Biard and S. Rommevaux (eds.), *Mathématiques et théorie du mouvement, XIVe–XVIe siècles* (Villeneuve d'Ascq: Presses Universitaires du Septentrion, 2008), 31–52; A. Harrison, "Blasius of Parma's critique of Bradwardine's *Tractatus de proportionibus*," in *Scienza e filosofia all'Università di Padova nel Quattrocento*, ed. Antonino Poppi (Padua / Trieste: Edizioni Lint, 1983), 19–69; Federici Vescovini, *Astrologia e scienza*, 323–365.

ging constantly, i.e., separated by an equal difference) corresponds to its mean degree, appear in Blasius' questions on the *Physics*. Surprisingly, they appear in unexpected contexts, such as when investigating whether bodies move to their natural place when obstacles preventing them doing so are removed.[66]

While Blasius frequently draws on the theory of intensities in his works, he explicitly elaborates on it at three distinct junctures: firstly, in a dedicated question specifically addressing the problem; secondly, in a separate treatise on the 'latitude of forms' where Blasius deals with the geometric treatment of intensities; and thirdly, in his tenth question (*secunda redactio*) on Book 5 of Aristotle's *Physics*.[67]

Several aspects of Blasius' perspective on the theory of intensities, which delves into the understanding of how properties can undergo intensive variations and the methods for quantifying these variations, have been examined by scholars.[68] Consequently, I shall confine my analysis here to examining the role of this theory within the specific context of his questions on the *Physics*, focusing on two aspects: the ontology of intensive properties associated with his utilization of this theory, and the range of its applications across the various fields and themes addressed in this work. These two points will make it possible to appreciate the central role of the theory of intensities for the unity of scientific knowledge, as this theory bridges distinct fields of knowledge by allowing the application of a common vocabulary to very different types of objects.

4.1 *What Are Intensive Properties?*

Discussions of intensive properties in the Latin philosophical tradition came from distinct sources. Some elements derived from medical writings on degrees of qualities, while others originated from theological discussions, most precisely from debates over the intensification of charity in human beings.[69] In

---

66 See for instance text 5 (IV, q. 2) in the present volume, n. 12. Several studies have been devoted to this important piece of medieval physics and its influence. Suffice it here to refer the interested reader to Sylla, "The Oxford Calculators' Middle Degree Theorem in Context."

67 Let us note that some elements relevant to this theme may also be found in Blasius' logical writings; cf. Blasius of Parma, *Questiones super tractatus logice magistri Petri Hispani*, III, q. 16, "Utrum termini de predicamento qualitatis sunt de se invicem predicabiles vel de suis inferioribus cum hiis adverbis 'magis' et 'minus'," 322–325.

68 Biard, "Blasius of Parma on the Calculation;" Di Liscia, "Biagio Pelacani da Parma's Geometrisation of Latitudes," Federici Vescovini, *Astrologia e scienza*, 205–238; Federici Vescovini, "La '*Quaestio de intensione et remissione formarum*'," 433–466.

69 A. Maier, "Das Problem der intensiven Grösse," in *Zwei Grundprobleme der scholastischen*

the fourteenth century, these elements turned into a sophisticated set of concepts which constituted what Murdoch called an 'analytical language.' Just like the theory of supposition and the theory of infinity and continuity, the analytical language of intensities was not one single theory in the sense of providing a unified account of what intensive properties are and how they should be measured. Rather, this language was a set of technical terms comprising notions like 'degree,' 'latitude,' 'intension,' 'remission,' and a few other related ones in which *different theories* could be formulated. These notions, in other words, could be defined, analyzed, and related to one another in quite different ways.

The language of intensities with all its associated concepts gained broader application in the fourteenth century, extending to different disciplines among which natural philosophy was probably the most important. In natural philosophy, this language served to describe and analyze velocities and motions in space. It was even employed to refer to changes pertaining to quantity like condensation and rarefaction, which at first glance have little to do with intensive properties strictly speaking.[70]

However, this vocabulary had been primarily elaborated in order to understand the nature and behavior of qualities. It is only at a later stage of the reflections on qualities that the terms originally designed to analyze them were gradually applied to local motion. This application stemmed from the fact that nothing prevents one from treating local motion as an intensive property, where the varying velocities of a body in motion represent the different 'degrees' of motion. This is why, while some authors were cautious to clearly differentiate the analysis of qualities from that of local motion, others tended to conflate these two types of changes because both could be viewed as two instances of intensive properties. For instance, the English philosopher and scientist Thomas Bradwardine (ca. 1300–1349), one of the initiators of the so-called 'Oxford Calculators' movement, formulated in his influential *Treatise of Proportions* (1328) conclusions about motion which were expressed in a general manner and could theoretically be applied to all types of changes, including qualitative changes.[71]

A consequence of this deep connection to the nature of qualities is that a preliminary issue related to the theory of intensities was to understand how

---

Naturphilosophie (Rome: Edizioni di storia et letteratura, 1968), 1–109; Solère, "Plus ou moins;" Sylla, "Medieval Concepts of the Latitude of Forms;" E.D. Sylla, "Medieval Quantifications of Qualities: The 'Merton School'," *Archive for the History of Exact Sciences* 8, 1971–1972, 9–39.

70 See, for instance, text 7 (IV, q. 7) in the present volume, nn. 31–33; ibid., nn. 41–42.

71 Thomas Bradwardine, *Tractatus de proportionibus*: see esp. c. 3, 112–114.

qualities can vary in intensity, from an ontological point of view. Four main types of theories regarding the nature of intensive variations had been proposed since the mid-thirteenth century, each of which could be traced back to theories formulated by ancient philosophers:

a. The theory of participation
b. The addition theory
c. The admixture theory
d. The succession theory

These theories should be viewed as *families* of theories rather than single well-defined theories. The description of these theories, which have been well studied, is beyond the scope of this paper.[72] Blasius mainly considers two of them as plausible: first, the addition theory, which had been originally designed by thirteenth-century Franciscans, before spreading outside of the Franciscan circle and being widely adopted in the fourteenth century; second, the succession theory, defended by Godfrey of Fontaines (ca. 1350–1306) and Walter Burley (ca. 1255–1344), among others. The addition theory of intense properties and the succession theory were strongly opposed on a central point: according to the addition theory, intensive properties are composed of 'degrees' which, together, constitute a quality. A variation of intensity, in this theory, is a change in the quantity of degrees present in a particular quality. The addition theory regards intensive variations as structurally similar to quantitative changes, in that it involves degrees of intensities rather than material parts.

By contrast, the succession theory held that no quality can be really composed of different degrees or parts. In this view, every quality is a simple element, insofar as it is a form, and that forms possess a nature essentially different from the partitional structure that characterizes matter. On this view, qualities should be identified with types of quantities. The succession theory nonetheless allowed one to speak of 'degrees' of qualities to describe their intensity. Unlike in the case of the addition theory, however, this reference to degrees in this framework did not designate a real composition of degrees internal to qualities. Rather, it was only a way to describe a quality which, in itself, was conceived as a metaphysically simple entity.

---

72 For comparative analyses of these theories, see Maier, "Das Problem der intensiven Grösse"; Sylla, *The Oxford Calculators and the Mathematics of Motion*; Sylla, "Medieval Concepts of the Latitudes of Forms"; Sylla, "Medieval Quantifications of Qualities;" E. Jung, "Intension and Remission of Forms," in H. Lagerlund (ed.), *Encyclopedia of Medieval Philosophy* (Berlin: Springer, 2011), 551–555; S. Roudaut, *La mesure de l'être. Le problème de la quantification des formes au Moyen Âge (ca. 1250–1370)* (Leiden / Boston: Brill, 2021).

The opposition between the different theories of how qualities vary involve, as can be seen, metaphysical questions over the nature of forms and the different types of forms that exist. As for other topics where he finds two opinions to be plausible, Blasius thinks that both the succession of theory and the addition theory have good arguments supporting them. Thus, he does not clearly choose between these two theories, although his preference seems to go the addition theory:

> And according to this, it must be held that when a later degree comes [to be], the preceding one is corrupted. Because such a thing does not sound good in the ears, it is conceded that intensification occurs by adding a degree to a degree, while the prior one remains. And when it is inferred that then in our hemisphere from sunrise to the third hour there was infinite light, etc., I say that many are forced by this argument to say that intensification does not occur by adding a degree to a degree as above. However, I will say in response to the argument that it does not follow from this that any agent produces an infinite light or an infinite heat. But it is well conceded only that with any given light, a more intense light will be induced, and with any given heat, a more intense one will be induced without reaching the extreme.[73]

Blasius refers in this passage to an argument inspired by the ideas of Walter Burley, the most famous proponent of the succession theory. Burley remarked that if the addition theory were true, then one and the same accident (a degree of heat) would migrate from one subject to another (different parts of the atmosphere).[74] Another argument of succession theorists was that, if the addition theory were true, then the air of our hemisphere would be infinitely hot

---

73 Blasius of Parma, *Quaestiones super VIII libros Physicorum* (secunda redactio), v, q. 10, 161^{ra-rb}: Et iuxta hoc est tenendum quod in adventum posterioris gradus precedens corrumpitur. Quia tale hoc non bene sonat in auribus, conceditur intensionem fieri per additionem gradus ad gradum, priori remanente. Et cum infertur quod tunc in emisperio nostro ab ortu ad horam tertiarum esset lumen infinitum etc., dico quod per hoc argumentum coguntur multi ad dicendum quod intensio non fiat per additionem gradus ad gradum ut supra. Dicam tamen ad argumentum quod ex hoc non concluditur aliquod agens producere lumen infinitum nec caliditatem infinitam. Sed bene conceditur tantum quod quocumque lumine dato inducetur lumen intensius et quacumque caliditate data citra summam inducetur intensior.

74 Walter Burley, *De intensione et remissione formarum* (Venice: Octavianus Scotus, 1496), 5^{ra-rb}, 14ra; cf. James of Forli, *De intensione et remissione formarum* (Venice: Octavianus Scotus, 1496), 18vb.

because the air would receive continuously (at each instant) new degrees of heat from the sun, which should be added to one another in the air. Given the infinity of instants following from the infinite divisibility of time, there is an infinity of instants within any given interval of time, and thus an infinity of degrees induced in the air over this interval. Blasius concedes that the addition theory does not entail infinite values even if one takes into account the infinity of instants during which degrees are added, without however explaining how that type of conclusion may be avoided.

4.2 *Powers of Intensive Properties and How to Measure Them*

Two additional features of Blasius' ontology of intensive properties deserve to be mentioned here. The first has to do with the types of powers associated with qualities. Blasius follows the Parisian tradition which distinguished two types of powers within a quality, assigning to them different degrees: a power of acting, on the one hand, and a power of resisting, on the other.

> Second difficulty: Does something hot as four [degrees] and something cold as four [degrees] act and react when applied to each other?
>
> To this, some said no, because then the action would result from a proportion of equality.
>
> But without many arguments, let it be said that they act and react to each other. For, in action, the active power of one thing should not be compared to the active power of another, but rather to its resistance, and vice versa. It stands, therefore, that something hot as four has a resistance of two, and something cold as four has a resistance of two. Because in this case both act on both.[75]

Distinguishing these two powers was a way Parisian Masters of the mid-fourteenth century had found to explain how processes of reaction between bodies of equal power are possible. According to the way medievals interpreted Aristotle's account of action, an action can only take place if the power of a body (the 'agent') overcomes or exceeds the power of another body on which it acts (the 'patient'). A difficult point was to understand how a process of reaction can take place when two bodies of the same power react to each other, and

75 See text 8 (v, q. 1) in the present volume, nn. 33–35. Let us note that, although Blasius distinguishes active power and resistance for heat and coldness, he attributes in this passage the same values to both qualities, a point on which Parisian authors (John Buridan, Marsilius of Inghen …) would disagree.

even more so how a body seemingly weaker than another can react to it (like when a hot iron bar plunged into cold water warms it a bit, despite itself being cooled).

John Buridan, Albert of Saxony, Nicole Oresme, and Marsilius of Inghen chose to explain this process by distinguishing two types of degrees present in qualities which are not equal.[76] While heat, for instance, is the most active primary quality, it is less resistive than other qualities. This distinction between two degrees allowed the explanation of how a cold body, despite being heated by a more powerful (hot) agent, can still react on this more powerful hot body. As evidenced by the passage just quoted, Blasius accepts this position, which was generally rejected by the Oxfordian authors he draws on elsewhere, namely William Heytesbury, John Dumbleton, and Richard Swineshead.[77]

Another noteworthy aspect of Blasius' ontology of intensive properties lies in what I have called elsewhere the 'new admixture theory' of addition, which combined elements of the addition theory with ideas of the admixture theory of intensive properties.[78] According to this last theory, intensive variations are due to the different proportions according to which opposite qualities are mixed within the same body. For instance, degrees of whiteness depend on the presence of blackness in the same subject: the less black is present, the whiter is the object. The main problem of this view was that it directly contradicts the axiom according to which contraries cannot be simultaneously present in the same subject. But the idea of a mixture responsible for intensive variations gained nonetheless prominence in the fourteenth century under a different form. It became more and more common to claim that, although qualities cannot be present together in the same subject, *degrees* of opposite qualities can be simultaneously present when the qualities are not at their highest degree. For instance, if the maximum of a quality is (conventionally) fixed at 8 degrees, a body can contain 5 degrees of heat and 3 degrees of coldness, although it cannot contain 8 degrees of heat plus any additional degrees of coldness.

---

76 On this theory, see S. Caroti, "Da Buridano a Marsilio di Inghen: la tradizione parigina della discussione 'De reactione'," *Medioevo* 15, 1989, 173–233.

77 On the Oxfordian approach to reaction, generally opposed to the Parisian one, see also S. Caroti, "Da Walter Burley al 'Tractatus de sex inconvenientibus'. La tradizione inglese della discussione medievale 'De reactione'," *Medioevo* 21, 1995, 257–374.

78 S. Roudaut, "Paul of Venice's Theory of Quantification and Measurement of Properties," *Noctua* 9(2), 2022, 104–158. On this evolution of the addition theory, cf. also S. Caroti, "Some Remarks on Buridan's Discussion on Intension and Remission," *Vivarium* 42(1), 2004, 58–85; M. Clagett, *Giovanni Marliani and Late Medieval Physics* (New York: Columbia University Press, 1941), 37–38.

Blasius accepts this theory, which can be viewed as a refinement of the addition theory, since it combines its epistemological virtues with a dynamic explanation of the admixture theory based on the theory of contrariety. Blasius seems to admit that the sum of degrees remains constant when primary qualities are concerned, meaning that the loss of one degree entails the acquisition of a degree of the opposite quality.[79] This principle entails that when one quality reaches its maximal degree, its opposite quality must conversely reach its minimal degree. From this, Blasius deduces consequences that directly echo paradoxes raised by Richard Swineshead in the first treatise of his *Liber calculationum*.[80] Blasius argues in particular that even when a quality is at its maximal degree, the opposite quality will still possess some degree of intensity. This is because if it did not, the more intense quality would have to be infinitely intense.

This paradoxical conclusion stems from the inverse proportional relations between degrees of opposite qualities present in a subject, given the assumption that the intensity of a quality is understood by reference to the non-degree. This last point means that the intensity of a quality depends on its distance from its absence, also called its non-degree. A quality is twice more intense than another if it is twice more distant from this non-degree, and it is twice less intense if it is twice closer to it. Blasius' conclusion that a quality at its maximal degree must be accompanied by its opposite quality still possessing some degree stems from the infinite divisibility of the continuum (the intensive degrees being conceived as continuous like other physical magnitudes, i.e., time, space, and motion).

To understand this point, let us consider two qualities (e.g., heat and coldness) present with the same intensity in the same subject. Given the principles just mentioned, these two qualities must be both at their mean degree. Let it be admitted, as a convention, that the maximum degree is equal to 8 degrees (which Blasius often does), in such a way that this heat and this coldness will be both at degree 4 at *t1*. Now, if one quality intensifies and reaches its maximal degree (degree 8) at *t2*, it will be twice more intense than before, being now twice farther from the non-degree than it was at *t1* (degree 4). Conversely,

79 Blasius of Parma, *Quaestio de intensione et remissione formarum*, 501. Blasius admits this principle of conservation but restricts it to primary qualities, while denying it for other properties like colors. In a second 'determination,' Blasius calls this principle into question but appears to accept it nonetheless (503).

80 Richard Swineshead, *Liber calculationum*, tr. 1, 2vb. Let us note that in this treatise, Swineshead does not draw exactly the same conclusion. His own conclusion does not concern two opposite qualities but rather the degree of intensification of a quality and its conversely proportional degree of attenuation or remission.

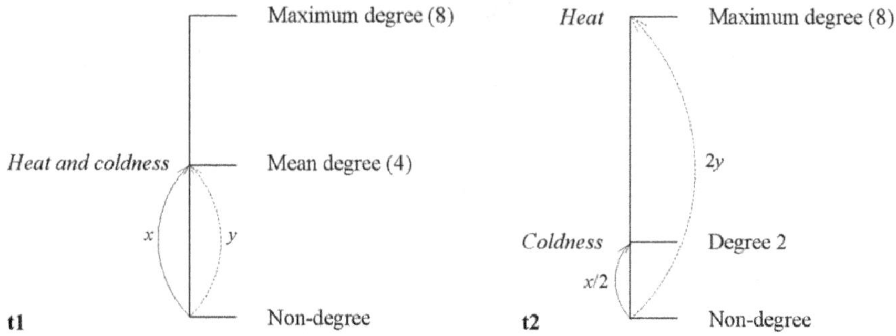

FIGURE 6.2 Representation of the inverse relations between heat and coldness in a subject

coldness will be now twice less intense, being now twice less distant from the non-degree. But one can see that the new state acquired at *t2* by the now less intense quality is not the non-degree itself, but only half the distance between the mean degree and the non-degree (see Fig. 6.2).

To reach the non-degree, this quality should be divided an infinite number of times, since the distance to the non-degree can always be divided without ever reaching it through a finite number of divisions, whereas the other quality is already at its maximal degree. It follows that, despite being inversely proportional, two opposite qualities are such that when one is at its maximal degree, the other still has some degree of intensity and is not completely absent. This point fits well with Blasius' inclination to integrate medicine in his physics, since the view that every body always contains a mixture of opposite qualities was key to the medical theory of mixtures.[81] These reflections further demonstrate the intimate connection between the theory of intensities and the analytical language of infinity and continuity.

Blasius' reflections on how to measure qualities go beyond these considerations about limits of qualities. Biard has shown that Blasius takes into account the matter in which a quality is instantiated to determine how to measure it.[82] The idea that the power of a quality must be measured by taking into account not only its intensity and its extension in a subject but also its matter (together with its density) was not new. It was already present, for instance, in Richard

81 On this point, see also Biard, *Hypothèse matérialiste*, 55, where he notes that Blasius accepts in his questions on *On Generation and Corruption*, just like in his questions on the *Physics* and in his question on the intensity of forms, that in a mixture heat and coldness are never pure but always mixed with each other: *Quaestiones circa libros de generatione et corruptione*, ms. Vat. Chig. O IV 41, q. 16, 30[ra]: Sexta conclusio est: cuiuslibet mixti caliditas compatitur secum aliquos gradus frigiditatis, et idem dico de frigiditate respectu caliditatis.

82 Biard, "Blasius of Parma on the Calculation," 239 and 243.

Swineshead's *Liber calculationum*. Blasius does not mention in his questions on the intensity of forms the notion of 'multitude of form' (*multitudo formae*) which served in Swineshead's work to define the power of a quality. However, his view that a quality must be measured by taking into account matter appears to depend on similar considerations.

In his study of Blasius' theory on how to measure qualities, Biard notes that the language of degrees and latitudes is mainly an *operational* or *practical* instrument employed by Blasius to ground a new approach to physics. This language does not constitute an ontological theory in which each concept can be associated with a real item in the world, nor a mere instrument for analyzing logical sophisms, but a practical device for investigating concrete physical phenomena.[83]

The preceding observations confirm this view, which aligns with Blasius' interest in the possibility of representing degrees and latitudes with geometric figures, a technique for analyzing and calculating intensities that began to develop in the mid-fourteenth century, to which Blasius devoted a special question. Although these figures do not exist as physical objects outside the mind, Blasius viewed geometry as providing a practical tool for calculating and measuring intensive properties through the demonstrative power of such representations.[84]

4.3 *Applications and Functions of the Theory of Intensities*

The theory of intensities in Blasius' questions on the *Physics* is not restricted to the study of sensible qualities or local motion; far from it. Blasius demonstrates a comprehensive use of this analytical language, which he is willing to apply to various types of objects. This explains why he sometimes formulates rules about intensities in purely general terms, without referring to what these degrees are to be applied:

> Sixth difficulty: In a given latitude uniformly difform ending at the non-degree, would the more intense half separated from the whole have a greater activity than the whole latitude?
>
> And it seems that not, by the determination of the question.
>
> And from another side it appears to the observer that it is so.

83 Biard, "Blasius of Parma on the Calculation," 246.

84 For more on this topic, refer to the detailed study by Di Liscia, "Biagio Pelacani da Parma's Geometrisation of Latitudes," where the author shows, however, that Blasius' interest in the effectiveness of geometric representations is tempered by his recognition of the inherent limitations of this method.

> The answer to the difficulty is yes. And the reason is that this half is more intense than the whole latitude. If the whole latitude is [intense] as four, it will be the case that the more intense half will be as six, as is clear to the intelligent. And here it is presupposed that a uniformly difform latitude corresponds to its mean degree.[85]

Blasius describes here the active power resulting from a certain distribution of intensities in a latitude, without indicating anything regarding the nature of the object possessing this latitude (whether it is a spatially extended or material object or, on the contrary, a purely temporal one). In this passage, the latitude starts from the non-degree up to a given degree, which must be the degree 8, as the reader can deduce. The more intense half of the latitude (starting from degree 4 to degree 8) is equal to its mean degree, namely the degree 6, whereas the whole latitude, starting from the non-degree to the degree 8, only equals 4. The interesting point in this case is that this property is formulated abstractly as a general property of latitudes, which can then be transposed to different sorts of objects, where this result will have different implications.

In Blasius' *Questions on the Physics*, one of the most extensive applications of the theory of intensive properties concerns the analysis of causal interactions between bodies through their primary qualities. The sophismatic character of thought-experiments inspired by the Oxford Calculators is undoubtedly present in Blasius' works but, more than in the works of these authors, Blasius employs the language of intensities to investigate problems related to concrete physical interactions. The examples illustrating this point in Blasius' *Questions on the Physics* are so numerous that quoting one, which is lengthy and typical of Blasius' practical use of this language, will suffice to give the reader a taste of what is to be found in these questions:

> Likewise, if any simple body were to have within itself the principle of its alteration, this would be most clearly seen in heated water, which, with the [source of] heat removed, appears to be altered from within toward coldness. But it is argued against this, because I will argue as it was argued before about fire. Because given some water uniformly hot throughout, either a part acts on a part adjacent to it, or on itself. And whichever is said, it follows that an action results from a proportion of equality. And with this, it follows that things similar in degree and form act and react on each other. If perhaps it is said that water, which is altered from within

85 See text 8 (V, q. 1) in the present volume, nn. 50–53.

> toward coldness, is difformly hot, let some given water be, for the sake of argument, uniformly difformly hot, ending exclusively at the highest degree of heat on the most intense side. I will argue then as before: Either a part acts on a part immediately adjacent to it, as the more intense part in heat on the less intense part—but this is not the case, because from this follows that conclusion, namely, 'the attenuated [part] intensifies the intense [part] faster than the intense [intensifies] the less intense.' Because suppose that there is a hotter part *A*, whose heat, for example, is of six [degrees]. And consequently, the coldness coextended with it will be of two [degrees]. And let there be another more attenuated part *B*, whose heat is of five [degrees]. And consequently, the coldness coextended with it will be of three [degrees]. As commonly admitted by those who talk about intension and remission, when discussing intensification and attenuation, one quality is more intense than another because it is less mixed with its contrary. Then I argue in this way: *B* acts faster on *A* than *A* on *B* after the reaction takes place; therefore, the attenuated [part] intensifies the intense [part] faster than the intense [intensifies] the attenuated. The antecedent is proven, because the proportion of the heat of this [part] *B* to the coldness of *A* is greater than the heat of *A* to the coldness of *B*; therefore, *B* acts faster, etc. The antecedent is obvious, because the heat of *B* to the coldness of A is as five to two, for instance, as the *dupla sesquialtera* [proportion].[86] But the heat of *A* to the coldness of *B* is precisely like the double [proportion], as is obvious; and consequently, I have the proposed conclusion. And beyond this, I have that *A* intensifies in heat, and likewise *B*. And in the exact same way for any other parts of the given water. Therefore, some given water does not revert to coldness as from an intrinsic principle.[87]

Several things must be noted here. The first is the intimate link between the language of intensities and the theory of proportions, the latter being employed here for understanding the mode of action proper to qualities existing in a body. We can also note that Blasius explicitly endorses here what I labeled the 'new admixture' theory, according to which contrary degrees of qualities are present in a body and make up a total sum of degrees that always remains constant (here, the total sum of degrees is again arbitrarily fixed at 8 degrees, which can be distributed in different ways, like 6 degrees of heat and 2 of cold, or 5

86 A *sesquialtera* proportion, in the Latin terminology of proportions, refers to the proportion 3:2. The *dupla sesquialtera* proportion corresponds to the proportion 5:2.

87 See text 2 (II, q. 2) in the present volume, n. 12.

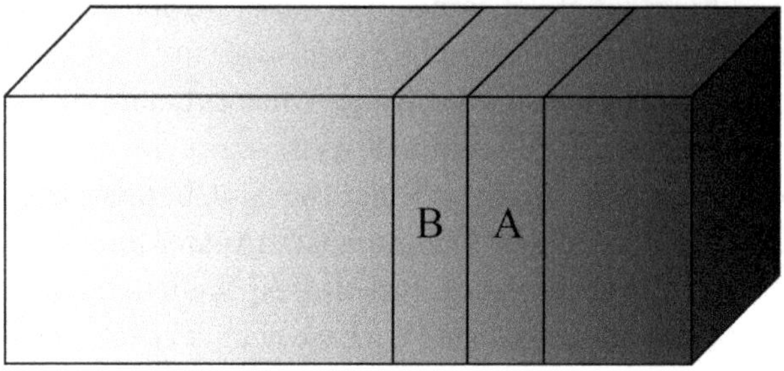

FIGURE 6.3 Reconstruction of a case of uniformly difform distribution of heat, as described by Blasius of Parma

degrees of heat and 3 of cold, etc.). Third, and more importantly, it can be seen with this passage that the language of intensive properties enables Blasius to investigate a quite concrete problem, namely whether bodies have in themselves an intrinsic principle of alteration.[88]

This problem is far from being a merely logical exercise or a sophism like many problems dealt with in the Oxford Calculators tradition. It does not correspond to a method of 'rule testing' regarding the application of predicates, as the works of the Oxford Calculators have sometimes been described.[89] The problem at stake here is relevant to natural philosophy understood in the strong sense as a study of nature. It also represents a classical problem of the medical tradition, due to its implications regarding the conditions necessary for an organism to maintain a specific state (health or disease).

More precisely, Blasius offers in this passage an argument showing that no intrinsic principle of alteration is possible. He does so by imagining a body qualified by heat in a uniformly difform way, i.e., such as the body represented in Figure 6.3, where the black color represents heat. By selecting two parts of this body, one of which is more intense (hotter) than another, Blasius shows that if an intrinsic principle of alteration were present, then a less intense part of such a body would intensify another more intense, and not the other way round, as one might expect. Indeed, if *A* denotes the more intense part having 6 degrees of heat and 2 of cold, and *B* denotes the less intense or attenuated (*remissum*) having 5 degrees of heat and 3 of cold, *B* will act faster on *A* than *A* on *B*, for the ratio between *B*'s heat over *A*'s cold is greater than the converse relation (5:2 > 6:3).[90]

88 See the essay on Blasius' theory of motion in this volume.

89 Murdoch, "From Social into Intellectual Factors," 289–290.

90 This diagram somewhat simplifies the case described by Blasius, as the different degrees

The language of intensities is here put to use to analyze thoroughly the way in which the different parts of a body, given their respective degrees of qualitative intensity, must act on each other. This passage illustrates well the general function of the theory of intensive degrees in Blasius' *Questions on the Physics*: it represents an analytical tool making it possible to infer conclusions about the natural effects of substances and to study the causal relations between bodies.

In Blasius' system, the analytical language of intensive properties is not only applied to elemental qualities like heat and cold, but also qualities existing in the mind. Blasius, for instance, attributes degrees to moral and intellectual values, something that had become common in fourteenth-century psychology and theology.[91] In particular, Blasius attributes degrees to evidence and to science conceived as an intellectual quality.[92] This point is of particular importance, since one of the most constant features of Blasius' approach to philosophical problems is providing different solutions or determinations to one question, to which he attributes different degrees of probability.

The inclusion of spiritual qualities like science, evidence, or moral virtues in the class of intensive properties was not, once again, unusual for this period. But Blasius goes further than anyone else before him by holding that science could not increase if it were not an intensive quality *inhering in matter*, this last point being one of the most striking claims of his theory of mental qualities.[93] Besides, let us note that considering spiritual qualities as intensive properties seems to go against the rule according to which qualities vary in a converse proportional relation, which can only be valid for sensible qualities that have a positive contrary, since properties like justice or charity do not have real contraries.

Blasius not only applies the technical language of intensities to accidents but also to substances and, more generally, to any object which can be considered in the universe of being. More precisely, Blasius engages in an interesting development of the analytical language of intensities in the fourteenth century, namely the application of this language to the order of species composing the world, often referred to at that time as controversies over the 'perfection of spe-

---

of the latitude are supposed to be 'uniformly difformly' distributed, i.e., in a continuous manner and not as units discretely ordered, as in this diagram.

91 See D.A. Di Liscia, "Kalkulierte Ethik: Vives und die 'Zerstörer' der Moralphilosophie (Le Maistre, Cranston und Almain)," in S. Ebbersmeyer and E. Kessler (eds.), *Ethik: Wissenschaft oder Lebenskunst? Modelle der Normenbegründung von der Antike bis zur frühen Neuzeit* (Münster: LIT, 2007), 75–105; Roudaut, *La mesure de l'être*, esp. c. 5, and the bibliography therein.

92 Blasius of Parma, *Quaestio*, 532.

93 Biard, *Hypothèse matérialiste*, 162.

cies.'[94] It had become more and more common in the fourteenth century to describe the total range of perfections represented by all possible species of the universe as the 'latitude of being.'[95] On this latitude of being, each species represents a certain 'degree' which is not a degree of intensity like in the case of qualities, but a degree of perfection.

Blasius dedicates the third question on Book 2 of the *Physics* to this theme. Blasius believes that all species (be they substances or accidents) in the world are located in what he calls the "perfectional latitude of beings" (*latitudo perfectionalis entium*), on which species are ordered like points.[96] On this latitude of being, two species can always be compared through a certain proportion, so that the perfectional distance between them may be calculated or quantified, at least in theory (although this calculation may not be accessible to our knowledge).[97]

The idea that all species may be viewed as different degrees on the latitude of being plays a key role in the theory of causality of Blasius, who contends that a being possessing a given degree of essential perfection cannot produce alone a being whose own degree of perfection is higher.[98] More generally, this way of applying a quantifying terminology to the whole range of beings composing the

94 Cf. J. Biard, "Natures et degrés de perfection. L'échelle des étants chez Blaise de Parme," in L. Bianchi, O. Grassi, and C. Panti (eds.), *Edizioni, traduzioni e tradizioni filosofiche (secoli XII–XVII). Studi per Pietro B. Rossi*, vol. 2 (Rome: Aracne, 2018), 397–410.

95 On this, see J.E. Murdoch, "*Subtilitates Anglicanae* in Fourteenth-Century Paris: John of Mirecourt and Peter Ceffons," in M.P. Cosman, B. Chandler (eds.), *Machaut's World. Science and Art in the Fourteenth Century*, (New York: New York Academy of Sciences, 1978), 51–86; J.E. Murdoch, "*Mathesis in philosophiam scholasticam introducta*. The Rise and Development of the Application of Mathematics in Fourteenth-Century Philosophy and Theology," in *Arts libéraux et philosophie au Moyen Âge. Actes du quatrième Congrès international de philosophie médiévale* (Montreal: Institut d'Études médiévales, 1969), 215–254; E.P. Mahoney, "Metaphysical Foundations of the Hierarchy of Being According to Some Late-Medieval and Renaissance Philosophers," in P. Morewedge (ed.), *Philosophies of Existence, Ancient and Medieval* (New York: Fordham University Press, 1982), 165–267, D.A. Di Liscia, "Perfections and Latitudes: The Development of the Calculators' Tradition and the Geometrisation of Metaphysics and Theology," in *Quantifying Aristotle*, 278–327; Roudaut, *La mesure de l'être*, 321–373.

96 Blasius of Parma, *Quaestiones super VIII libros Physicorum* (secunda redactio) [transcr. Biard, provisional edition], II, q. 3: 89vb: Primum ⟨corollarium⟩: cuiuslibet entis producti et producibilis est perfectio essentialis respectu latitudine perfectionali entium assignanda ut punctus et non ut linea.

97 Blasius of Parma, *Quaestiones super VIII libros Physicorum* (secunda redactio) [transcr. Biard, provisional edition], II, q. 3, 90rb: Quarta conclusio: cuiuslibet rei naturalis ad aliam rem naturalem, sive eiusdem specie fuerit sive alterius, est secundum perfectionem essentialem certa proportio.

98 Biard, *Hypothèse matérialiste*, 95–108, and Biard's essay in this volume.

universe goes along with a new approach to the natural world—typical of the fourteenth century—according to which it can be described through the use of mathematics, insofar as the general architecture of the world, seen through the lens of degrees and latitudes applied to species, can be described in the language of mathematical proportions. This possibility of applying mathematical techniques to the order of species composing the natural world departs significantly from the assumption, established in the Aristotelian tradition from the principle of equivocity of being, that only things of the same genus may be compared.

The practical significance of applying mathematical analyses to different objects through the terminology of degrees and latitudes becomes even more evident when considering the diverse disciplines that Blasius integrates in his physics. Let us briefly summarize this point by looking at different examples of scientific disciplines incorporated into Blasius' *Questions on the Physics*. Since light is an intensive property, Blasius often employs the terminology of latitude, intension, and remission when it comes to dealing with optical problems and more generally phenomena related to light, as in this passage:

> First argument: If half of the sun is taken away from the whole, the remaining half would also illuminate as intensely and extensively as it does now. That it would do so intensively is evident because, since the body of the sun is uniformly luminous, there is as much luminosity in half of it as in the whole; and consequently the half could induce as much latitude intensively as at the beginning of the light, and therefore, the removed half did not contribute [to illuminate]. That it would do so extensively is also evident because the base of the pyramid of light will be in this case the whole diameter of the sun. And before the removal [of the half], it could not be greater; therefore, etc.[99]

The same goes for the science of weights, which is particularly important in Blasius' physics. Although heaviness and lightness are not intensive properties properly speaking, it had become common in the fourteenth century, largely due to the Oxford Calculators tradition, to describe weights in intensive terms, just like condensation and rarefaction. When it comes to the science of weights, Blasius attributes numerical values to weights and degrees of intensity to the forces, heaviness and lightness, involved.[100]

---

99 See text 8 (v, q. 1), in the present volume, n. 3.

100 See the essay on Blasius' science of weights in this volume.

Finally, in line with the traditional use of numerical values in medicine for denoting intensities of substances (e.g., for the composition of medicines), Blasius in his *Physics* employs the language of intensities when it incorporates elements of medical theories, which most of the time involve interactions of qualities, as the texts quoted in this section show.

These many uses of the language of intensities demonstrate the main function attributed by Murdoch to the 'new analytical languages,' namely to achieve the unity of science through the use of a common language. The language of intensities can be applied to different disciplines due to the extreme flexibility of this language. Primarily aimed at quantifying qualities, as they appear in different sorts of disciplines (natural philosophy, psychology, medicine) as central concepts of the medieval philosophical framework, the language of intensities may even be extended to processes and objects that are not qualities strictly speaking, like local motion, or more abstract objects, like species.

5 Conclusions

What Murdoch called the new "analytical languages" of the fourteenth century play an important role in Blasius' *Questions on the Physics*. From Murdoch's differentiation of these languages into three main layers, we can distinguish different levels of functions proper to these languages. At the risk of oversimplification, we may say that these levels roughly correspond to the generality of the questions in which they are employed. The theory of supposition, seen as the most basic analytical language by Murdoch, mainly serves in Blasius' physics the purpose of defining the most basic physical terms such as motion, principle, substance, generation, and the like. We have seen, however, that pieces of purely semantic analysis can be also detected in more applied questions, like the problem of explaining the trajectories of falling bodies.

Furthermore, it appears that the theory of supposition, which Blasius mainly inherits from the Parisian tradition, serves to counterbalance potentially problematic ontological implications of his methods of philosophical investigation. Blasius' methods, indeed, are based on appeals to thought-experiments in which entities like points, instants, the void, and infinite magnitudes, as well as fire capable of freezing other bodies, appear frequently.[101] The theory of supposition also allows Blasius to constrain the proliferation of entities by offering a means to reduce them to ways of referring to real beings.

101 On a fire capable of freezing other bodies, see text 4 (II, q. 8) in the present volume, n. 26.

The language of continuity and infinity appears much more often and concerns a wider range of topics. This language represents indeed one of the main operators involved in the series of *pro et contra* arguments placed at the beginning of his questions and in the arguments put forward for their determinations. The paradoxes involving infinity and continuity serve to demonstrate the inconsistency of hypotheses under examination and, when combined with the theory of proportions, provide the relevant degree of certainty associated with mathematical reasoning.

The language of intension and remission, or what I called here the 'language of intensive properties,' is employed as a common language for analyzing different types of objects, including objects that extend far beyond the realm of qualities properly speaking. This language represents a universal conceptual grid through which the most basic language of proportions may be applied to virtually anything to which language may refer.

To this extent, the new analytical languages enable Blasius to ground his new approach to physics, ranging from the analysis of the most basic furniture of the world to the study of laws and behaviors proper to natural bodies. Murdoch's claim that one of their primary purposes was to make possible a certain unity of sciences fits perfectly with the case of Blasius' *Questions on the Physics*, one of the most striking features of which is the endeavor to incorporate all the sciences available at his time into a unified framework marked both by a more empiricist approach to natural phenomena and a mathematical analysis of them.

EDITION

Blasius Pelacanus de Parma

Quaestiones super VIII libros Physicorum

Description of the Manuscripts

Robert Andrews

The *Questiones disputate super octo libros Physicorum*, secunda redactio, of Blasius Pelacanus de Parma is available in two manuscripts: Vaticano, Biblioteca Apostolica Vaticana, Vat. lat. 2159 ff. 60^{v}–229^{v} and Vaticano, Biblioteca Apostolica Vaticana, Vat. lat. 3012 ff. 2^{v}–110^{v} and 136^{v}.[1] The manuscript Chigi O IV. 41 ff. 219^{r}–272^{v}, incomplete, contains the *redactio prima* and so will not be considered here. We have not ourselves examined the manuscripts.

The editions published herein are from the two manuscripts, Vatican lat. 2159 (V1) and Vatican lat. 3012 (V2). V1 is preferred because it is the only complete manuscript. Even when both versions are available, V2 shows its unreliability, as in homoeoteleuta in *Quaestiones super* VIII *libris Physicorum* II q. 2 f. 28vb ("ut … quietis") and *Quaestiones super* VIII *libris Physicorum* V q. 1 (not edited) f. 63rb ("Et … distantiam"). In one case, Book II, q. 7, a text refered to as *Federici Vescovini* is used. It can be found in: Graziella Federici Vescovini, "Su alcune testimonianze dell'influenza di Alberto Magno come 'metafisico', scienziato e 'astrologo' nella filosofia padovana del cadere del secolo XIV: Angelo di Fossombrone e Biagio Pelacani da Parma," in *Albert der Große: Seine Zeit, sein Werk, seine Wirkung*, ed. Albert Zimmermann (Berlin: De Gruyter, 1981) 155–176.

1 V1 = Vaticano, Biblioteca Apostolica Vaticana, Vat. lat. 2159. Written 1397, paper, mm. 290 × 215, ff. I, 230, coll. 2.

Vaticano, Biblioteca Apostolica Vaticana, Vat. lat. 2159 was first described by Ireneo Affò.[2] It was copied in Pavia by a student of Blasius, the physician Bernardus a Campanea de Verona, in 1397. Bernardus has garnered his own scholarly attention, as a scribe and as the one-time owner of this manuscript as well as of the codices Vat. lat. 2160, 2185, and 3066, and Cesena, B. Malatestiana S.VII.5 and S.VIII.2.[3] Our manuscript provides the first biographical data

1 Images of these manuscripts are available online: Vat.lat.2159 | DigiVatLib and Vat.lat.3012 | DigiVatLib.

2 Ireneo Affò, *Memorie degli scrittori et letterati parmigiani* II (1791) II 121–122.

3 Johannes M.M.H. Thijssen "The Book-collection of Bernardus a Campanea de Verona" (1986) 11.

 | DOI:10.1163/9789004734968_009

about Bernardus—that he wrote in 1397 *in studio Papiensi* (Pavia).[4] The codex has been described by Anneliese Maier,[5] and Graziella Federici Vescovini has provided a list of questions.[6]

The manuscript contains only two works, both by Blasius Pelacanus de Parma.

1. *Conclusiones physicorum*, redactio secunda (or: *Expositio per conclusiones*). ff. 1ʳ–59ᵛ.

In summo f. 1ʳ inscriptio (manu librarii): "Terminus atque via petentibus esto Maria."

Titulus: "Incipit expositio primi libri *Physicorum* per conclusiones secundum serenissimum artium illustratorem magistrum Blaxium de Parma."
Incipit: "Quoniam quidem intelligere et scire contingit circa omnes scientias quarum sunt principia."
Explicit f. 49ᵛᵃ: "et in hoc cum dei laude finitum (sic) sententia octavi et ultimi libri *Physicorum* secundum solemnissimum virum artium illustratorem praeclarissimum Blasium de Parma, Expliciunt conclusiones octavi libri et ultimi *Physicorum* secundum Blasium de Parma qui 'subtilis doctor' appellatur, die Jovis XVIIIIa mensis Julii 1397 in studio papiensi scriptae per me Bernardum a Campanea de Verona hora tertia noctis."

2. *Quaestiones super VIII libros Physicorum*, redactio secunda, ff. 60ᵛ–229ᵛ.

Incipit f. 60ᵛ: "Gratia regis caelorum qui totius orbi elementalis summus est Imperator in laudem et gloriam Serenissimi Duci Mediolani ... librum *Physicorum* existimavi exponere infra scriptis quaestionibus, quarum prima iuxta textum in hac forma proponatur: Utrum scientifica notitia sit nobis de rebus naturalibus possibilis?"
Explicit f. 229ᵛ: "... scilicet rerum non naturalium sex. Et haec de tertio articulo. Pro quarto articulo patet solutio ad rationes in oppositum. Per ea quae dicta sunt. Et haec de quaestione ista. Et sic est hic finis quaestionum omnium VIII librorum *Physicorum*."

---

4 Thijssen "The Book-collection of Bernardus a Campanea de Verona" (1986) 300; cf. Graziella Federici Vescovini, "A propos de la diffusion des oeuvres de Jean Buridan en Italie du XIVe au XVIe siècle," in *The Logic of John Buridan*, ed. Jan Pinborg. Copenhagen: Museum Tusculanum Press. 1976, pp. 31–32.

5 Anneliese Maier and Marco Vattasso *Codices Vaticani Latini*, Codices 2118–2192 (1961) 105–111.

6 Federici Vescovini *Astrologia e scienza* (1979) 433–439.

Colophon f. 229^{v}: "Expliciunt questiones super 1, 2, 3, 4, 5, 6, 7, et 8 libris *Physicorum* Aristotelis disputate et in scriptis tradite in civitate Papie per perspicuum doctorem Blaxium de Parma anno domini 1397."

John Murdoch, *Album of Science*, page 14, provides an image from the *Quaestiones* (f. 85^{r}) as an example of the divisions of a medieval text.[7]

2 V2 = Vaticano, Biblioteca Apostolica Vaticana, Vat. lat. 3012. Written in 1405 and 1406. The manuscript has been described by Piacentini,[8] from whom is adopted the following description.

Vat. lat. 3012, written 1405–1406. Numbered up to f. 164 (but ff. 111–135 are lacking). Paper (ff. 1–110, 136) and parchment (ff. 137–164), mm. 360 × 255 (first part) and 350 × 230 (second part). Illuminated initial and coat of arms at f. 5^{r}. On f. 1^{r} a *titulus* from the fifteenth century, according to Piacentini, from the hand of Vigili, i.e. Fabio Vigili, who composed an incomplete inventory of Vatican manuscripts (between 1508–1513), to be found in Vat. lat. 7135, 7134, and 7136. The large-format codex consists of two parts copied and assembled by a student of Blasius, Fr. Petrus de Raymundis de Cumis, O.P. (Pietro Raimondi of Como); cf. f. 136ra.

The manuscript contains two works, both by Blasius Pelacanus de Parma.

1. Blasius Pelacanus de Parma *Quaestiones super VIII libros Physicorum* redactio secunda, ff. 1^{r}–110^{v}, 136^{r}.

Titulus f. 1^{r}: "*Questiones Iacobi de Parma super libros Phisicorum.*"
Prologus f. 5ra: "Gratia regis caelorum qui totius orbi elementalis summus est imperator in laudem et gloriam serenissimi Duci Mediolani …"
Incipit f. 5ra: "Utrum scientifica notitia sit nobis a rebus naturalibus possibilis."
Explicit f. 110vb: "… eius est conveniens formae."
Colophon f. 136^{v}: "Iste questiones Blaxii super libros *Physicorum* sunt fratris Petri de Raymundis de Cumis Ord. Predic. quas scribi fecit anno domini 1409 anno scilicet primo quo fuit magister studii conventis Papiensis quas et sub ipso magistro Blaxio audivit."

7 John E. Murdoch, *Album of Science: Antiquity and Middle Ages* (1984) 14.
8 Paola Piacentini, "Ricerche sugli antichi inventari della Biblioteca Vaticana," in *Un pontificato ed una città: Atti del convegno, Roma, 3–7 dicembre 1984*. Edited by Massimo Miglio and Francesca Nuttia. Roma (1986) I pp. 115–178.

2. Blasius Pelacanus de Parma *Questiones circa tractatum proportionum magistri Thome Braduardini*, ff. 137ra–163rb. Partially revised by both Blasius and Pierre de Raymond de Cumes (cf. f. 163rb). Described by Federici Vescovini;[9] edited by Biard and Rommevaux.[10]

Incipit f. 137ra: "Queritur circa tractatum proportionum Magistri Thome Berduerdini ab exemio artium doctore monarchaque domino magistro Blaxio de Parma, utrum contingat omnem motum alteri in velocitate et tarditate proportionari."

Explicit f. 163rb: "Ratio Magistri non concluderet, quia procedit ex ignotis, falsis, et dubiis. Deo gratias, amen, etc."

Colophon f. 163rb: "Expliciunt questiones super tractatum de proportionibus secundum gloriosum ac famosum doctorem artium quod monarchantur (?) dominum magistrum Blaxii de Parma etc. Istas questiones super tractatum de proportionibus ego frater Petrus de Raymundis de Cumis emi a magistro Iacobo de Panisalibus artium et medicine doctore anno domini 1406 die 29 augusti, et ipsas pro parte correxit magister Blaxius de Parma huius operis compilator, in residua autem pars ego correcxi."

Abbreviationum Interpretatio

⟨…⟩	addendum censemus
(?)	lectio incerta
V1 117ra	incipit folio 117 recto codicis V1 prima columna

add.	addidit
adnot.	adnotavit
a.c.	ante correxit
cap.	caput
cf.	confer, conferatur
cod.	codex
col.	columna
corr.	correxit
del.	delevit
exp.	expunxit

9 Federici Vescovini *Astrologia e scienza* (1979) 430–431.

10 Blasius Pelacanus de Parma *Questiones circa tractatum proportionum magistri Thome Braduardini*, ed. Joël Biard et Sabine Rommevaux (2005).

f.	folium
ff.	folia
hom.	homoeoteleuton
litt.	littera
mg.	in margine
om.	omisit
ord. inv.	ordinem invertit
q.	quaestio
rep.	repetivit
scrip.	scripsit
s.l.	sub linea, supra lineam
s.l.	supra linea, sub linea
var.	variatio orthographica

1

Liber II q. 1: Utrum domificator in faciendo domum faciat aliquid rebus naturalibus condistinctum

V1 = Vat. lat. 2159 ff. 84vb–86ra; V2 = Vat. lat. 3012 ff. 27rb–28ra

V1 84vb V2 27rb [1] Circa secundum librum *Physicorum*[1] primo quaeritur utrum domificator in faciendo domum faciat aliquid rebus naturalibus condistinctum.

[2] Et sic ista quaestio duo quaerit. Primum utrum faciens domum faciat aliquid. Secundum est utrum illud artificiale sit distinctum a rebus naturalibus. Et arguo primo per primum affirmative. Secundo contra secundum.

[3] Primo[2] sic. Ista consequentia valet: 'tu facis domum, ergo tu facis aliquid'. Tenet argumentum, quia arguitur ab inferiori ad superius affirmative.[3] Ergo omnis domus est aliquid vel aliqua.

[4] Secundo[4] sic: faciens domum agit. Aut igitur agit aliquid, aut nihil. Si nihil agit, ergo non agit. Ergo non agit domum. Si aliquid, habetur propositum quod 'faciens domum facit aliquid'.

[5] Tertio[5] sic: faciens domum praemiatur a prudenti viro, cuius non est dare praemium alicui sine causa. Et patet quod non praemiatur nisi quia facit domum. Et per consequens, quia facit aliquid; alioquin praemiaretur in nihil faciendo.

---

1 consequentia] argumentum V2 3 ista quaestio duo] in ista quaestio quaeritur duo V2 4 rebus naturalibus] re naturali V2 5 per primum affirmative. Secundo] affirmative contra primum deinde V2 7 Tenet argumentum] Argumentum probatur V2 8 Ergo] eo quod V2 | vel aliqua *om.* V2 | aliqua] alica *var.* 10 agit] Et ultra *add.* V2 13 sine] absque V2 | quod] quia V2

---

1 Resp. Aristoteles *Physica* II,1 192b27–31: Similiter autem est et aliorum unumquodque que fiunt; nullum enim ipsorum habet principium in se ipso factionis, sed alia quidem in aliis et ab exteriori, ut domus et aliorum manu incisorum unumquodque.

2 Adn. mg. V1: Primum argumentum.

3 Cf. Gualterus Burlaeus *De puritate artis logicae* (1955) 86: consequentia ab inferiori ad superius affirmando est formalis, et tamen tenet ratione terminorum.

4 Adn. mg. V1: Secundum argumentum.

5 Adn. mg. V1: Tertium argumentum.—Cf. Iohannes de Celaya *Expositio magistri Ioannis de Celaya Valentini in octo libros Physicorum* II q. 2 (1517) f. 39va: Contra istam solutionem ⟨Nominalium⟩ arguitur: sequeretur quod latomus non esset praemiandum quantumcumque faceret pulcherrimas et fortes domus. Consequens est manifeste falsum, igitur. Sequela probatur: latomus nihil facit; ergo non est praemiandus.

 | DOI:10.1163/9789004734968_010

[6] Quarto[6] sic: faciens domum alicui facit illum sibi debitorem sic quod sub iusto iudice cogi potest ad dandum praemium pro opere; ergo, etc.

[7] Arguo[7] ad secundam partem. Et probo quod domus et omne aliud artificiale sit rebus naturalibus condistinctum; quia domificator in faciendo domum non facit ligna nec lapides nec aliud naturale.

[8] Item,[8] si artificialia non distinguerentur a naturalibus, tunc domus non esset aliud quam lapides et ligna.

[9] Sed contra: quia tunc libere posses facere ligna et lapides, quod est falsum. Et tenet consequentia sic arguendo: 'tu facis B domum, et B domus est lapides et ligna, ergo tu facis lapides et ligna'.

[10] Confirmatur:[9] quia si artificialia non distinguerentur etc., sequeretur quod nullus domificator posset domificare, nec alius artifex in sua arte posset aliquid operari. Probatur argumentum sic: quia nullus dominificator potest facere lapides nec ligna; ergo nullus domificator potest facere domum. Tenet consequentia postquam[10] domus non est aliud nec alia quam lapides etc.

[11] Contra arguitur pro parte negativa. Et[11] primo contra primum, ut quod domificator faciens domum non faciat aliquid nec aliqua, ut arguitur sic: 'domus non est aliud nec alia quam suae partes ut dictum est in primo[12] huius; sed partes domus sunt lapides, ligna, et tabula quae fuerunt antequam esset domus; | ergo sequitur quod domificator in domificando non facit aliquid nec aliqua, quia tunc faceret illa quae sunt et fuerunt magnis temporibus'. V1 85ra

[12] Item,[13] si domificator in faciendo domum facit aliquid, quaero: quid est illud? Quia primo nec substantia nec accidens, quia quaererem: quae substan-

1 sic *om.* V2 3 secundam] aliam V2 5 naturale] et facit aliquid ergo illud est artificiale *add.* V2 8 ligna et lapides] lapides et ligna V2 11 sequeretur] sequitur V2 12 alius] allius *var.* V1 13 sic *om.* V2 15 consequentia] argumentum V2 | postquam] per adversarium *add.* V2 | etc.] et ligna V2; ergo tu facis lapides et ligna *add. mg.* V2 16 Contra arguitur pro parte negativa] Pro parte negativa arguitur V2 18 dictum] conclusum V2 19 lapides] et *add.* V2 | tabula] fabula V2 21 sunt et fuerunt] fuerunt et sunt V2 23 Quia *om.* V2

6 Adn. mg. V1: Quartum argumentum.

7 Adn. mg. V1: Primum argumentum.

8 Adn. mg. V1: Secundum.

9 Adn. mg. V1: Confirmatur.

10 Cf. infra n. 20.

11 Adn. mg. V1: Primum argumentum

12 Cf. Averroës *In Phys.* I com. 17 (1562) f. 16M; *Auctoritates Aristotelis* 144 (43): Totum nihil aliud est nisi omnes partes simul sumptae.

13 Adn. mg. V1: Secundum argumentum.

tia fit a domificatore? Si[14] tu dicas quod sit B domus, contra: sit illud instans praesens primum instans in quo nunc primo est B domus. Arguitur sic: isti lapides et haec ligna fuerunt antequam esset iste domificator; sed isti lapides et haec ligna sunt B domus; ergo B domus fuit antequam fuit iste domificator, ergo ipse non fecit B domum. Et hoc argumentum probat quod non sit possibile domificatorem domificare, nec alius artifex secundum suam artem aliquid producere potest. Patet etiam quod artificialia non sunt naturalibus condistincta, quia ut constat securis non est aliud quam ferrum et lignum, quae sunt naturalia, et sic de aliis.

V2 27va [13] In hac quaestione erunt quattuor | articuli. In primo problema. In secundo determinatio quaestionis. Et in tertio, difficultates. In quarto, ad rationes in oppositum.

[14] Quantum[15] ad primum sit problema propter quid est. Quod data aequilibri aequalium brachiorum, cuius aequilibris una cordarum appensa uni brachio sit circumvolubilis, alia vero corda appensa alteri sit affixa eidem secundum angulum rectum invariabilem. Stante isto casu, propter quid est quod, ponderibus aeque gravibus positis, pondus pendens in corda fixa secundum angulum rectum fit levius, aliud vero gravius?

[15] Respondetur[16] quod huius non est alia causa nisi quod brachium cui corda est affixa secundum angulum rectum describit partem arcus magis cur-

---

2 primum instans *om.* V2 4 fuit iste] fuerit ille V2 5 ipse] ille domificator V2 6–7 alius artifex ... potest] alium artificem secundum suam artem aliquid producere V2 12 oppositum] contrarium V2 13 sit] haec *add.* V2 | propter quid est *om.* V2 14 aequalium brachiorum *ord. inv.* V2 17 positis] apensis V2

---

14 Adn. mg. V1: *Manicula.*

15 Adn. mg. V1: Primus articulus. *Manicula.*—Cf. Iordanus Nemorarius *Elementa Jordani* propositio E.7 (1952) 136: Equis ponderibus in equilibri appensis, si equalia sint appendicula, alterum autem circumvolubile et alterum secundum rectum angulum fixtum, quod in circumvolubili appenditur gravius est secundum situm. Cf. Iordanus Nemorarius *Liber Jordani de ponderibus* prop. 7 (1952) 158. Cf. Blasius Pelacanus de Parma *De ponderibus* II prop. 6 (1952) 254: Quare manifestum est quod si eque gravia in equilibri pendeant ita ut unum brachio sit affixum, reliquum circumvolubili pendeat, fixum descendet, aliud in contrarium movebitur. Cf. Blasius de Parma *Questiones super tractatum de ponderibus* q. 3 (1967) 528: Quintum dubium. Utrum equalitas appendiculorum, quorum unum est circumvolubile, reliquum fixum ad angulum rectum cum eius brachio, sit causa diversitatis ponderum in pondere. Quod sic per sextam propositionem Iordani. Cf. Blasius de Parma *Questiones super tractatum de ponderibus* q. 3 (1967) 530: Sextum dubium. Utrum, equis ponderibus in equilibri appensis, uno in appendiculo circumvolubili brachio pendente reliquo alteri brachio infixo, talis situato, sit causa diversitatis in ponderibus. Non arguo.

16 Cf. Blasius Pelacanus de Parma *De ponderibus* II prop. 6 (1952) 256: Subiungitur aliud corollarium in hac forma: Cum equilibris equalium brachiorum fuerit unum appendiculum

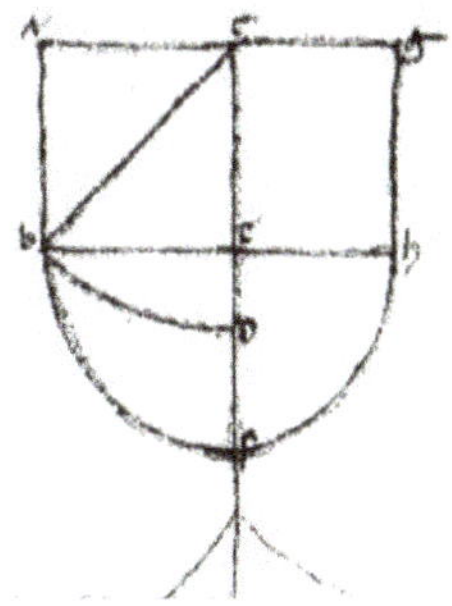

FIGURA 1A
Vat. lat. 2159 f. 85[ra]

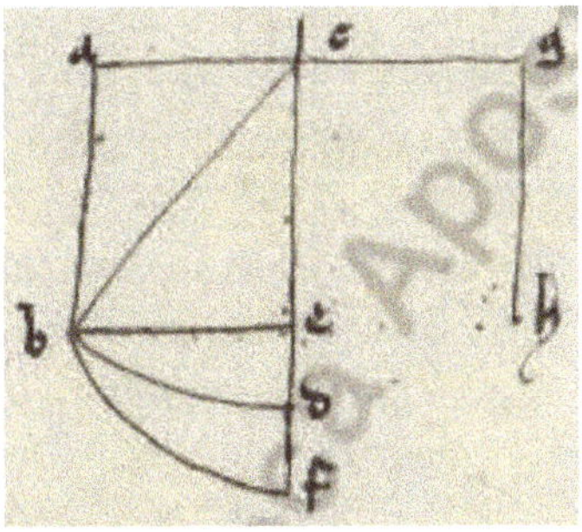

FIGURA 1B
Vat. lat. 3012 f. 27[va]

vam quam pondus pendens in corda circumvolubili. Pro cuius declaratione fiat descriptio in hac forma. Unde sit aequilibris cuius brachia sint aequalia; et sit A, B corda fixa cum brachio secundum rectum angulum. Et protrahatur linea C, B, ad cuius motum describetur arcus B, D, quae linea movetur una cum corda A, B. Sit autem alia corda G, H. Tunc pondus H describet unam quartam circuli rectiorem quam describat pondus apensum in corda fixa. Et per hoc erit pondus H gravius quam pondus B. Et quod pondus H rectius moveatur quam pondus B patet, quia si corda A, B erat circumvolubilis ut est corda G, H, tunc pondus B describeret quartam B, C, F, cui quartae correspondent E, F de directo, quae linea est maior quam C, D, correspondens incessui per quem descenderet B in corda fixa. Et hoc de primo. | V1 85[rb]

[16] Transeo ad[17] secundum articulum, in quo erit duo processus. In primo determinabitur quaestio pro una parte. In secundo pro alia.

---

6 erit] est V2   11 fixa] fisa *var.* V1   12 Transeo] Quantum V2 | articulum *om.* V2

---

fixum ad angulum rectum et alterum circumvolubile, eque gravibus appensis, gravius est quod in fixo pendet. Et si in circumvolubili pendens brachio infingatur, in hoc situ equaliter ponderabunt. Prima pars declaratur: Sint brachia AB et BC, appendicula AD et CE. Et faciat appendiculum AD angulum rectum cum suo brachio, et sit fixum cum eo. Aliud appendiculum cum CE sit circumvolubile in suo brachio. Et per consequens facit cum eo angulum acutum eo quod grave pendens in eo facit ipsum inclinare versus locum gravium. Quare describit quartam minoris circuli alio appendiculo, et per consequens pondus pendens in appendiculo fixo ad angulum rectum cum brachio sit gravius alio, quod est prima pars. Secunda pars patet, quia cum AB et BC brachia sint equalia, erunt semidiametri equalium circulorum. Cum igitur *e* pondus fuerit affixum cum brachio secundum certum modum, ut tangetur in posterium, per primam conclusionem equaliter ponderabit cum *d* eo quod huiusmodi pondera respiciunt quartas equalium circulorum. Ergo et cetera.

17 Adn. mg. V1: Secundus articulus.

[17] Quantum[18] ad primum sit prima[19] propositio de praeambulis: quibuscumque duabus rebus datis, quarum neutra est de essentia alterius, potest unam deus ab alia separare, et separatim conservare. Patet, quia sic fieri contradictionem non includit.

[18] Secunda[20] propositio: motus localis non est aliud nec alia quam mobile quod movetur. Probatur, quia nisi esset alia entitas superaddita ipsi mobili, cum non sit de essentia ipsius, per[21] primam[22] propositionem, potest deus hunc motum separare a mobili, et separatim conservare. Faciat igitur haec deus. Et tunc sequitur quod motus est, et nihil movetur illo motu. Hoc patet.

[19] Tertia[23] propositio: figura non est aliud nec alia quam res figurata. Probatur, quia nisi esset alia entitas superaddita rei figuratae, separet igitur deus figuram a re figurata, et conservet utrumque in esse. Et[24] sit figura separata A, et res quae erat figurata illa figura sit B. Et quaeratur an B sit termino vel terminis clausum, an ne. Si sic, ergo B est figuratum. Tenet argumentum, quia omne clausum termino vel terminis est figuratum.[25] Et sic sequitur quod deus non subtraxit figuram a re figurata. Si vero B non sit figuratum, sequitur quod B est

---

3 conservare] detinere V2 6–7 cum non] quin V2 7–8 hunc motum separare] separare hunc motum V2 8 separatim] utrumque *add.* V2 12 conservet utrumque] utrumque illorum conservet V2 | figura] figurata *add. sed del.* V2 16 Si vero B *rep.* V2 | est] sit V2

---

18 Adn. mg. V1: Primus processus.

19 Adn. mg. V1: Prima propositio.—Cf. Blasius Pelacanus de Parma *Questiones super tractatus logice magistri Petri Hispani* De praedicamentis III,5 (2001) 264: est enim principium in *Sacra theologia* quod quibuscumque duabus rebus distinctis datis, deus potest unam ab alia separare et, si sibi placuerit, unam ab alia separatim conservare. Cf. *Chartularium Universitatis Parisiensis*, Opiniones damnatae n. 141 (1899) I, 551: Quod Deus non potest facere accidens esse sine subjecto, nec plures dimensiones simul esse.

20 Adn. mg. V1: Secunda propositio.—Cf. Ioannes Buridanus *Quaestiones super Physicorum* III q. 7 (2016) 75: posteriores moderni propter rationes praedictas posuerunt quod motus non sit alia res a mobili. Cf. Blasius Pelacanus de Parma *Questiones circa tractatum proportionum magistri Thome Braduardini* q. 1 (2005) 55: Quarto noto quod probabile est, et multi tenent, quod motus est ipsum mobile quod movetur.

21 Adn. mg. V1: *Manicula.*

22 Cf. supra n. 17.

23 Adn. mg. V1: ⟨⟨Tertia propositio.⟩⟩—Cf. Albertus de Saxonia (attrib.) *Quaestiones in Aristotelis Physicam* II q. 3 (1999) 317: Ex hoc potest poni una conclusio quod figura et esse figuratum non distinguuntur realiter.

24 Adn. mg. V1: *Manicula.*

25 Cf. Euclides *Elementa* I def. 14 (ed. H. Busard) 33: Figura est que termino vel terminis continetur. Cf. Ioannes Buridanus *Quaestiones super Physicorum* II q. 3 (2015) 257: figuratum est quod termino vel terminis clauditur.

infinitum. Tenet argumentum, quia illud est 'infinitum' quod est extensum absque termino.[26]

[20] Hiis[27] praemissis, sit prima conclusio: domus non est aliud nec alia quam lapides et ligna, etc. Probatur, quia si non esset, ergo domus ⟨esset⟩ aliqua entitas superaddita hiis rebus, et tunc illa entitas maximum esset figura facta ab artifice, concurrente motu | locali lapidis super lapidem. Sed hoc non V2 27vb molestat conclusionem, quia per tertiam[28] propositionem, figura non est aliud nec alia quam res figurata. Nec motus est aliud nec alia quam illa quae moventur. Et haec duo solum concurrunt ut figura et motus.

[21] Secunda[29] conclusio: omnis res artificialis est res naturalis, una vel plures. Patet de domo, per primam conclusionem.[30] Et sic de pari ⟨ratione⟩ dicendum est de qualibet re artificiali.

[22] Confirmatur[31] conclusio, quia quaelibet res artificialis habet in se principium motus et quietis,[32] ergo quaelibet artificialis est res naturalis, una vel plures. Antecedens patet.

[23] Tertia[33] conclusio: faciens domum non facit aliquid nec aliqua. Probatur, quia domus non est aliud nec alia quam lapides et ligna, ut dicit conclusio praecedens prima,[34] quia domus non est res superaddita rebus naturalibus. Sed faciens domum non facit lapides nec ligna; ergo non facit aliquid nec aliqua—quae conclusio est responsalis.

---

1 est] dicitur V2 5 esset] figurata *add. sed del.* V2 11 primam conclusionem] conclusionem praecedentem V2 12 qualibet] alia *add.* V2 14 quaelibet] res *add.* V2 15 Antecedens] Assumptum V1 17–18 conclusio praecedens prima] praecedens conclusio V2 18 res] aliqua entitas V2 | Sed] ut patet de plano *add.* V2 19 ergo] faciens domum *add. mg.* V2 20 conclusio est *ord. inv.* V2

---

26 Cf. Ioannes Buridanus *Quaestiones super Physicorum* III q. 14 (2016) 124: Licet igitur multis modis dicatur infinitum, ut secundum vigorem, secundum durationem, secundum divisionem etc., tamen hic solum intenditur de infinito secundum magnitudinis extensionem, prout infinitum diceretur corpus extensum sine terminis. Et non curamus hic de 'infinito' syncategorematice accipiendo, sed categorematice.

27 Adn. mg. V1: Prima conclusio.

28 Cf. supra n. 19.

29 Adn. mg. V1: Secunda conclusio.

30 Cf. supra n. 20.

31 Adn. mg. V1: Confirmatur.

32 Cf. Aristoteles *Physica* II,1 192b14–5: Que quidem enim natura sunt omnia videntur habentia in se ipsis principium motus et status; *Auctoritates Aristotelis* 144 (50): Natura est principium et causa motus et quietis in quo est primo per se et non per accidens.

33 Adn. mg. V1: Tertia conclusio.

34 Cf. supra n. 20.

[24] Quarta[35] conclusio: haec domus, civitas, vel castrum est facta vel factum a natura naturaliter operante. Probatur: isti lapides et haec ligna sunt facta a natura etc.; et isti non sunt aliud nec alia quam haec domus, civitas; ergo, etc.

V1 85va [25] Quinta[36] conclusio: | clamidem meam fecerunt pecudes in Anglia naturaliter operantes. Probatur: clamis mea non est aliud nec alia quam ista lana quam produxerunt pecudes in Anglia.

[26] Corollarium:[37] Papiae est facta ista clamis hodierna die quae clamis facta fuit Angliae centum annis retroactis. Patet conclusio intelligenti.

[27] Secundum[38] corollarium: hoc argumentum non valet: 'B domus est aliquid vel aliqua; ergo qui fecit B domum fecit aliquid vel aliqua'. Patet, quia antecedens est verum et consequens falsum. Et si quaeratur defectus huius consequentiae, respondetur: causa est quia arguere ab inferiori ad superius cum verbo importante inchoationem temporis, fallit argumentum propter negationem inclusam tali verbo, sicut sunt haec verba 'facit', 'fit', 'incipit', 'generat', etc. Et ideo non valet argumentum 'facio Sortem album, igitur facio Sortem coloratum'. Et sic non valet 'facio securim, igitur facio aliquid'.

[28] Aliud[39] corollarium: nunc primo facta est B domus, quae domus fuit ante domificatorem qui fecit illam. Prima pars potest patere ex casu, quem casum pone ad tui libitum. Secunda pars patet, quia isti lapides et haec ligna fuerunt antequam fuerit domificator qui fecit illam domum; sed isti lapides et haec ligna sunt ista domus; igitur, etc.

[29] Aliud corollarium:[40] nunc primo est facta ista domus quae facta est ab aeterno. Patet, quia capio unum lapidem qui fuerit ab aeterno, gratia exempli, et capio artificem qui concavet dictam lapidem, faciendo cameras, coquinam, et cacatorium. Et sit hoc instans praesens in quo ista sint perfecte completa.

---

1 facta vel *om.* V2 3 etc.] naturaliter operante V2 | non sunt aliud nec alia quam haec] lapides et haec ligna sunt ista V2 | civitas *om.* V2 5 mea *om.* V2 | ista *om.* V2 13 inchoationem] incoationem *var.* V1 14 inclusam] in *add.* V2 | incipit *om.* V2 15 Et *om.* V2 | valet] hoc *add.* V2 15–16 coloratum] ut bene scis *add.* V2 16 Et sic] Ergo V2 18 illam] domum *add.* V2 21 sunt … etc. *mg.* V1; fuerunt antequam esset tale domificator *rep. sed del.* V1 22 Aliud corollarium] Corollarium quartum V2 | est facta *ord. inv.* V2 23 quia] quoniam V2 24 coquinam] cochinam *var.* V2 25 cacatorium] cachatorium *var.* V2 | ista] haec omnia V2 | perfecte completa] perfecta et V2

---

35 Adn. mg. V1: Quarta conclusio.
36 Adn. mg. V1: Quinta conclusio.
37 Adn. mg. V1: Corollarium primum. *Manicula.*
38 Adn. mg. V1: Corollarium secundum.
39 Adn. mg. V1: Corollarium tertium.
40 Adn. mg. V1: Corollarium quartum.

Tunc patet corollarium primo: haec domus fuit ab aeterno, quia haec lapis fuit ab aeterno; et haec lapis est ista domus; igitur, etc. Patet alia pars intelligenti.

[30] Et iuxta hoc potest concedi quod aliquis novus effectus crastina die, hora tertiarum, potest produci ab aeterno, capiendo li 'aeternum' nominaliter. Arguendo[41] sic, et pono in casu quod deus crastina die, hora tertiarum, producat unum lapidem qui est B. Tunc patet quod B crastina die, hora tertiarum, producetur. Sed quod ab aeterno patet, quia producetur a deo, et deus est quoddam aeternum; ergo, etc. Ex quo patet quod hoc non valet argumentum: 'B productum est ab aeterno, igitur B non | habet principium ex parte temporis'. V2 28ra
Et haec de primo processu.

[31] Nunc[42] pro secundo processu huius articuli subiungo aliam determinationem; et sit prima[43] conclusio. Ad B domum fiendam requiruntur lapides et ligna vel eis aequivalentia. Et dico 'eis aequivalentia' quia bene potest fieri domus absque lignis et lapidibus. Patet conclusio.

[32] Secunda[44] conclusio: haec domus vel illa non est lapides et ligna qualitercumque disponi possunt. Patet, quia protractis lignis et lapidibus versus terram, non est B domus. | V1 85vb

[33] Ex[45] praedictis, tertia conclusio: ad hoc quod B domus sit, vel alia domus, requiritur quod lapides et ligna vel eis aequivalentia taliter se habeant merito artificis qualiter ante B domum non se habuerunt. Patet ista conclusio per praecedentem.[46] Quia postquam ad B domum fiendam requiruntur ista, ut dicit prima[47] conclusio. Nec sufficit quod qualitercumque possint se habere, se habeant ut B domus sit; requiratur igitur quod ista sic vel sic se habeant. Et notanter dico 'merito artificis', quia domus ut sic est res artificialis, et per consequens est opus artificis.

1 corollarium *om.* V2 | aeterno] ex casu *add.* V2 2 Patet alia pars] Prima pars. Secunda patet V2 5 aeternum] aeterno V2 8 et] ergo ab aeterno quia V2 9 ergo, etc. *om.* V2 | hoc non valet argumentum] argumentum istud non valet V2 12 huius articuli *om.* V2 19 vel alia *om.* V2 23 Nec] Non V2 24 B domus sit] sit B domus V2

41 Adn. mg. V1: *Manicula.*
42 Adn. mg. V1: Secundus processus.
43 Adn. mg. V1: Prima conclusio.
44 Adn. mg. V1: Secunda conclusio.
45 Adn. mg. V1: Tertia conclusio.
46 Cf. supra n. 32.
47 Cf. supra n. 31.

[34] Quarta[48] conclusio: ad domum fiendam, ultra partes materiales quae sunt lapides et ligna, requiritur taliter se habere vel taliter quod bene lingua proferri non potest, ut quae res sit taliter vel taliter se habere.

[35] Ex[49] quo corollarie sequitur quod nec ista domus nec illa est lapides et ligna. Patet inspecta secunda conclusione et tertia.[50] Et si quaeratur 'quid est lapides et ligna taliter se habere?', respondetur quod ad talem interrogationem factam per 'quid' non est respondendum de proposito, quia per interrogationem factam proprie quaeritur de substantia vel de termino de praedicamento substantiae. Sed praedicta 'taliter se habere merito artificis' non est substantia nec terminus de praedicamento substantiae. Et hoc est simile huic, si quaeratur 'quid est Sortes currere?'. Unde quando sic quaeritur, non potest proprie responderi. Sed si quaeratur 'qualiter est Sortem currere?', respondendum est quod est Sortes aliqualiter sive velociter moveri. Ex quo sequitur quod faciens domum non facit aliquid nec aliqua, sed facit aliqualiter esse. Et haec de secundo processu huius articuli.

[36] Quantum ad tertium articulum sit prima[51] difficultas: utrum sit concedendum quod artificialia sint naturalia.

[37] Respondetur quod non, iuxta secundam[52] determinationem, quia nullum artificiale est aliquid nec aliqua, sed taliter esse vel taliter. Iuxta primam[53] determinationem dicendum est quod sic. Et iuxta illam est convenienter concedendum quod artificialia habent in se principium motus et quietis, ut patet postquam[54] naturalia sunt.

[38] Secunda[55] difficultas: utrum naturalia possint fieri artificialia.

[39] Respondetur quod sic, iuxta primam[56] determinationem et iuxta secundam.[57] Tamen nota quod iuxta primam, tu potes concedere hanc de prae-

---

2 ligna] etc. *add.* V2 3 potest] nec calamo scribi *add.* V2 8 proprie] per propter quid V2 13 aliqualiter sive *om.* V2 16 articulum *om.* V2 19 Iuxta] vero *add.* V2

---

48 Adn. mg. V1: Quarta conclusio.
49 Adn. mg. V1: Corollarium.
50 Cf. supra n. 32–33.
51 Adn. mg. V1: Prima difficultas.
52 Cf. supra n. 31–35.
53 Cf. supra n. 17–30.
54 Cf. infra II q. 2.
55 Adn. mg. V1: Secunda difficultas.
56 Cf. supra n. 17–30.
57 Cf. supra n. 31–35.

senti naturalia sunt artificialia. Iuxta vero secundam tu habes negare hoc, quia sicut domus non est lapides et ligna, sic per conversionem lapides et ligna non sunt domus.

[40] Tertia[58] difficultas: utrum a natura possit produci res artificialis.

[41] Respondetur quod non, quia res artificialis denominatur ab artifice qui est artifex per habitum.[59] Et haec de tertio articulo.

[42] Transeo ad quartum articulum, | ad rationes in oppositum. Et quia rationes V1 86ra
sunt logicales, logicis dimittantur. Et haec de quarto articulo, etc. | V2 28rb

---

2 per conversionem *om.* V2 6 articulo *om.* V2 7 Transeo *om.* V2 | articulum *om.* V2 8 logicales] loicales *var.* V1; loycales *var.* V2 | logicis] loicis *var.* V1; loycis *var.* V2 | Et ... etc. *om.* V2

---

58 Adn. mg. V1: Tertia difficultas.

59 Cf. Aristoteles *Ethica ad Nicomachum* VI,5 1140a20–21: Ars quidem igitur, ut dictum est, habitus quidam cum racione vera factivus est. Cf. Petrus de Bruxellis *Argutissime subtiles et fecunde questiones phisicales* II q. 1 art. 2 (1510) (s.p.): Item sexto *Ethicorum* ars est habitus factivus.

Liber II q. 2: Utrum quaelibet ens naturale habeat in se principium motus et quietis

V1 = Vat. lat. 2159 ff. 86ra–89rb; V2 = Vat. lat. 3012 ff. 28rb–31ra

V2 28rb [1] Secundo quaeritur utrum quodlibet ens naturale habeat in se principium motus et quietis.[1]

[2] Arguitur[2] quod non. Et primo de motu locali. Materia prima est ens naturale quae, privata omni forma, non habebit in se principium motus etc. Patet, quia ubicumque ipsa poneretur, ipsa ibi staret; nec ad aliquam differentiam loci posset violentari. Quia si violentaretur ad motum sursum, ipsa naturaliter moveretur deorsum, quod non est dicendum.

[3] Item,[3] si corpus solis esset in media regione aëris, nec moveretur sursum nec deorsum; igitur corpus solis non habet in se principium motus etc. Antecedens patet, quia si moveretur sursum, igitur esset corpus leve; et si deorsum, esset grave. Modo conclusum est primo *De caelo*[4] quod “caelum non est grave nec leve nec aliqua pars eius”.

[4] Et ubi tu non concederes primum casum de materia prima, nec secundum de sole, arguo de terra. Et sit A una portio terrae privata omni gravitate et similiter levitate, quae terra non videtur habere in se principium motus. Quia ubicumque ponatur, ibi stabit, sic quod si in concavo caeli ponatur, non descendet sed ibi stabit; quare etc.

---

1 Secundo quaeritur] *ord. inv.* V2 | quodlibet ens naturale] quaelibet res naturalis V2 3 ens] locale *add.* V2 4 quae] non *add. sed del.* V1 5 ipsa[2] *om.* V2 | aliquam] alicam *var.* V1 9 motus *om.* V2 16 ponatur] ipsa *add.* V2

---

1 Resp. Aristoteles *Physica* II,1 192b20–3; cf. *Auctoritates Aristotelis* 142 (50): Natura est principium et causa motus et quietis in quo est primo per se et non per accidens. Cf. Blasius Pelacanus de Parma *Questiones disputate super octo libros Physicorum* prima redactio II q. 4 (ed. Biard): In oppositum est Aristoteles in hoc II° ⟨libro⟩, capitulo primo. Diffiniens naturam, dicit enim quod natura est principium motus et status.

2 Adn. mg. V1: Primum argumentum.

3 Adn. mg. V1: Secundum argumentum.

4 Aristoteles *De caelo* III,3 296b18–9: Quoniam autem hec quidem supponuntur, hec autem ex dictis ostensa sunt, manifestum quia neque levitatem neque gravitatem habet corpus totum.

 | DOI:10.1163/9789004734968_011

[5] Item[5] capio unam parvam sphaeram ignis, et ponatur in centro mundi, vel sic situetur quod centrum eius sit centrum mundi. Tunc[6] arguo sic: remoto omni impedimento extrinseco, iste ignis distans a loco eius naturali non movetur ad ipsum; ergo iste ignis non habet in se principium motus nec quietis. Valet argumentum, quia si haberet principium etc., moveretur sursum. Et antecedens patet, quia aut moveretur per unam semidiametrum aut per omnes. Non primum, quia quaereretur statim quare magis movetur per unam semidiametrum quam per aliam. Nec per omnes, eo quod oporteret istum ignem rarefieri ad duplum, et ultra centuplum, quod non contingit.

[6] Item,[7] nutrimentum naturaliter movetur ad membra, et tamen non habet in se principium motus ad membra. Patet antecedens, quia si nutrimentum moveretur violenter et non naturaliter ad membra, tunc omnis mutatio nutrimenti in membra esset laboriosa, quod est falsum. Patet igitur quod naturaliter movetur ad membra. Sed quod nutrimentum non habeat in se principium huius motus patet, eo quod nutrimentum est de genere gravium. Sed cum movetur ad membra, movetur ergo sursum et secundum omnem differentiam locorum, ut dextrum, sinistrum et ante, retro.

[7] Similiter[8] ferrum movetur ad magnetem naturaliter. Quod patet, quia quando ferrum movetur ad magnetem semper velocius movetur et velocius, quod est de ratione | motus naturalis. Et quod non habeat in se principium V1 86rb huius motus patet, quia tunc ferrum haberet naturalem inclinationem ad loca directe contraria. Patet, quia primo enim ferrum removetur a centro mundi, et ascendit ut coniugantur magneti. Similiter remoto magnete, movetur idem ferrum ad centrum mundi ut coniugnatur centro. Et isti sunt motus oppositi, qui procedere non possunt ab eodem principio.

[8] Item,[9] grave | hora qua movetur sursum per violentiam non habet in se V2 28va principium motus nec quietis. Primo non habet in se principium quietis quia non quiescit. Nec habet in se principium motus, quia maxime haberet in se principium motus quo movetur; sed hoc non, quia motus quo movetur est violentus repugnans eius principio intrinseco.

1 sphaeram ignis *ord. inv.* V2 2 mundi] mondi *var.* V2 5 argumentum *rep.* V2 | haberet] in se *add.* V2 6 semidiametrum] tantum *add.* V2 9 centuplum] centecuplum *var.* 10 tamen] nutrimentum *add.* V2 13 in] ad V2 15 quod] omne *add.* V2 16 omnem differentiam] omnes differentias V2 17 ut] ad *add.* V2; sursum deor *add. sed del.* V1 19 quando] hora qua V2 24 qui] motus *add.* V2

5 Adn. mg. V1: Tertium argumentum.
6 Adn. mg. V1: *Manicula.*
7 Adn. mg. V1: Quartum argumentum.
8 Adn. mg. V1: Quintum argumentum.
9 Adn. mg. V1: Sextum argumentum.

[9] Nunc[10] probatur de alteratione, ut quod nullum corpus naturale habeat in se principium suae alterationis.

[10] Et[11] primo arguitur de simplici. Primo sic: ignis in sua dispositione convenienti exsistens non videtur posse alterari ab intrinseco; ergo non videtur habere in se principium suae alterationis. Argumentum stat. Et antecedens patet, quia si posset alterari, alteraretur igitur ad caliditatem. Aut igitur alteratur ad caliditatem intensiorem quam habeat; et hoc non, quia ignis datus postquam est in sua convenienti dispositione est summe calidus, est sic non potest fieri calidior. Nec ad caliditatem remissiorem potest alterari, quia talis alteratio non fieret nisi a contrario; quare, etc.

[11] Item,[12] aut una pars ignis dati ageret in aliam sibi propinquam; et hoc non, quia tunc simile in gradu et forma mutuo se alterarent, aut una pars ageret in seipsam. Et hoc non, propter eandem causam.

[12] Item,[13] si aliquod corpus simplex haberet in se principium suae alterationis, hoc maxime videretur de aqua calefacta,[14] quae remoto calefaciente videtur alterari ab intrinseco ad frigiditatem. Sed contra arguitur, quia arguam ut prius[15] arguebatur de igne. Quia data aqua uniformiter calida per totum, aut pars agit in partem sibi propinquam, aut in seipsam. Et quocumque dicatur, sequitur quod actio fiet a proportione aequalitatis. Et cum hoc sequitur quod similia in gradu et forma invicem agant et reagant. Si[16] forte dicatur quod aqua quae alteratur ab intrinseco versus frigiditatem sit difformiter calida, sit ergo gratia argumenti aqua data uniformiter difformiter[17] calida terminata in

---

4 intrinseco] extrinseco V2 12 gradu] materia V2 16 arguitur] hoc quaeritur V2 19 actio] accio *var.* V1; acio *var.* V2 | aequalitatis] aequaliter V2 20 similia] simile V2 | gradu] materia *ante corr. mg.* V2 | reagant] Sed *add.* V2

---

10 Adn. mg. V1: De alteratione.

11 Adn. mg. V1: Primum argumentum.

12 Adn. mg. V1: Confirmatur.

13 Adn. mg. V1: Secundum argumentum.

14 Cf. Ioannes Buridanus *Quaestiones super octo libros Physicorum Aristotelis* (*secundum ultimam lecturam*) II q. 5 (2015–2016) I 271: prout multi dicunt, aqua calefacta refrigeraret se remoto calefaciente vel impediente.

15 Cf. supra n. 10.

16 Adn. mg. V1: *Manicula.*

17 Cf. Guillelmus Heytesbury *Termini naturales* n. 57 (2017) 95: Qualitas uniformiter difformis est illa cuius omnes partes immediate secundum extensionem sunt immediate secundum intensionem, et terminatur ad eundem gradum. Cf. Blasius Pelacanus de Parma *Quaestiones musicae* q. 3 (1976) 132: Alia est diffinitio vel descriptio latitudinis uniformiter difformis, videlicet: latitudo uniformiter difformis est latitudo difformis cuius quibuscumque tribus partibus datis equaliter ad invicem distantibus, proportio excessus intensioris ad medium ad excessus (!) medii ad extremum est equalis proportio distantie intensioris ad

extremo intensiori ad summum gradum caliditatis exclusive. Arguam tunc ut prius: aut pars agit in partem sibi immediatam, sicut pars intensior in caliditatem in partem remissiorem—sed hoc non, quia ex isto sequitur illa conclusio,[18] scilicet 'remissum velocius intendit intensum quam intensum remissum'. Quia sit pars calidior A, cuius caliditas gratia exempli sit ut sex. Et per consequens | V1 86va frigiditas[19] coextensa cum illa erit ut duo. Et sit alia pars remissior B, cuius caliditas sit ut quinque. Et per consequens, frigiditas coextensa cum illa sit tria. Sicut communiter volunt loquentes[20] de intensione et remissione, ut quod una qualitas intensior est alia quia est minus permixta cum suo contrario. Tunc arguo sic: B velocius agit in A quam A in B postquam reactio habet locum; ergo remissum velocius intendit intensum quam intensum remissum. Probatur antecedens, quia maior est proportio caliditatis ipsius B ad frigiditatem A quam caliditas A ad frigiditatem B; ergo B velocius agit, etc. Patet antecedens, quia caliditas B ad frigiditatem A est sicut proportio quinque ad duo, ut puta dupla sesquialtera. Sed caliditas ipsius A ad frigiditatem B est praecise sicut dupla, ut patet; et per consequens habeo conclusionem propositam. Et ultra hoc habeo quod A intendetur in caliditatem, et B similiter. Et pariformiter de quibuscumque aliis partibus aquae datae. Ergo data aqua non | reducet se ad V2 28vb frigiditatem tamquam a principio intrinseco.

[13] De[21] mixto potest argui, quia dato mixto, patet quod ipsum appetit permanere et conservari in esse; igitur mixtum non habebit in se principium suae alterationis. Et si habuerit in se principium suae alterationis, hoc erit versus dispositionem sibi convenientem. Et ubi hoc concedatur, quilibet homo distans a temperamento deberet alterari ab intrinseco ad temperamentum, et sic quili-

---

1 Arguam tunc] Arguitur V2 6 erit *om.* V2 7 quinque] duae(?) *ante corr. mg.* V1 | coextensa cum illa sit] extensa cum erunt V1 8 communiter volunt *ord. inv.* V2 9 permixta] permista *var.* V1 10 velocius] fortius V2; fortius *ante corr.* V1 12 frigiditatem] ipsius *add.* V2 13 caliditas] ipsius *add.* V2 | frigiditatem] ipsius *add.* V2 14 caliditas] ipsius *add.* V2 | frigiditatem] ipsius *add.* V2 15 sesquialtera] sexquialtera *var.* | frigiditatem] ipsius *add.* V2 21 suae *om.* V2 22 principium suae alterationis *om.* V2 24 ad] versus V2

---

medium sicut medii ad extremum. Hane puto formam bonam et optimam ut apparebit bene et subtile consideranti particulas eius.

18 Cf. Blasius Pelacanus de Parma *Questiones disputate super octo libros Physicorum* VII q. 6 (2017) 53: et per consequens remissum velocius intendit intensum quam intensum remissum. Et quia hoc est impossibile…

19 Adn. mg. V1: *Manicula.*

20 Resp. Aristoteles *Topica* III,5 119a27–28; *Auctoritates Aristotelis* 325 (55): Illud est albius quod nigro est impermixtius. Cf. Gregorius Ariminensis *Lectura super Sententiarum* I d. 17 q. 3 (1979) 327: Unde quanto forma est permixtior formae contrariae, tanto est remissior; quanto autem est purior vel minus permixta contrariae, tanto est intensior.

21 Adn. mg. V1: Tertium argumentum.

bet homo sanus deberet esse summe temperatus et eiusdem complexionis cum alio.

[14] In oppositum est Aristoteles in principio[22] huius secundi, dicens hoc esse manifestum de quolibet ente naturali. Et per hoc ut constat ponit[23] differentiam inter naturalia et artificialia. Patet etiam ex alio, quia quodlibet ens naturale habet in se naturam, ut puta formam substantialem, quae est principium motus et quietis, ut vult definitio data[24] de natura, quae est haec: "natura est principium motus et quietis eius in quo est."

[15] In hac quaestione erunt quattuor articuli. In primo problema. In secundo determinatio quaestionis. In tertio difficultates. In quarto ad rationes in oppositum.

[16] Quantum[25] ad primum articulum sit hoc problema. Propter quod est quod, data aequilibri cuius unum brachium est quadruplum in longitudine ad aliud, assumptis duobus ponderibus quorum unum est alio gravius in quadruplo, si gravius ponitur in brachio breviori et reliquum pondus in alio, ista pondera in hac dispositione aequaliter ponderabunt?

[17] Respondetur quod causa huius est quia pondus pendens in brachio longiori est gravius gravitate situali alio secundum proportionem brachii longioris AB brevius, eo quod sicut brachium est quadruplum ad aliud, sic circumferentia descripta per brachium longius est in quadruplo rectior vel minus curva
V1 86vb quam circumferentia | brachii brevioris.

[18] Et quia sic est, ideo pondus in brachio longiori alio erit in quadruplo gravius alio, ut vult una suppositio[26] in hac facultate de ponderibus, quae est haec: 'gravius aliud alio quanto motus eius ad medium rectior'.

---

3 Aristoteles] Philosophus V2 | in principio] ad principium V2 6 quae] forma *add.* V2 7–8 ut … quietis *om. hom.* V2 8 est] etc. *add.* V2 9 articuli *om.* V2 10–11 in oppositum *om.* V2 12 articulum *om.* V2 13 quod *om.* V2 15 reliquum] relicum *var.* V1 16 ponderabunt] Ad problema *add.* V2 19 eo quod *rep.* V1

---

22 Aristoteles *Physica* II,1 192b20–3; cf. supra n. 1.

23 Aristoteles *Physica* II,1 192b27–31: Similiter autem est et aliorum unumquodque que fiunt; nullum enim ipsorum habet principium in se ipso factionis, sed alia quidem in aliis et ab exteriori, ut domus et aliorum manu incisorum unumquodque.

24 Aristoteles *Physica* II,1 192b20–3; cf. supra n. 1.

25 Adn. mg. V1: Primus articulus. *Manicula sicut canem.*

26 Iordanus Nemorarius *Elementa Jordani* suppositio 3 (1952) 128: Gravius esse in descendendo, quando eiusdem motus ad medium rectior est. Cf. Blasius de Parma *Questiones super tractatum de ponderibus* q. 3 (1967) 429: Utrum gravioris scilicet ponderis motus ad medium sit rectior.

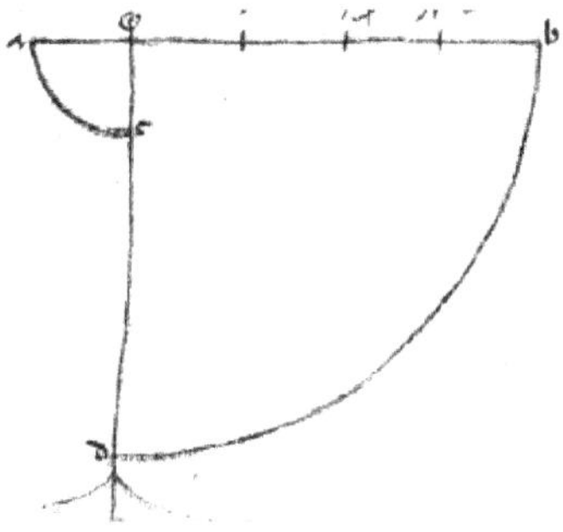

FIGURA 1
in ima col. f. 86va V1

[19] Sed[27] hic occurrit grandis difficultas. Quia ex isto videtur sequi idem grave velocius moveri per lineam curvam quam si per lineam rectam moveretur, quod est contra veritatem et intentionem omnium[28] philosophorum. Unde Aristoteles in quarto[29] huius dicet "cuiuslibet corporis extra locum eius exsistentis est movere ad locum naturalem per lineam brevissimam, quae recta est". Sed probatur consequentia, quia sit A pondus pendens in aequilibri cuius velocitas sui motus in deorsum sit ut quattuor. Deinde illud pondus ponatur in concavo orbis lunae, et derelinquatur suae naturae cuius est moveri per lineam rectam. Tunc quaero: aut hoc grave descendet infinite velocius per lineam rectam quam descendebat in brachio aequilibris, aut finite velocius. Non potest dari primum, quia sic grave moveretur velocitate infinita. Si dicatur quod finite velocius, detur igitur proportio velocitatis ad velocitatem, quae si fuerit quadrupla, tunc ponam datum pondus in brachio longiori in quadruplo quam prius. Si vero in decuplo velocius moveatur, tunc ponam datum pondus in brachio longiori in decuplo. Et sic patet propositum.

[20] Ad[30] hoc argumentum sic fortiter coloratum est respondendum.

2 lineam² *om.* V2 3 et] contra *add.* V2 4 huius *om.* V2 | eius] naturalem V2 5 movere] moveri V2 | naturalem] eius V2

27 Adn. mg. V1: Dubium. *Manicula.*

28 Cf., e.g., Blasius de Parma *Questiones super tractatum de ponderibus* q. 3 (1967) 445: Igitur velocius descendit grave per lineam rectam quam si per lineam curvam descenderit ceteris paribus. Cf. Blasius Pelacanus de Parma *Questiones disputate super octo libros Physicorum* secunda redactio I q. 5 (ed. Biard): sciendum quod omne corpus naturale extra locum eius existens inclinatur moveri ad locum eius secundum lineam rectam, et velocius movetur per illam quam si per curvam moveret.

29 Cf. Aristoteles *Physica* IV,1 208b8–11. Cf. Blasius Pelacanus de Parma *Questiones disputate super octo libros Physicorum* VII q. 5 (2017) 48: Et per hoc scriptum est IV° *Physicorum* quod omne corpus naturale, cum fuerit extra eius locum naturalem, appetit moveri ad ipsum per lineam brevissimam que recta est.

30 Adn. mg. V1: Respondetur.

[21] Prima[31] responsio posset esse ista, quod “motus rectus et circularis non sunt invicem comparabiles,” ut scribitur septimo *Physicorum*,[32] ad textum.

V2 29ra [22] Ista[33] responsio nulla est, quia | licet tales motus non sint invicem comparabiles ratione suorum spatiorum, quia linea recta non potest commensurare curvam nec econtra, cum una non possit reliquam tangere nisi in puncto, tamen sunt invicem comparabiles penes velocitatem et tarditatem, ut dicitur quarto huius.[34]

[23] Et[35] ideo dicendum est dupliciter. Una solutio sit ista, quod semper grave movetur velocius per rectam quam si per curvam moveretur, servata identitate medii in raritate et spissitudine. Et ad argumentum[36] in oppositum, ‘vel grave velocius movetur per rectam’ etc. ‘finite vel infinite’, dico quod finite. Et cum dicitur ‘sit hoc in decuplo’, dico quod hoc admisso per elongationem brachii quantumlibet non velocitabitur motus gravis in aequilibri ad decuplum.

[24] Unde est sciendum quod convenit brachium aequilibris tantum elongari quod grave positum in ipso non descendet, quia contingeret ipsum ascendere.

[25] Pro quo intelligas trabem rectam in superficie terrae orthogonaliter. Tunc ubi huic trabi applicetur unum brachium aequale se in diametro terrae
V1 87ra in cuius extremitate | ponatur unum pondus, in hoc casu illud grave movebitur deorsum per curvam usque quo directe applicabitur centro mundi. Ubi vero illud brachium fieret longius ad decuplum ut quod adaequaretur semidiametro mundi, hoc non moveretur sic quod per eius motum describeret quartam circuli. Quoniam sic ascenderet in tali motu et nedum quod illud brachium adaequatur se in diametro mundi; immo ubi solum adaequaretur diametro terrae. Nec in hoc casu moveretur describens quartam circuli. Ergo ex dictis patet propositionem esse modificandam sic quod licet pondus fiat gravius in brachio longiori et velocius descendat, non tamen velocius descenderet in brachio quamlibet longiori, quod potest patere intelligenti. Et haec de primo articulo.

---

2 textum] testum *var.* V1 5 reliquam] relicam *var.* V1 6 dicitur] scribitur V2 8 dupliciter. Una] aliter et V2 9–10 identitate] idemtitate *var.* V1 12 elongationem] longationem V2 14 aequilibris] in *add.* V2 14–15 elongari] longari V2 17 orthogonaliter] ortogonaliter *var.* 21 ut] gratia exempli *add.* V2 22 quod *om.* V2 28 articulo *om.* V2

---

31 Adn. mg. V1: Prima responsio.

32 Aristoteles *Physica* VII,4 248b5–6: At vero si sunt comparabilia, accidit modo dictum, equalem rectum esse circulo. Sed non comparabilia sunt.

33 Adn. mg. V1: Improbatur responsio.

34 Aristoteles *Physica* VII,4 249a12–13: Quare si que in equali tempore ducta equali longitudine eque velocia erunt, equalis rectus et circularis.

35 Adn. mg. V1: Alia responsio.

36 Cf. supra n. 18.

[26] Transeo ad secundum articulum, in quo erunt duo processus. Primo ponam unam determinationem, secundo aliam. Et in primo loquar prius de motu locali, deinde de alteratione. Circa motum localem, propter aliqua argumenta, ponam aliquas conclusiones quoad imaginationem.

[27] Et[37] sit haec prima data a portione materiae, omni forma privata, ista materia in hac dispositione nec movetur naturaliter nec quiescit. Patet primum, quia si naturaliter moveretur, aut sursum aut deorsum. Et si sic, sequitur quod habeat gravitatem vel levitatem; ergo non esset privata omni forma. Patet secundum, quia si quiesceret naturaliter, ergo ille locus esset ei naturalis, qui locus esset sursum vel deorsum vel in medio. Ergo ubi distaret a loco dato, tunc remoto omni impedimento extrinseco, naturaliter moveretur ad ipsum. Sed omnis motus naturalis fit a principio intrinseco.[38] Et sic ista materia haberet in se principium activum intrinsecum, et sic non esset privata forma.

[28] Secunda[39] conclusio: data materia sic disposita, illa non movetur nec quiescit violenter. Patet statim, quia si moveretur violenter, ergo naturaliter inclinaretur ad motum oppositum. Et idem dico de quiete.

[29] Tertia[40] conclusio: contradictionem formaliter non includit aliquod ens naturale sic disponi quod in hac dispositione exsistens non potest moveri nec quiescere. Patet ex dictis, quia data portione materiae sic disposita, illa non potest moveri | nec quiescere. V2 29[rb]

[30] Sequitur[41] quarta conclusio: quod inter entia naturalia contradictionem non includit aliquid esse quod non potest violentari. Patet statim, quia materia prima est primo infinitae passibilitatis,[42] et per consequens non est alicuius resistentiae; quare conclusio vera.

[31] Sit quinta[43] conclusio: data terra privata omni qualitate motiva, illa exsistens in concavo caeli, remoto omni impedimento extrinseco, non movebitur

1 Transeo ad secundum articulum] Ad secundum articulum transeo V2 3 Circa] contra V2 4 quoad] ad V2 | imaginationem] ymagionem V2; ymaginationem *var.* V1 9 ei] locus eius V2 13 intrinsecum] extrinsecum V2 21 quiescere] Et *add.* V2 24 est primo] ita est V2 26 Sit *om.* V2 27 concavo] concavolo V2

37 Adn. mg. V1: Prima conclusio. *Manicula.*

38 Cf. *Quaestiones Cracovienses super octo libros Physicorum Aristotelis* q. 107 (1969) 185: Pro responsione sciendum, quod differentia est inter motum naturalem et violentum, quia motus naturalis fit a principio intrinseco conferente vim passo.

39 Adn. mg. V1: Secunda conclusio.

40 Adn. mg. V1: Tertia conclusio.

41 Adn. mg. V1: Quarta conclusio.

42 Cf. Marsilius de Inghen *Quaestiones super libros De generatione* II q. 6 (1567) 232b: materia est infinitae passibilitatis.

43 Adn. mg. V1: Quinta conclusio. *Manicula.*

deorsum. Patet, quia si moveretur deorsum, capio tunc unam aliam portionem terrae exsistentem in sua convenienti dispositione, quam pono in concavo caeli V1 87rb iuxta aliam. Et quaero: | aut[44] istae movebuntur aeque velociter; et hoc non, quia tunc duo corpora quorum unum esset grave, reliquum non, aequaliter moveretur per idem medium. Nec una movebitur velocius alia infinite, quia tunc illa moveretur velocitate infinita. Nec finite, quia sit hoc in quadruplo velocius. Ergo ubi remitteretur gravitas ad subquadruplum, moveretur aeque velociter cum alia; et hoc est falsum. Quia tunc grave et non-grave aequali velocitate moverentur.

[32] Sit sexta[45] conclusio: data una portione materiae, gravitate informata absque forma substantiali in concavo caeli exsistente, imaginandum est, remoto omni impedimento extrinseco, istam materiam moveri deorsum a gravitate intrinseca. Patet istud. Et assumo exemplum de eucharistia. Unde[46] data hostia consecrata, ibi non est forma substantialis; immo solum remanet ibi albedo, dulcedo, extensio, cum tanta gravitate. Et evidens est quod si dimittatur suae naturae, ipsa descendet per sphaeram aëris.

[33] Patet etiam ex alio. Quia corpus naturale non movetur deorsum ratione qua terra, sed ratione qua grave. Patet, quia contingit saepe aquam in forma aquae moveri sursum, ut vapores levitate inducta. Et sic hoc vel illud movetur deorsum ratione qua grave. Igitur ubi aliqua materia absque forma substantiali esset informata gravitate, ipsa deorsum moveretur. Ex quo infertur quod forma substantialis non videtur debere dici principium intrinsecum motus localis.

[34] Septima[47] conclusio: non inconvenit per totam istam horam hanc terram naturaliter moveri sursum. Declaro quia sit A una portio terrae terminata ad superficiem ipsius terrae. Et volo quod per totam istam horam augeatur terra, sic quod eius superficies veniat ad concavam ignis. Et in hoc casu patet conclusio. Quia cum ista augmentatio sit naturalis, quae augmentatio realiter

---

6 Nec] *mg.* V2 | finite] infinite V2 10 Sit *om.* V2 | portione] proportione *ante corr.* V1 13 eucharistia] heucaristia *var.* V1; eucaristia *var.* V2 16 sphaeram] speram *var.* 18 in] sub V2 20 grave] Patet quia contingit saepe *add. seu rep.* V2 23 conclusio] illa conclusio *add.* V2 24 Declaro] de caelo V2 27 quae] quia V2

---

44 Adn. mg. V1: *Manicula cum oculo.*

45 Adn. mg. V1: Sexta conclusio. *Manicula.*

46 Cf. Ioannes Buridanus *Quaestiones super octo libros Physicorum Aristotelis* (*secundum ultimam lecturam*) II q. 5 (2015–2016) I 273: Item magnum argumentum ad propositum est ex eo quod fide credimus de sacramento altaris, videlicet quod ibi accidentia nobis apparentia non habent substantiam sibi subiectam et tamen ita agunt et movent sicut si haberent, scilicet gravitas movendo deorsum, color et sapor movendo sensum, caliditas vel frigiditas calefaciendo vel frigefaciendo.

47 Adn. mg. V1: Septima conclusio. *Manicula.*

non distinguitur a motu locali, patet portionem terrae datam continue moveri sursum naturaliter.

[35] Potest conclusio etiam probari per aliud medium se habens conformiter ad praedicta, quia potest terra levificari a sole vel aqua, per cuius levitatem inductam terra vel aqua movebitur sursum ultra mediam regionem, qui motus videtur dici naturalis. Postquam nihil descendit ratione qua terra nec aliquid ascendit ratione qua ignis, sed potius ratione qua grave vel leve.

[36] Sit[48] ultima conclusio: omne corpus naturale cuius forma educta est de potentia materiae habet in se principium activum motus et quietis. Patet conclusio,[49] quia omne tale corpus naturale est de genere gravium vel levium; quare conclusio vera. Et dico in conclusione 'forma cuius educta est de potentia materiae', quia caelum non habet in se principium motus quo movetur. Quare, ut probabitur octavo huius,[50] "primus motor non est in magnitudine, | V2 29va sed separatus ab omni magnitudine". Et idem est putandum de aliis Intelligentiis | V1 87va moventibus orbes. Et dico 'principium activum' quia dato lapide exsistente in summitate terris, iste lapis non habet in se principium motus, quia non movetur; nec quietis, quia licet quiescat, violenter quiescit. Bene tamen habet principium activum per quod est aptum natum moveri etc. Et hoc de motu locali.

[37] Nunc loquamur de alteratione. Et primo loquamur de simplicibus. Et[51] pro isto sit prima conclusio: nullum elementum simplex substantialiter et qualitative habet in se principium suae alterationis. Ista conclusio probata est per

---

3 etiam *om.* V2   6 nihil] nisi V2   8 est] sit V2   21 elementum] ellementum *var.* V2

---

48 Adn. mg. V1: Octava conclusio.—Cf. Nicolaus Oresme *Questiones super Physicam* II q. 1 (2013) 170: Tunc pono conclusiones. Prima est illa quod omne ens naturale compositum habet in se principium passivum motus et quietis.

49 Cf. Ioannes Buridanus *Quaestiones super octo libros Physicorum Aristotelis* (*secundum ultimam lecturam*) II q. 2 (2015–2016) I 252: De quibus ponitur tertia conclusio, scilicet quod omne ens naturale sub orbe lunae existens habet in se principium activum sui motus localis naturalis, quia omne tale est grave vel leve.

50 Aristoteles *Physica* VIII,10 266b25–6: Quod quidem igitur non contingit infinitam esse potentiam in finita magnitudine. Cf. Guillelmus de Ockham *Quodlibeta septem* VII q. 16 (OTh IX 1980) 762: Quia, VIII *Physicorum* in fine, probatur quod primus motor non est in magnitudine finita, quia virtus infinita non potest esse in magnitudine finita.

51 Adn. mg. V1: Prima conclusio.—Cf. Nicolaus Oresme *Questiones super Physicam* II q. 2 (2013) 176: Prima est quod nullum corpus simplex primo modo, scilicet substantialiter et qualitative, alteratur ex se. Cf. ps.-Marsilius de Inghen (pseudo-Ioannes Duns Scotus) *Quaestiones in VIII libros Physicorum Aristotelis* II q. 2 (1891) 513: Prima conclusio. Nullum elementum simplex existens in sua naturali dispositione, secundum qualitates et secundum locum, habet in se ipso principium activum sui motus.

unum argumentum: quia aut pars ageret in partem sibi propinquam, aut in se ipsam. Et quomodocumque dicatur, deveniemus ad illud quod similia in gradu et forma agunt invicem et reagunt.

[38] Secunda[52] conclusio: quodlibet elementum alteratum ad contrarium, remoto alternante, habet in se principium alterationis eius. Patet[53] de aqua calefacta quae, remoto calefaciente, alterat se ad dispositionem sibi convenientem tamquam ad frigiditatem pristinam.

[39] Tertia[54] conclusio: nullum elementum, sive simplex sit qualitative sive alteratum, habet in se principium alterationis ad dispositionem sibi disconvenientem. Patet conclusio ex duabus praecedentibus.

[40] Quarta[55] conclusio: ubi corpus simplex substantialiter et qualitative in vacuo poneretur, illud numquam tenderet ad non-esse. Patet conclusio, quia omne quod tendit ad non-esse tendit ab intrinseco vel ab extrinseco. Modo ignis datus non habet in se causam suae alterationis ad disconveniens. Nec ab extrinseco, cum positum sit in vacuum.

[41] Contra[56] quam conclusionem est obiectio in hac forma: mundus inferior et quaelibet eius pars dependentiam habet a superis in esse et conservari, ut scribitur primo[57] *Meteororum*, cum dicitur quod "necesse est hunc mundum esse continuum lationibus superioribus, ut omnis virtus inde gubernetur". Ergo ubi aliquod simplex poneretur in vacuo, videretur statim tendere ad non-esse.

[42] Dicendum[58] est hoc argumentum esse insolubile. Et pro isto tutius est negare conclusionem positam quam concedere. Sed ad argumentum, cum

2 dicatur] ad hoc *add.* V2 | ad illud *om.* V2 | gradu] materia V2 13 Modo] Non ab intrinseco quia V2 (*et melius*) 18 Meteororum] Methaurorum *var.* V2 22 est] quod V2 23 Sed] tamen V2

52 Adn. mg. V1: Secunda conclusio.

53 Cf. Blasius Pelacanus de Parma *Questiones circa libros de generatione et corruptione* I q. 1 (ed. Biard): aqua calefacta, remoto calefaciente (calefante *ed.*), reducit se ad frigiditatem pristinam.

54 Adn. mg. V1: Tertia conclusio.—Cf. Ioannes Buridanus *Quaestiones super octo libros Physicorum Aristotelis (secundum ultimam lecturam)* II q. 5 (2015–2016) I 277: Ex notatis videtur mihi dicendum primo quod forma substantialis non est principium per se activum ad mutationes et quietes sibi disconvenientes.

55 Adn. mg. V1: Quarta conclusio. *Manicula.*

56 Adn. mg. V1: Contra quartam conclusionem.—Cf., supra II q. 7 n. 16.

57 Aristoteles *Meteora* I,2 339a21–23: Est autem ex necessitate continuus iste superioribus lationibus, ut omnis ipsius uirtus gubernetur inde; *Auctoritates Aristotelis* 171 (2): Totus mundus sensibilis contiguus est lationibus superioribus caeli sensibilis ut inde tota ejus virtus gubernetur. Cf. infra II q. 7.

58 Adn. mg. V1: Respondetur.

dicebatur 'si illud simplex corpus positum in vacuo corrumperetur, aut ab extrinseco aut ab intrinseco', respondetur quod corrumperetur per solam absentiam conservantis, quem modum corruptionis innuit Philosophus secundo huius[59] ad textum dum dicit quod "absentia nautae est causa submersionis navis." Et sic per absentiam luminosi corrumpitur lumen in hemisphaerio nostro.

[43] De[60] mixtis autem ponatur ista conclusio: quodlibet mixtum habet in se principium suae alterationis. Patet conclusio, quia in omni mixto aut sunt formaliter elementa aut sunt qualitates elementorum contrariae. Et per consequens in omni mixto est continua actio et reactio cum tendentia ad non-esse.

[44] Alia[61] conclusio: nullius mixti principium suae alterationis ad disconveniens est forma substantialis. Patet, quia illa appetit conservare totum in esse cuius est forma, et materia appetit informari per illam formam. Relinquitur igitur quod causa alterationis in mixto sunt elementa aut qualitates elementorum contrariae. Et hoc de isto processu.

[45] Nunc[62] loquamur pro secundo processu aliter quam in praedictis. | V1 87vb

[46] Sit prima[63] conclusio: | omne corpus naturale, sive simplex sive mixtum, certam habet latitudinem caliditatis. Patet conclusio in simplicibus, in quibus est magis dubium. Et patet hoc de igne et de aëre. Patet idem de aqua. Quia ut constat, aqua dependentiam habet a caelo et ab astris, et specialiter a luna in tantum quod, cum luna fuerit in quinta decima, omnes aquae augentur V2 29vb

3 quem modum] quemadmodum *ante corr.* V2 5 hemisphaerio] emisperio *var.* 7 mixtis] mistis *var.* V1 12 substantialis] neque materia *add.* V2 14 elementorum] eorum V2 15 hoc de isto processu] haec sit tertia conclusio V2 16 pro] in V2 | quam *om.* V1 | praedictis] Et *add.* V2 17–19 Sit prima ... aëre] Quarta ... aëre *scrip. ante* Nunc (n. 44) V2 17 simplex] fuerit *add.* V2 18 in[2]] de V2 19 patet hoc] primo V2

59 Aristoteles *Physica* II,3 195a12–13: que enim presens causa huius est, hanc et absentem causamur aliquando contrarii, ut absentiam gubernatoris navis eversionis, cuius erat presentia causa salutis.

60 Adn. mg. V1: Alia conclusio.—Cf. Nicolaus Oresme *Questiones super Physicam* II q. 1 (2013) 177: Nullum elementum simplex substantialiter et qualitative habet in se principium suae alterationis. Cf. Albertus de Saxonia *Quaestiones in Aristotelis De caelo* I q. 16 (2008) 198: Tertia conclusio: substantia mixta heterogenea bene in se habet principium suae alterationis et corruptionis. Cf. ps.-Marsilius de Inghen (pseudo-Ioannes Duns Scotus) *Quaestiones in VIII libros Physicorum Aristotelis* II q. 2 (1891) 514: Quodlibet mixtum heterogeneum habet in se principium activum suae alterationis.

61 Adn. mg. V1: Alia conclusio.

62 Adn. mg. V1: Secundus processus.

63 Adn. mg. V1: Prima conclusio.

et in meliori dispositione exsistunt quam prius, eo quod melius conservatur a corpore lunari dum est plenum lumine, sicut est in quinta decima, quam alio tempore. Quod lumen lunare non distinguitur a lumine solari nisi sicut lumen incidens a reflexo; hoc autem lumen est calefactivum et conservativum aquarum per eius caliditatem. Est etiam imaginandum quod aqua dependet ab aliis astris. Et putandum est omnia astra esse calefactiva, quia omnia astra recipiunt lumen a sole; deinde reflectunt illud lumen ad nos.

[47] Idem patet de terra. Et potest argui sic: terra quantum est de mente philosophantium non ponitur summe frigida. Et per consequens cum frigiditate terrae coextensa est certa latitudo caliditatis.

[48] Item est videre quod in profundum terrae reperiuntur multa mineralia calefactiva et multum translucentia. Et per consequens calefactiva quod non esset nisi cum frigiditate terrae esset certa latitudo caliditatis coextensa. Scribunt[64] insuper sapientes quod "frigiditas absque caliditate non ingreditur opus naturae". Patet idem de mixtis, quia postquam in quolibet mixto sunt qualitates elementales sub esse intenso aut refracto, in quolibet mixto erit certa latitudo caliditatis.

[49] Item,[65] in omni mixto est certa permixtio miscibilium. "Permixtio autem non fit nisi a caelo", ut scribitur quarto *Meteororum*.[66]

[50] Secunda[67] conclusio: non convenit caliditatem conservari absque humido. Patet: primo enim videmus hoc in omnibus viventibus, de quibus dicit Aristoteles libro *De longitudine vitae*[68] quod "vita constat per calidum et humidum, calido agente in humidum". Tu vides idem in aliis. Unde deficiente

---

5 imaginandum] ymaginandum *var.* 6 Et *om.* V2 8 potest argui] potes arguare V2 10 terrae] terra V1; illa V2 12 calefactiva] calefractiva *ante corr.* V2 18 Permixtio] Mixtio V2 22 Aristoteles] Philosophus V2

---

64 Adn. mg. V1: *Manicula.*—Cf. Beda (attrib.) *Axiomata* (1605) 106: Frigiditas non ingreditur opus naturae. Cf. Ludovicus Buccaferreus *Lectiones super primum librum Meteorologicorum Aristotelis* (1590) 40^{b}: Frigiditas enim ex Galeni et Avicennae sententia, non ingreditur opus naturae.

65 Adn. mg. V1: *Manicula.*

66 Non invenimus. Cf. Thomas de Aquino *Quaestiones disputatae de potentia* v a. 10 arg. 2: Praeterea, omnis mixtio elementorum ex motu caeli causatur. Cf. Blasius Pelacanus de Parma *Questiones disputate super octo libros Physicorum* secunda redactio II q. 7 (ed. Biard): Prima propositio: ab astris qualitates elementales hic inferius assidue causantur, et alie consequenter de secundis.

67 Adn. mg. V1: Secunda conclusio.

68 Aristoteles *De longitudine et brevitate vitae* 5 466a18–19: Intelligendum enim est animal natura humidum calidumque esse, cuiusmodi vita quoque ipsa est. Cf. Blasius Pelacanus

humido in licino,[69] statim ignis extinguitur. Et idem contingit in quocumque alio combustibili. Et per hoc cineres privatae omni gradu humiditatis inflammari non possunt.

[51] Idem habes arguere in simplicibus. Videtur quod in igne est certa latitudo humiditatis quam forma substantialis ignis sibi determinat, ad cuius consumptionem corrumpitur forma substantialis ignis.

[52] Idem[70] patet de intentione Philosophi secundo *De generatione*,[71] dicentis ignem esse primo calidum. Volunt namque sapientes[72] in naturalibus quod “ignis quantumcumque in sua convenienti dispositione exsistat non sit summe siccus”; immo quod habeat unam humiditatem caliditati eius proportionatam. Et ideo dicebat Philosophus quarto *Meteororum* et primo[73] quod “non datur humidum absque sicco neque siccum absque humidum.”

[53] Idem patet de aëre; quod aër habeat humiditatem nullus negat. Et magis de aqua patet quod sit de genere humidorum. Sed suspectum est hoc de terra, cum ipsa ponatur summe sicca. Sed quod ipsa habeat | humiditatem sibi natu- V1 88ra
ralem patet, quia ubi esset absque humiditate, tunc ipsa esset divisa in omnem partem possibilem dividi, quod est impossibile. Et patet argumentum, quia ubi esset aliqua pars terrae continua, hoc esset solum ex principia humiditatis continuantis partem cum parte. Et hoc operatur humiditas in naturalibus, tam simplicibus quam mixtis.

---

10 proportionatam] proportionatum V1 13 nullus negat] nec hoc negatur V2 14 suspectum est hoc de terra] de terra est suspectum V2 15 sibi *om.* V2

---

de Parma *Questiones circa libros Metaurorum Aristotelis* I q. 1 (ed. Biard): vita constat per calidum et humidum, ut habemus in libro *De longitudine et brevitate vite*.

69 Id est, ellychnium candelae vel lucernae. Cf. Ioannes Buridanus *Quaestiones super octo libros Physicorum Aristotelis* (*secundum ultimam lecturam*) II q. 2 (2015–2016) I 250: candela similiter principium suae combustionis lichini. Cf. Buenaventura de San Agustín *Artium cursus: Libros de generatione, coelo mundo et anima* De generatione et corruptione II, q. ult., art. ult., § I, n. 244 (1739) III 203b: Unde sicut deficiente humido in lucerna, deficit flamma, ita deficiente humido in vivente, deficit etiam calor, ut apparet in senibus.

70 Adn. mg. V1: *Manicula*.

71 Aristoteles *De gen. et cor.* II,3 331a3–7: Sed tamen simpliciter quatuor entia unius unumquodque, terra quidem enim sicci magis quam frigidi, aqua autem frigidi magis quam humidi, aër autem humidi magis quam calidi, ignis autem calidi magis quam sicci.

72 Cf. Marsilius de Inghen *Quaestiones super libros De generatione* II q. 6 (1567) 232a: Corollarium: ignis non habet siccitatem in summo.

73 Aristoteles *Meteora* II,4 359b32–3: Est autem neque humidum sine sicco, neque siccum sine humido, sed omnia hec dicuntur secundum excessum. Cf. Aristoteles *De sensu et sensato* 4 441b26: nec humidum absque sicco, nec siccum absque humido.

[54] Tertia[74] conclusio: quodlibet corpus naturale, tam simplex quam mixtum, | V2 30ra sive in naturali eius dispositione exsistat sive non, habet in se principium suae alterationis activum ad disconveniens. Patet, quia omne tale habet caliditatem, quae caliditas continue agit in humiditatem tandem consumendo illud; ergo conclusio vera.

[55] Alia conclusio:[75] omne corpus etc. habet in se principium activum suae corruptionis. Patet, quia consumpto humido per actionem caloris naturalis, necesse est caliditatem extingui et corruptionem illius sequi.

[56] Et[76] sequitur alia conclusio: impossibile est aliquod corpus naturale, remoto omni extrinseco corrumpente, aeternari. Patet consequens: quocumque corpore naturali dato, non potest impediri actio calidi in humidum; igitur conclusio vera. Et hoc est quod dicit Avicenna capitulo De complexione pueri et iuvenis.[77] Dicit namque quod "si virtus nutrimenti corporis infinita foret et semper restauraret humidum tale quale deperditum est, adhuc vita finiretur propter continue fieri humidum impurius et impurius." Et ex hoc apparet quod Ecclesia[78] scribit dum dicit "cinis es et in cinerem reverteris." Quia consumpto humido per actionem caloris naturalis, totum resolvitur in cinericium, sicut carbo fit cinis.

[57] Octava conclusio: in omni corpore, tam simplici quam mixto, est certa latitudo frigiditatis. Patet, quia in omni corpore etc. est certa latitudo caliditatis. Et quia omnis actio in praedictis est ratione contrarietatis, patet propositum. Nec alio quin calidus ageret in humidum.

[58] Quibus sic visis, facile est respondere ad motiva quibus videntur stare conclusiones quae positae sunt in primo processu.

---

4 consumendo] et exterminando *add.* V2 6 Alia] Sexta V2 7 actionem] acionem *var.* 9 alia] quinta V2 10 consequens] quoniam V2 11 igitur] quare V2 13 nutrimenti] s.l. V2 15 impurius] Et hoc tam in simplicibus quam in mixtis postquam calidum agit continue in humidum et in omni corpore simplici et mixto est caliditas, et per consequens humiditas *add.* V2 19–22 Octava ... humidum *om.* V1 23 sic *om.* V2

---

74 Adn. mg. V1: Tertia conclusio.

75 Adn. mg. V1: Quarta conclusio.

76 Adn. mg. V1: Quinta conclusio.

77 Avicenna *Canon medicinae* I fen. 1 doc. 3 cap. 3 (1595) 18b: Amplius, si haec virtus infinita foret et semper permutationem in corpore faceret eius, quod dissolvitur aequaliter et uno modo; dissolutio tamen uno modo non esset, immo semper quotidie augeretur, et corpus dissolutionem tolerare non posset, et dissolutio humiditatem finiret.

78 Genesis 3, 19: pulvis es et in pulverem reverteris. Cf. Albertus Magnus *Summa theologia* II tr. 17 q. 113; *Opera omnia* XXXIII (1894) 327a: accipitur ex hoc verbo Domini, Genesis III, 19: *Quia pulvis es, et in pulverem reverteris*; vel ut habet alia translatio: *Cinis es, et in cinerem reverteris.*

[59] Et[79] subiungo huic tertium processum. Dando tibi modum per quem potes sustentare nihil corrumpi ab intrinseco, nec aliquid habere in se principium suae alterationis ad disconveniens, sive illud sit simplex sive mixtum. Pro quo noto quod aliquid habere principium intrinsecum motus est illud naturaliter inclinari ad motum. Et idem dico de alteratione, et consimiliter de corruptione.

[60] Praemitto secundo, quod omne ens naturale appetit permanere et in esse conservari.[80] Istud patet ex ultima[81] quaestione primi huius. Et patet etiam per rationem, quia quocumque ente naturali, dato forma substantialis illius, appetit informare materiam quam informat; et materia appetit informari illa forma.

[61] Dico tertio quod omne ens naturale hora suae generationis producitur a generante mediantibus dispositionibus qualitativis facientibus pro conservatione formae in materia et totius compositi. Istud relinquitur tamquam notum. Et patet per rationem, quoniam primo, quantum est ex parte materiae, ipsa non est nata recipere successive hanc formam et illam nisi fuerit aliter disposita et aliter, ita quod alia et alia praeparatio arguit aliam et aliam formam. Ex quo patet quod omnis forma substantialis in materia pro conservatione eius determinaret sibi certam latitudinem vel proportionem qualitatum activarum ad passivas.

[62] Et tunc sequitur quod dicendum ad | propositum, ut sit illa conclusio V1 88rb
generalis:[82] nullum ens naturale, sive simplex fuerit sive mixtum, habet in se principium activum suae alterationis versus dispositionem sibi disconvenientem. Patet conclusio, quia si conclusio non stat, sit igitur A corpus simplex vel mixtum habens in se principium activum etc., de quo quaero: aut istud corpus est in sua convenienti dispositione, aut non. Si sic, cum hoc ens appetat permanere et in esse conservari, et maxime in | dispositione convenienti, sequi- V2 30rb
tur quod hoc non alterabit se ad dispositionem disconvenientem. Quia ubi sic, appeteret eius corruptionem.

1–2 potes sustentare] potest substentari V2 5 consimiliter] similiter V2 6 quod] s.l. V2 15 et] vel V2 20 dicendum] est *add.* V2 | ad *rep.* V1

---

79 Adn. mg. V1: Tertius processus.

80 Cf. Averroës *In Phys.* I com. 81 (1562) 46rb: Omne enim ens, ut dictum est, diligit se permanere. Cf. Ioannes Buridanus *Quaestiones super octo libros Physicorum Aristotelis (secundum ultimam lecturam)* II q. 5 (2015–2016) I 277–278: Omne enim ens naturaliter diligit se permanere, ut dicit Commentator primo huius.

81 Blaesius de Parma *Quaestiones super VIII libris Physicorum* I q. 10, "Utrum quodlibet ens naturale naturaliter appetat permanere."

82 Adn. mg. V1: Prima conclusio.

[63] Item, postquam hoc corpus est in sua convenienti dispositione, sequitur quod ipsum habet determinatam latitudinem proportionum qualitatum activarum ad passivas per quas forma substantialis conservatur in materia. Ergo forma substantialis non appetit corrumpere huiusmodi qualitates sic proportionatas dictam formam conservantes in materia. Nec etiam illae qualitates invicem se alterant postquam invicem sunt debitae proportionatae. Similiter istae qualitates sic se habentes appetunt conservari in esse postquam quodlibet ens, tam substantiale quam accidentale, appetit permanere.

[64] Si vero dicatur quod A corpus naturale non est exsistens in dispositione eius convenienti, sicut dicimus de aqua calida, et idem etiam de animali febriente, adhuc patet propositum. Quoniam primo: de aqua calida non dicimus quod habeat in se principium activum per quod alteretur ad caliditatem intensiorem, sed bene ad frigiditatem tamquam ad sibi dispositionem convenientem. Quod idem putandum est de mixto tam animato quam inanimato.

[65] Ex[83] quo est dicendum corollarie quod, licet principalia membra dicantur mutuo contraoperari, illa contraoperantia est conservativa dictorum membrorum et totius compositi.

[66] Secunda[84] conclusio: nullum corpus naturale, sive simplex fuerit sive mixtum, habet in se originaliter principium activum suae corruptionis. Patet conclusio, quia nullum tale habet in se originaliter principium suae alterationis versus disconveniens; igitur nec corruptionis. Tenet argumentum, quia non est corruptio sine praevia dispositione. Et dico in conclusione 'originaliter', quia bene contingit quod ex occursu sex rerum non-naturalium,[85] quae dicuntur somnus, vigilia; motus, quies; etc., inducta sit mala dispositio habituata per quam insequitur mors vel corruptio mixti.

---

2 qualitatum] *rep.* V2 22 disconveniens] dispositionem disconvenientem V2 | corruptionis] corruptionem V2 23 dispositione] indispositione V2 *forte melius* 25 somnus] sopnus *var.* V1; sompnus *var.* V2 26 insequitur] sequitur V2

---

83 Adn. mg. V1: Corollarium. *Manicula.*

84 Adn. mg. V1: Secunda conclusio.

85 Cf. Galenus *Ars medica* (1821) 367: His igitur expositis, in singulis eorum, quae necessario corpus immutant, proprium salubrium causarum genus inveniemus. Unum quidem ex ambientis aeris contactu, alterum ex motu et quiete, tam corporis universi, tam ejus partium. Tertium ex somno et vigilia. Quartum ex his, quae assumuntur. Quintum ex his quae exernuntur et retinentur. Sextum ex animi affectibus. Cf. Rodolphus Goclenius *Disquisitiones philosophicae* XVII (1599) 63: … a medicis vocantur non naturales, ut aër, somnus, vigilia, motus, quies, etc.

[67] Alia conclusio:[86] omne quod corrumpitur ab extrinseco corrumpitur tamquam aliqua primitiva, quae postmodum uti causa coniuncta efficitur. Patet ex dictis postquam nihil corrumpitur ab intrinseco.

[68] Forte[87] hic dicetur contra hoc sic: contingit aliquem mori morte naturali quae fit a principio intrinseco, conservata semper proportione vel latitudine proportionum qualitatum activarum ad passivas, sicut contingit de hiis qui moriuntur per resolutionem, stante semper eadem proportione calidi ad humidum et aliarum qualitatum.

[69] Hic[88] respondetur quod nulla est mors naturalis, ubi per 'mortem naturalem' tu excludas omnia extrinseca active concurrentia ad alterationem disconvenientem. | Primo namque ex parte loci qui numquam est sic conformis V1 88va
locato quin alteret illud, et ex parte constellationis ac temporis. De[89] tempore namque dicit Philosophus quarto huius[90] quod "est causa corruptionis de per se et totius mali". De astris loquuntur astronomi[91] volentes reducere tales effectus in causas universales ex parte astrorum. Et per hoc volunt[92] quod cuilibet

1 Alia] Tertia V2 3 intrinseco] extrinseco V2 5 quae] mors naturalis *add.* V2 10 tu *om.* V2 14 loquuntur] locuntur *var.* V2 15 quod cuilibet] unicuique V2

86 Adn. mg. V1: Alia conclusio.

87 Adn. mg. V1: Contra.

88 Adn. mg. V1: Respondetur.

89 Adn. mg. V1: *Manicula.*

90 Aristoteles *Physica* IV,3 222b17: in tempore autem omnia fiunt vel corrumpuntur. Cf. Blasius Pelacanus de Parma *Questiones disputate super octo libros Physicorum* prima redactio II q. 4 (ed. Biard): hoc est quod dicitur IV° *Physicorum* ad testum, ubi dicitur quod tempus (terminus *ed.*) est causa corruptionis de per se et totius mali, et dicitur quod tempus (terminus *ed.*) non est nisi primum mobile cum suis astris.

91 Cf. Blasius Pelacanus de Parma *Questiones disputate super octo libros Physicorum* secunda redactio II q. 7 (ed. Biard): Pro primo premitto quod causarum efficientium quedam dicitur causa universalis, non quod per causam universalem intelligatur aliqua res universalis, sed quod universaliter videtur concurrere ad quemcumque effectum, ut celum et astri eorum motu et lumine, primo motore non excluso.

92 Cf. Blasius Pelacanus de Parma *Questiones disputate super octo libros Physicorum* prima redactio II q. 4 (ed. Biard): Et per consequens apparet quod omnis generatio et corruptio fiat a constellatione, et de ratione constellationis est respectu cuiuslibet individui certa periodus sue durationis, ut scribitur II° *De generatione*. Cf. Averroës *In De caelo* II com. 57 (1562) 386vb: tempus iuventutis sit aequale tempori senectutis, et ista duo tempora determinantur per revolutiones orbium declivis. Cf. Albertus Magnus *Super Dionysium De divinis nominibus* (1971) 155: quamvis superiora determinent inferiora, haec tamen non consequuntur determinationem illam necessario, ut dictum est, nec illa sunt principia istorum propinqua et essentialia, nec iterum est mensura quam necesse sit sequi suum mensuratum, quia quamvis vita alicuius sit determinata ad determinatum tempus secundum propriam periodum, potest tamen plus vel minus vivere, secundum quod disponit se ad hoc vel illud.

individuo est certa et determinata periodus impensa ultra quam praeterire non licet; abbreviari vero potest. Haec Avicenna[93] ubi de hoc.

[70] Verumtamen de motu locali potest dici periformiter ad praedicta, ut valet quod omne corpus naturale ad intra circulum lunae exsistens naturaliter appetit quiescere. Probatur: omne corpus naturale appetit esse in suo loco naturali propter conservari et in eo quiescere. Et sequitur consequenter quod nullum corpus naturale ad intra circulum lunae exsistens appetit localiter moveri. Patet, quia omnis appetitus naturalis determinatus est sic ad hoc quod V2 30va non ad oppositum. Non enim appetitus naturalis est idem contrariorum, ut | est potentia rationalis quae potest in opposita. Modo clarum est quod appetitu naturali quodlibet ens naturale appetit quiescere, ergo non moveri localiter. Et sequitur consequenter quod omnis motus localis ad intra circulum lunae est violentus. Patet postquam est contra inclinationem corporis naturalis.

[71] Si quaeras: igitur a quo movetur grave cum fuerit extra locum eius naturale non detentum? Dico quod inchoative movetur a loco. Habet namque locus naturalis virtutem attractivam locati sicuti magnes ferri. Prosecutive autem locus et impetus qui acquiritur per motum Intelligentiis superioribus concurrentibus, de quibus scribit Commentator duodecimo *Metaphysicae*,[94] "unumquodque ens naturale dirigi in locum naturalem ab Intelligentia in qua non est error". Nec valet hoc argumentum per motum gravis: 'in deorsum grave acquirit locum eius naturalem; ergo talis motus est naturalis'. Nec valet 'grave appetit esse in loco naturali, ergo appetit moveri versus eius locum naturalem'. Quia sic probaretur quod 'tu assumis libenter reubarbarum aut turbith[95], quia tu appetis sanitatem; ergo appetis sumere reubarbarum' etc. Non enim appetens

2 ubi] *praem.* et Philosophus V2 | hoc] Et hoc de secundo articulo *add. sed del.* V1 7 circulum lunae] lunam V1 9 ad] hoc *add.* V2 11 localiter *om.* V2 12 localis] corporis *add.* V2 13 corporis naturalis] naturalem V2 15 inchoative] incoative *var.* V? 16 naturalis] unam *add.* V2 | Prosecutive] prosequtive *var.* V2 17 superioribus] ad hoc *add.* V2 23 turbith] turbit *var.* 24 etc. *om.* V2

93 Avicenna *De generatione et corruptione* c. 15 (1987) 147: Et quaelibet potentiarum corporis et quaelibet materia habet [necesse vel] necessario certum terminum ultra quem non est transitus, dato quod inveniantur earum causae secundum quod indigent: et haec est mors naturalis. Cf. Blasius Pelacanus de Parma *Questiones disputate super octo libros Physicorum* prima redactio II q. 4 (ed. Biard): Et idem Avicenna fatetur prima primi dicens quod unicuique individuo destinata est certa mensura ultra quam progredi non contingit.

94 Adn. mg. V1: *Manicula.*—Averroës *In Metaph.* XII com. 44 (1562) 327vb: Et ideo omnes intendunt hunc motum diurnum, qui est primi motoris actio et principium aliarum actionum. Apparet enim quod alii motus proprii sunt propter istum motum. Cf. Averroës *In Metaph.* XII com. 18 (1562) 305rb: et ideo forma apud ipsum non habet generationem neque corruptionem nisi accidentaliter, scilicet per generationem et corruptionem compositi, et cum hoc conservaverit hoc fundamentum, nullus error accidet ei.

95 Turbith, id est Operculina turpethum.

finem naturaliter appetit naturaliter requisita vel media quibus acquiritur finis. Et ideo scribitur[96] quod "hic vult pisces, et non piscari". Et haec de secundo articulo quaestionis.

[72] Transeo[97] ad tertium articulum. Et quaero difficultates. Et primo circa materiam de motu locali.

[73] Inter[98] quas sit haec prima: utrum in aliquo casu | possit[99] contin- V1 88vb
gere quod, data aequilibri brachiorum inaequalium, ut problema[100] praesentis quaestionis proponit, scilicet quod unum brachium sit quadruplum ad aliud; et sint duo pondera habentia eandem proportionem; et gravius ponatur in longiori et aliud in alio—quaeritur an in hoc casu possibile sit brachia aequilibris ab aequalitate non recedere?

[74] Respondetur quod sic. Unus casus est ubi idem sit centrum mundi et centrum circa quod praedicta brachia sit revolubilia, quod centrum dicitur centrum motus. Et patet quod in hoc casu, brachia aequilibris ab aequalitate non recedent. Dico tamen quod gravius descendet motu recto, et tantum movebitur quod centrum mundi et punctus aequaliter distans ab extremis brachiorum est idem centrum.

[75] Alius casus est iste, quia data dicta aequilibri brachiorum inaequalium, et datis duobus ponderibus aeque gravibus positis in aequilibri, non fiet motus ubi extrema circumvolubilium cordarum a linea directionis aeque distent. Patet, quia pondera in hiis locis respiciunt quartas eiusdem circuli, ut fiat sic figura:

---

1 naturaliter *om.* V2 2 non] vult *add.* V2 2–3 articulo quaestionis *om.* V2 4 articulum *om.* V2 | primo] circha *add. mg.* V2 11 ab aequalitate non recedere] *marg.* V2; ad aequalitatem non descendere *a.c.* 13 dicitur] appetitur V2 16 aequaliter] inaequaliter *a.c.* V2 20 a linea] lineae V2 21 circuli] aequales *add.* V2 21–22 ut fiat sic figura *om.* V2

---

96 Cf. Odo de Ceritona *Fabulae* LXXI De ciconia et catto (1896) IV 243: Quod videns Catus qui libenter comedit pisces, licet non velit humectare pedes. Cf. *Proverbialia Dicteria*, "Labor, diligentia, studium" (1575) 58: Cattus amat pisces, sed non vult tangere flumen. Cf. Albericus de Rosciate *Dictionarium iuris quam civilis quam canonici*, "Memoria" (1601) 190va: Catinus vult pisces, sed non piscari.

97 Adn. mg. V1: Tertius articulus.

98 Adn. mg. V1: Prima difficultas.—Cf. Iordanus Nemorarius *Liber Jordani de ponderibus* prop. 8 (1952) 160: Octava propositio: Si fuerint brachia libre proportionalia ponderibus appensorum, ita ut in breviori gravius appendatur, eque gravia erunt secundum situm appensa. Cf. Blasius Pelacanus de Parma *De ponderibus* propositio X (1952) 258: Brachiis libre inequalibus, si fuerint pondera proportionalia, non fiet in hac dispositione motus.

99 Adn. mg. V1: *Manicula.*

100 Cf. supra n. 16.

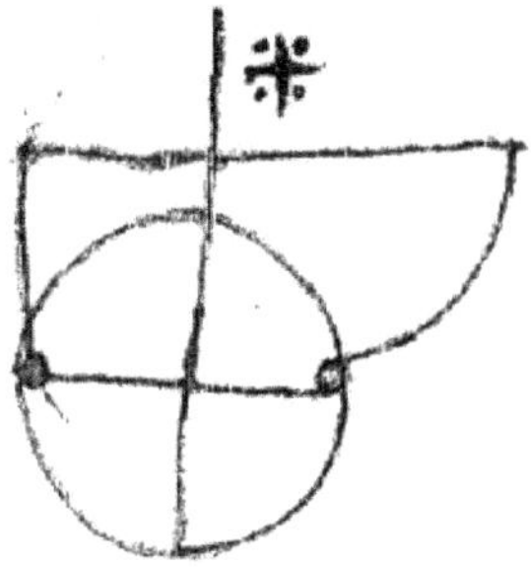

FIGURA 2
in ima pag. f. 88vb V1

[76] Secunda[101] difficultas: utrum ferrum naturaliter moveatur ad magnetem.

[77] Respondetur quod non. Quod probatur, quia ubi ferrum naturaliter moveretur ad magnetem, ferrum quamcumque magnum moveretur ad magnetem; quod est falsum. Apparet enim quod modicum ferrum movetur ad magnetem. Et ubi ponatur magnum frustum ferri illud non movetur, cum tamen deberet velocius moveri. Et hoc non nisi quia motus ferri ad magnetem non est nisi motus tractus sic quod magnes attrahens parvum ferrum non potest ad se trahere quodlibet maius.

[78] Dico iuxta hoc[102] quod nutrimentum movetur ad membra, non tamen quod membra fatigentur in attrahendo nutrimentum. Sed sic est quod sicut ferrum recipit quandam dispositionem a magnete, qua introducta sub certa latitudine movetur ad magnetem et non a principio principali intrinseco ut a forma substantiali, sic nutrimentum recipit dispositionem a determinato
V2 30vb membro quod | nutrimentum sic dispositum movetur ad illud membrum, et aliud ad aliud membrum secundum aliam et aliam dispositionem. Non tamen a principio principali intrinseco, cum nutrimentum semper sit de genere gravium.

[79] Tertia[103] difficultas: quia ita est quod data hostia consecrata, et posita in sphaera aëris, ipsa movetur deorsum. Nunc quaeritur an ipsa movetur naturaliter,
V1 89ra an violenter? Et arguitur quod nullo modo. | Quia in ea non est forma substantialis quae dicitur principium motus et quietis, ut scribitur secundo huius.[104] Nec violenter, quia non videtur quod sit violentans in proposito.

4–5 magnetem; quod] ipsum et hoc V2   8 attrahens] trahens V2   20 movetur[2]] sic *add.* V2

---

101 Adn. mg. V1: Secunda difficultas. *Manicula.*—Cf. supra n. 7.

102 Cf. supra n. 6.

103 Adn. mg. V1: Tertia difficultas.

104 Aristoteles *Physica* II,1 192b14–5: Que quidem enim natura sunt omnia videntur habentia

[80] Ad quod respondetur quod iste motus est naturalis, et sic naturaliter movetur deorsum. Sufficit enim in proposito quod ibi sit gravitas. Et proportionabiliter dixi quod ubi esset una materia privata omni forma substantiali, informata tamen gravitate, quod illa materia cum sua gravitate naturaliter moveretur deorsum. Et[105] sic de hostia consecrata, licet ibi non sit materia.

[81] Et si quaeras quod in hoc videtur accidens transire de subiecto in subiectum,[106] quia ibi est gravitas per quam nihil est grave; et sic cum movetur localiter modo est in una parte aëris, modo in alia:

[82] Ad quod respondetur quod ex hoc non transit accidens de subiecto ad subiectum, eo quod aër non subicitur sibi, nec est sicut subiectatio huius ponderositatis. Unde si tu transiveris de ecclesia ad forum, albedo informans superficiem faciei tuae, licet sit immediata aëri exteriori ita bene sicuti superficiei faciei tuae, non tamen accidens istud transit de subiecto in subiectum. Et sic dico quod licet gravitas sit immediata aëri, et modo in uno loco, modo in alio, illa gravitas non informat illum aërem nec aliquod corpus naturale.

[83] Transeo[107] ad materiam de alteratione. Et sit haec prima difficultas: utrum simplex ut ignis positum in vacuo perpetuaretur ibi aut corrumperetur.

[84] Respondetur ut supra[108] ut quod statim tenderet ad non-esse per absentiam conservantis.

[85] Sed[109] difficultas sequitur in proposito magis ardua. Quia si iste ignis corrumpetur, quaeritur an materia informata forma ignis recipiat aliam formam, aut nullam. Si nullam, dabitur ergo materia sine forma. Si aliquam, ergo fiet generatio in vacuo. Et tunc quaero: quod generabitur, et a quo generabi-

---

3 esset] quod *add.* V2 4 illa] ibi V2 10 sicut] sicuti V2 17 ibi] ibidem V2 18 ut^2 *om.* V2 22 dabitur ergo] relinquitur V2

---

in se ipsis principium motus et status; *Auctoritates Aristotelis* 144 (50): Natura est principium et causa motus et quietis in quo est primo per se et non per accidens. Cf. Blaesius de Parma *Quaestiones super VIII libris Physicorum* II q. 1 n. 22.

105 Adn. mg. V1: *Manicula.*—Cf. Blasius Pelacanus de Parma *Questiones disputate super octo libros Physicorum* prima redactio II q. 4 (ed. Biard): Tertium argumentum ad principale, et suppono fidem nostram esse veram et solidam, tunc accipio ostiam consacratam, et clarum est quod ibi datur ponderositas et sapor sine subiecto.

106 Cf. Aristoteles *De gen. et cor.* I,10 327b22; *Auctoritates Aristotelis* 167 (6): Passiones non sunt separabiles a subiectis. Cf. Ioannes Buridanus *Quaestiones De anima* II q. 2 n. 11: Et iterum communiter dicitur quod impossibile est accidens transire de subiecto in subiectum. Cf. Paulus Veneti *Logica magna* pars 2 fasc. 6 (1978) 22: Consequens est falsum, quia tunc unum accidens migraret de subiecto in subiectum—quod est contra Philosophum in primo *De generatione*.

107 Adn. mg. V1: Alia difficultas.

108 Cf. supra n. 41.

109 Adn. mg. V1: Dubium.

tur? Cum ibi non sit generans nec aliquis influxus caelestis vigore, cuius saepe dicimus[110] aliquid generari ex putrefactione.

[86] Respondetur[111] hic quia, ut scriptum est primo[112] huius, "dato uno inconvenienti, multa sequuntur". Et ideo expedit quod dicatur istam materiam annihilari. Nec mirum, quia sicut iste mundus dependet a superiori, sic et materia istorum inferiorum dependet a superiori in esse et in conservari. Et quod annihiletur non est contra Philosophum. Stat bene simul quod materia ista annihilabitur, et quod ipsa sit incorruptibilis, sicut diceremus de anima vel de una alia Intelligentia. Deus enim potest annihilare animam tuam, quae est V1 89rb incorruptibilis. |

[87] Posset aliter dici ut quod dicta materia subito sic rarefieret quod applicaretur corpori ut concavo caeli vel alteri corpori intermedio, et tamen reciperet novam formam.

[88] In vacuo potest fieri generatio, quia deus potest creare homines, ut virum et mulierem, et eosdem conservare per tempus sufficiens; et tunc isti duo suis naturalibus poterunt in vacuo generare.

[89] Et similiter in data materia posset deus creare intellectiva; et sic erat homo quem deus potest conservare.

[90] Et haec nobis sufficiant de tertio articulo.

[91] Quantum ad quartum articulum patet solutio inspectis determinationibus V2 31ra positis in quaestione et difficultatibus. Et hoc de quarto articulo. |

---

1 influxus] influsus *var.* V1 5 annihilari] anichilari *var.* 6 inferiorum *om.* V2 14–15 In … potest conservare *om.* V1 19 nobis sufficiant *om.* V2 | tertio articulo] secundo V2 20 articulum *om.* V2 | determinationibus] et hoc de quarto *add. sed del.* V1 21 quarto articulo] tota quaestione V2

---

110 Cf., e.g., Blasius Pelacanus de Parma *Questiones circa libros de generatione et corruptione* II q. 8 (ed. Biard): Patet antecedens, quia ex solo aspectu corporum supercelestium multa animalia generantur ex putrefactione. Cf. Thomas de Aquino *Summa theologiae* I q. 71 ad 1: loco cuius virtutis, in iis quae ex putrefactione generantur, est virtus caelestis corporis. Cf. Ioannes Buridanus *Quaestiones super octo libros Physicorum Aristotelis* (*secundum ultimam lecturam*) II q. 5 (2015–2016) I 274: in generatione multorum animalium ex putrefactione corpora supercaelestia sunt agentia principalia.

111 Adn. mg. V1: Respondetur.

112 Aristoteles *Physica* I,2 185a11; cf. *Auctoritates Aristotelis* 140 (7): Uno inconvenienti dato, plura contingunt.

3

Liber II q. 7: Utrum ad cuiuscumque rei naturalis generationem praeter particulare agens requiratur influxus causae universalis, quae causa universalis dicitur sol

V1 = Vat. lat. 2159 ff. 96ra–98rb; V2 = Vat. lat. 3012 ff. 35vb–37rb

Federici Vescovini = Graziella Federici Vescovini, "Su alcune testimonianze dell'influenza di Alberto Magno come 'metafisico', scienziato e 'astrologo' nella filosofia padovana del cadere del secolo XIV: Angelo di Fossombrone e Biagio Pelacani da Parma," in *Albert der Große: Seine Zeit, sein Werk, seine Wirkung*, ed. Albert Zimmermann (Berlin: De Gruyter, 1981) 155–176.

[1] Quaeritur septimo utrum ad cuiuscumque rei naturalis generationem praeter particulare agens requiratur influxus causae universalis, quae causa universalis dicitur sol, quia secundo huius[1] dicebatur quod homo generat hominem, et sol. Et ita intelligitur de aliis planetis.[2]

[2] Arguitur negative.

[3] Primum[3] argumentum: Sole et luna, exsistentibus sub circulo nostri hemisphaerii, fiunt multorum generationes in hemisphaerio nostro, ut constat

1 cuiuscumque] cuiuslibet V2 4 intelligitur] intelligatur V2 5 negative] ad quaestionem *add.* V2 7 nostro] et *add.* V2

1 Aristoteles *Physica* II,2 194b13: Homo generat hominem et sol.

2 Cf. Iacobus de Ascoli *Quodlibet* I q. 14 (cod. Vat. lat. 1086, ff. 73^{b}–75^{a}) "Utrum agens particulare habeat per se aliquem effectum." Resp. Averroës *In Metaph.* VII com. 31 (in VII,9 1034a30–1034b7) (1562) f. 181rbE: In generabilibus autem non ex se, id est casu, apparebit quod corpora coelestia sunt illa quae dant istis aliquid loco seminum, et virtutum quae sunt in seminibus, et in eis quae generantur ex semine. Cf. Federici Vescovini (1981) 155–176. Ed. Joël Biard. Cf. Blasius Pelacanus de Parma *Questiones disputate super octo libros Physicorum* prima redactio, II q. 4, "Utrum qualitas in vigore proprio possit producere formam substantialem" (ed. Biard).

3 Adn. mg. V1: Primum argumentum.

 | DOI:10.1163/9789004734968_012

V1 96rb primo.[4] Tunc sol et luna non concurrunt, propter terram abscindere | radios eorum. Confirmatur, quia in camera tua bene obstructa fiunt generationes, etc.

[4] Secundo,[5] in generatione rei naturalis sufficit agens particulare; igitur, etc. Antecedens patet, quia quacumque hora diei et noctis, si applicaveris ignem straminibus, generabitur ignis.

[5] Tertio,[6] homo potest generare hominem, et asinus asinum, et hoc est eis naturalissimum, ut scribitur secundo *De anima*[7] in hac forma, "naturalissimum est unumquodque generare tale quale ipsum est, ut animal animal et planta plantam." Confirmatur, quia homo vult generare hominem et nihil sibi resistit. Idem dico de asino. Igitur homo, quandocumque vult, potest generare sine concursu causae universalis. Antecedens patet, quia ubi aliquid resisteret, esset passum ut materia, quae non esset apta recipere formam nec dispositionem agentis. Sed hoc non, quia materia ponitur infinitae passibilitatis et nullius resistentiae.[8] Immo, ut dicitur primo *Physicorum*,[9] "semper materia appetit formam, quaecumque sit illa," quia per illam perficitur et actuatur.

---

1 abscindere] ascindere *var.* V2 4 rei] particularis *add. sed del.* V2 5 noctis] data *add.* V2 7 et *om.* V2 9 unumquodque] unumquemque V2 | generare] sibi *add.* V2 | et *om.* V2 10 planta] plancta *var.* V2 11 potest *rep.* V2 12 sine] absque V2; sed *add. sed del.* V2 | Antecedens *om.* V2 15 dicitur] scribitur V2

---

4 Cf. Aristoteles *De gen. et cor.* II,10 336a29–b2; cf. Ioannes de Sacrobosco *Sphaera* (1478) f. 84: Ab Aristotele vero in libro *De generatione et corruptione* dicitur circulus obliquus, ubi dicit quod secundum accessum et recessum solis in circulo obliquo fiunt generationes et corruptiones in rebus inferioribus.

5 Adn. mg. V1: Secundum argumentum.

6 Adn. mg. V1: Tertium argumentum.

7 Adn. mg. V1: *Manicula.*—Cf. Aristoteles *De anima* III,4 415a: Naturalis enim viventibus operibus est quaecumque perfecta et non imperfecta, quam generationem per se vanam habere, facere alterum sicut et ipsum est. Cf. Blasius Pelacanus de Parma *Questiones super perspectiva communi* prima pars, q. 10 (2009) 155: dicebat Philosophus secundo *De anima* ad textum: 'naturalissimum est unumquodque producere tale quale ipsum est'. Cf. Blasius Pelacanus de Parma *Questiones disputate super octo libros Physicorum* prima redactio II q. 9 (ed. Biard): Consimilia verba scribuntur II° *De anima*, dum dicitur quod naturalissimum est unumquodque generare sibi tale quale, ut animal animal.

8 Cf. Marsilius de Inghen *Quaestiones super libros De generatione* II q. 6 (1567) 232b: materia est infinitae passibilitatis. Cf. supra Blasius Pelacanus de Parma *Questiones disputate super octo libros Physicorum* VII q. 2 n. 29: materia prima est primo infinitae passibilitatis, et per consequens non est alicuius resistentiae.

9 Aristoteles *Physica* I,9 192a22–23; *Auctoritates Aristotelis* 142 (32): Materia appetit formam sicut femina virum et turpe bonum.

[6] Quarto[10] sic: a media regione saepe ceciderunt ranae, pisces, viperae, et vitulus, ut narrat Avicenna in eius *Metaphysica*,[11] et clarum est quod ista omnia fuerunt genita in media regione sine concursu agentis universalis, quia in media regione non videtur sol influere, quia media regio semper est frigida et humida, sol autem calidus et siccus virtualiter.

[7] Quinto[12] sic: nulla solaris qualitas nec alterius astri potest ad nos pertingere; | igitur, etc. Patet antecedens, quia ubi esset identitas medii ab astris usque V2 36ra
ad nos, radii solis et sideris ad nos pertingerent. Sed quia non est ita, eo quod magna est difformitas medii in raritate et densitate, sequitur quod radii divertant se a recto incessu; ergo, etc. Quod autem sit magna difformitas ex parte medii, patet inspectis pluribus elementis ut igne, aëre, aqua, etc.

[8] In[13] oppositum est Philosophus in hoc secundo,[14] volens quod ad quemcumque effectum ultra causam particularem concurrat causa universalis, quae causa universalis per ipsum in textu dicitur sol.

[9] In hac quaestione erunt quattuor articuli. In primo problema; in secundo de quaesito; in tertio difficultates; in quarto etc.

---

1 sic *om.* V2 2–3 ista omnia fuerunt genita] illa genita fuerunt V2 3 sine] absque V2 | agentis] naturalis *add. sed exp.* V2 5 siccus] est *add.* V2 6 nulla] caliditas *add. sed del.* V1 8 solis et sideris] solares et siderei V2 9 in] etiam V2 10 ex *rep.* V2 14 in textu *om.* V2 16 de quaesito *om.* V2 | in quarto *om.* V2

---

10 Adn. mg. V1: Quartum argumentum.

11 Cf. immo pseudo-Avicenna *De diluviis* II, 6 (1949) 307: Et non est inopinabile secundum quosdam quos tu scis, ut corrumpantur et plante et multa genera eorum et post generentur per generationem, non gignitionem, quoniam multa alia fiunt per generationem et gignitionem. Et similiter plante. Et ex capillis fiunt serpentes, ex ficubus scorpiones et mus de terra et ranae de pluvia. Et omnia huiusmodi gignuntur. Et quando haec gignitur abscinditur, in multis annis possibile est ut veniat secundum quosdam constellatio et aliqua praeparatio elementorum quae faciat ea generari. Immo dicimus quod quaelibet species quae sit per complexionem elementorum in quantitatibus scitis dum elementa fuerunt, et sua divisio secundum illas quantitates et congregatio earum fuerit possibilis, congregatio earum erit possibilis. Et si prima complexio non sufficit, haec generantur ex secunda et tertia. Quoniam quemadmodum generatur animal ex complexione humorum post complexionem elementorum, non est inopinabile quod fiat complexio secunda sine semine et sine spermate. Cf. Blasius Pelacanus de Parma *Quaestiones disputatae De anima* I q. 8 (1974) 78: in media regione, saepe vigore alicuius astri vel plurium, generata sunt animalia diversarum specierum; visa a nobis, cadent deorsum, ut sunt ranae, pisces, boves etc.

12 Adn. mg. V1: Quintum argumentum.

13 Adn. mg. V1: In oppositum.

14 Aristoteles *Physica* II,2 194b13; cf. supra n. 1.

[10] Quantum[15] ad primum articulum sit problema: propter quid est quod in specie humana reperiuntur plures difformitates ex parte corporis quam in aliis animalibus? Videmus enim quod in specie humana reperiuntur multae dissimilitudines et impuritates filiorum ad parentes. Unde saepe pulcher homo generat filium claudicantem, nigrum cum pater sit album strumosum, et aliquando contingit quod homo turpis producit filios pulchros bene formatos, et sic de aliis. | In brutis vero animalibus hoc raro contingit. (V1 96va)

[11] Ad problema uno modo responderet philosophus naturalis et aliter astronomus et aliter theologus.

[12] Philosophice loquendo, causa est quia hora coitus contingit quod mulier habet saepe fortem estimationem ad alium quam ad virum suum. Et, ut dixi in problemate praecedentis quaestionis,[16] fortes estimationes et motus animae immutant corpora. Hinc est quod huius dissimilitudines in specie humana reperiuntur. In brutis vero non, quia animal brutum non habet estimationem nec huiusmodi praevaricationem in virtutibus cognoscitivis nisi forte ad advenam. Et ob hoc saepe viri habent mulieres eorum odio, nescientes causam huius veritatis, sed potius estimantes quod eorum mulieres habuerint agere cum aliis. Possetis etiam introducere pro argumenti complemento ea quae dixi[17] in conclusionibus huius secundi *Physicorum* in materia de casu et fortuna, ut quod hora coitus in specie humana mulier aliquando turpiter movetur et agitatur, et ex isto motu inepto semen fluit ad loca et sic causantur ista.

[13] Astronomus[18] vero diceret hoc esse a constellatione. Et in hac parte praefero sententiam philosophorum sermonibus astronomorum, quia ita con-

---

1 articulum *om* V2 | sit] hoc *add.* V2 3 Videmus] Vides V2 4 et impuritates *om.* V2 5 strumosum] stromosum *var.* V2 9 astronomus] astrologus V2 11 estimationem] extimationem *var.* 12 praecedentis quaestionis] priori V2 13 immutant] mutant V2 | huius dissimilitudines] huiusmodi difformitates V2 14 non habet estimationem] estimationem non habet V2 15–16 advenam] avenam V1 16 hoc] patet *add. sed del.* V1 18 complemento] ad problema *add.* V2 22 Astronomus] Astrologus V2 | diceret] dicet V2 23 astronomorum] astrologorum V2 | quia] quoniam V2

---

15 Adn. mg. V1: *Manicula.*

16 Blasius Pelacanus de Parma *Quaestiones super VIII libris Physicorum* II q. 6 (cod. Vat. lat. 2159) f. 94vb: tunc dominus facit sibi fortem estimationem, et in hac forti estimatione sanguis et spiritus movetur de centro ad partes corporis oppilatas et repletas malis humoribus.

17 Cf. Blasius Pelacanus de Parma *Quaestiones super VIII libris Physicorum* II q. 9, "Utrum asserentes omnia de necessitate evenire, et nihil a casu vel a fortuna possint eorum positionem substentare" (ed. Biard). Cf. Albertus Magnus *Physica* II tract. 3 cap. 3; *Opera omnia* 4,1 (1987) 137b: sunt enim quaedam mulieres et quaedam animalium, quae multum in coitu delectantur, et in delectatione illa movetur matrix, cum sperma infunditur nervis sensibilibus eius, et in motu illo dividitur sperma.

18 Cf., e.g., pseudo-Ptolomeus *De nativitatibus* cap. 53 "De aegritudinis nati" (1501) f. 9va:

tingeret in brutis sicut in specie humana eo quod generationes brutorum ita dependent a constellatione sicut etc.

[14] Et haec de primo articulo.

[15] Transeo ad secundum articulum in quo erunt duo processus. In primo breviter dicam de quaesito; in secundo dicam an ad cuiuslibet rei substantialis productionem ultra agens universale requiratur agens particulare.

[16] Quantum[19] ad primum, sit prima suppositio: mundo corruptibilium distante orbiculariter a superis, mundus corruptibilium tenderet ad non-esse. Declaratur conclusio primo auctoritate Aristotelis primo *Meteororum*[20] ubi de hoc dicit namque quod "necesse est hunc mundum inferiorem esse continuum lationibus superioribus ut omnis virtus inde gubernetur." Declaratur per rationem, quoniam, lateribus folliculi se mutuo tangentibus et obstructis foraminibus dicti folliculi diligenter, non possunt latera | dicti folliculi ab invicem separari. Et huius non est alia causa nisi quia quodlibet corpus naturale appetit caelo copulari aut saltem corpori tangenti illud. Videmus[21] etiam quod si V2 36rb

1 generationes] generationem V1 | ita] de *add. sed del.* V1 3 articulo] questionis *add.* V2 4 articulum *om.* V2 5 de *om.* V2 | an] ante *ante corr.* V1 7 suppositio] propositio V2 8 superis] tendit *add. sed del.* V1; tendit *add.* V2 | tenderet *om.* V2 9 Declaratur conclusio *om.* V2 | auctoritate] *corr.* s.l.; auto *a.c.* V2 9–10 ubi de hoc *om.* V2 10 inferiorem *om.* V2 10–11 continuum] contiguum V2 11 superioribus] *mg.* V2 | ut *rep.* V2 12 tangentibus] instans *add.* V2 13 dicti] dati V2 14 huius] haec V2 | causa *om.* V2 15 copulari] compulari *var.* V2 | saltem] saltim *var.* V2

Quando sol in nativitate diuna ad aliqua infortunatarum impeditus fuerit, natus aegritudinem in oculo dextro habebit, aut forte luscus aut strabo erit. Et si sole sic disposito nativitas nocturna fuerit, natus aegritudinem in stomacho vel cerebro habebit. Cf. Albertus Magnus *Physica* II tract. 2 cap. 18; *Opera omnia* 4,1 (1987) 125b: Et ideo secundum veritatem quam tradidit Ptolemaeus in libro *De nativitatibus*, nativitates tales occasionatae sunt in causa particulariter movente, quae est vis formativa seminis et praeparatio materiae ad formam, sed tamen sunt intentae in causa universaliter movente, quae est qualitas proveniens ex situ et respectu stellarum in hora, qua semen cadit in matricem.

19 Adn. mg. V1: Prima suppositio. *Manicula.*—Cf. Blasius Pelacanus de Parma *Questiones disputate super octo libros Physicorum* secunda redactio IV q. 4 (ed. Biard): Et tunc premitto hanc propositionem preambulam: absque orbium influxu in hunc alterabilem mundum, hic alterabilis mundus statim tenderet ad non esse.

20 Aristoteles *Meteora* I,2 339a21–23: Est autem ex necessitate continuus iste superioribus lationibus, ut omnis ipsius uirtus gubernetur inde; *Auctoritates Aristotelis* 171 (2): Totus mundus sensibilis contiguus est lationibus superioribus caeli sensibilis ut inde tota ejus virtus gubernetur. Cf. supra II q. 2 n. 40.

21 Cf. Blasius Pelacanus de Parma *Questiones disputate super octo libros Physicorum* secunda redactio IV q. 4 (ed. Biard): Patet hoc multis experientiis. Unde ut hoc concludatur, in natura videmus aquam moveri sursum de loco primo ut si posueris lignum perforatum

V1 96vb posuerimus canam perforatam secundum | unum extremum eius ad liquorem, et mediante hoc ore tu attrahas aërem in cana contentum, subito liquor ascendit et sursum movetur. Et huius non est alia causa nisi quia quodlibet corpus appetit copulari alteri etc., ut inde recipiat influxum a caelo. Ex quo patet quod non minus iste mundus dependet a superis in conservari quam radius solaris a sole, ut dicam quod, quam cito supervenit corpus opacum inter solem et lumen diffusum ab ipso per cameram nostram, statim lumen deperditur; sic etiam iste mundus subito tenderet ad non-esse ubi orbiculariter distaret a superis. Et hoc intellige esse verum quantum est per naturam, quia non est dubium quod ita deus posset conservare istum mundum distantem a superis sicut non distantem.

[17] Prima[22] conclusio et responsalis: ad cuiuslibet rei substantialis productionem, ultra agens particulare requiritur agens universale. Ista conclusio patet inspecta propositione praecedente[23] cum eius probationibus.

[18] Et sequitur corollarium,[24] quod nullum agens particulare se solo est alicuius effectus productivum. Patet postquam ad cuiuslibet rei productionem requiritur causa universalis. Et hoc de isto processu.

[19] Transeo ad secundum in quo quaeram utrum ad cuiuslibet rei substantialis productionem ultra agens universale requiratur agens particulare.

[20] Et circa hoc erunt duo modi dicendi.

[21] Pro primo praemitto quod causarum efficientium quaedam dicitur causa universalis, non quod per 'causam universalem' intelligatur aliqua res universalis, sed quia universaliter videtur concurrere ad quemcumque effectum. Quaedam vero dicitur causa particularis, et dicitur 'causa particularis' quia sic limitata est ad istum effectum producendum, quod non ad quemcumque.

---

1 posuerimus] posueris V2 | perforatam] foratam V1 | extremum eius] *ord. inv.* V2 2 hoc *om.* V2 3 huius] haec V2 4 copulari] corpori V1 5 mundus] mondus *var.* V2 8 ubi] nisi V2 9 ita *om.* V2 18 secundum] processum *add.* V2 20 circa] contra V2 23 effectum] ut caelum et astra eorum motu et lumine, primo motore non excluso *add.* V2 25 limitata] limitare *Federici-Vescovini* | istum] hunc V2

---

secundum unum eius extremum in aqua vel in aliquo liquore, reliquum autem extremum ponas in ore tuo, tunc attracto aere existente in ligno perforato subito est aqua in ore tuo. … Et ratio huius non est alia causa nisi quia ubi aqua non copularetur superiori mundo vel alteri corpori tangenti illud, statim tenderet ad non esse, quia absque influxu superiorum hic inferius iste mundus tendit ad non esse.

22 Adn. mg. V1: Prima conclusio.

23 Cf. supra n. 16.

24 Adn. mg. V1: Corollarium.

Unde licet possit indifferenter ignis agere in quodlibet combustibile generando ignem, causa universali concurrente, non tamen causa particularis concurrit generando lapidem vel asinum ex combustibili.

[22] Hac[25] evidentia praemissa, sit prima propositio: non contingit plures formas substantiales ultimatas informare eandem partem materiae. Patet de se.

[23] Prima[26] conclusio: non contingit causam universalem se sola formam substantialem hominis producere. Patet, quia causa universalis non disponit materiam plus ad hanc formam quam ad illam, ut potest haberi ex evidentia praemissa[27] et quia, ut dicit | (V1 97ra) propositio praemissa,[28] non est possibile plures formas substantiales informare eandem partem materiae. Sequitur quod causa universalis se sola nullam formam inducet substantialem; alioquin causa universalis determinaret se plus ad unam formam quam ad aliam, et sic esset sicut causa particularis. Et intelligatis in proposito per 'causam universalem' caelum cum suis astris sic influens. Ubi vero per 'causam universalem' intelligeretis ipsum deum, non est dubium quod ipse se solo, nullo alio active concurrente, posset producere formam substantialem hominis et creare novum caelum. Adhuc in proposito non excludetis ipsum deum, eo quod ubi deus non ageret sed omnem eius actionem suspenderet, caelum nihil | (V2 36va) inferius influeret. Et ergo per 'causam universalem' intelligatis caelum et eius influxum, Prima causa generaliter concurrente, nec solum generaliter concurrente pro generatione hominis, sed multum etiam particulariter pro quanto homo habet animam a deo creatam.

[24] Secunda[29] conclusio: ad cuiuslibet rei substantialis generationem, caelum videtur concurrere, et natura. Et per 'naturam' in praesenti intelligo particulare.

---

1 licet] agens *add. sed del.* V1 | possit indifferenter ignis] ignis possit indifferenter V2 3 lapidem vel asinum ex combustibili] ignem ex combustibili vel asinum V2 4 contingit] absolute est concedendum V2 5 ultimatas] vel terminatas *Federici-Vescovini* | Patet] propositio *add.* V2 7 causam universalem] caelum V2 | sola] solo V2 13 plus] magis V2 | sic] si sic tunc V2 | sicut *om.* V2 14 universalem *rep.* V2 16 solo] quia *add.* V2 | concurrente] per actum creationis *add.* V2 18 excludetis] excludas V2 21 generaliter] influente *add. sed exp.* V2 | concurrente] *om.* V2 22 etiam paticulariter] in particulari V2 25 intelligo] agens *add.* V2 25–26 particulare] patet conclusio ex primo processu *add.* V2

---

25 Adn. mg. V1: Prima propositio.

26 Adn. mg. V1: Prima conclusio.

27 Cf. supra n. 21.

28 Cf. supra n. 22.

29 Adn. mg. V1: Secunda conclusio.

[25] Tertia[30] conclusio: licet iste homo vel ille totalis totaliter generetur a causa universali, ut a deo, astris concurrentibus, tamen nec ille homo vel alius producitur a deo totaliter. Ista conclusio habet duas partes. Pro eius probatione oportet intelligere quod in generatione hominis, licet causa universalis et particularis concurrant, non est intelligendum unam istarum causarum producere unam particularem, et causam particularem aliam. Sed ille homo totus, totaliter noviter productus, producitur a deo, et totaliter producitur a causa particulari. Sed a neutro istorum tamquam a causa totali, quia tunc alterum istorum superflueret. Et sic homo se solo posset producere hominem, et asinus asinum, vel causa universalis se sola posset generare asinum vel hominem, quod non est dicendum.

[26] Nunc subiungo alium modum dicendi qui imputatur Avicennae in sua *Metaphysica*,[31] et similiter Alberto Magno,[32] et qui in lumine naturali non posset potenter improbari. Sed quia falsus est in se, et fidei obviat catholicae, ideo narrabo hunc modum solummodo ad hunc finem quod ubi infidelis talia diceret, eidem sciatis respondere. Hoc praemisso, condescendo ad propositum.

[27] Prima propositio:[33] ab astris qualitates elementales hic inferius assidue causantur, et aliae consequenter de secundis. Ista propositio videtur patere per hoc quia sol hic inferius caliditatem efficit, et tandem ignem, sicut in partibus Cypri, sole exsistente in Cancro vel in Leone, inflammantur nemora. In partibus vero nostris septentrionalibus saepe hoc fit mediante speculo[34] concavo
V1 97rb aut | vase[35] vitreo aqua pleno. Saturnus[36] autem frigiditatem inducit, luna vero

---

1 generetur] produceretur V2 2 vel] nec V2 3 a deo totaliter] totaliter a deo V2 6 particularem] eius partem V2 7 noviter *om.* V2 8 totali] producitur totaliter *add.* V2 10–11 vel … dicendum] *mg.* V2 10 asinum vel hominem] hominem et asinum V2 11 non est dicendum] est falsum V2 12 dicendi qui *rep. sed del.* V1 13 et *om.* V2 | et *om.* V2 13–14 non posset potenter *rep. sed del.* V1 14 falsus] falsius *Federici-Vescovini* 16 eidem] eisdem V2 17 Prima propositio *om.* V2 20 Cypri] Cipri *var.* | in² *om.* V2

---

30 Adn. mg. V1: Tertia conclusio. *Manicula.*

31 Cf. supra n. 6.

32 Cf. supra n. 6.

33 Adn. mg. V1: Prima propositio.

34 Cf. infra II q. 8 n. 3. Cf. Blasius Pelacanus de Parma *Questiones circa tractatum proportionum magistri Thome Braduardini*, redactio prima, q. 12 (2005) 230: Docet enim hoc experientia quod, exposito speculo concavo radiis solaribus, generatur ignis in combustibili. Cf. Marsilius de Inghen *Quaestiones super libros De generatione* I q. 16 (1567) 179^{a-b}: radii a speculo concavo reflexi prope speculum de per se deberent generare intensiorem calorem.

35 Cf. infra II q. 8 n. 3.

36 Cf. Robertus Grosseteste *De impressionibus aeris seu de prognosticatione* (1912) 42: Naturae

humiditatem, Mars autem siccitatem, et sic de aliis, quae qualitates sunt eiusdem rationis cum qualitatibus quae per elementa hic inferius causantur.

[28] Secunda[37] propositio: ab astris hic inferius omnis proportio elementalium qualitatum et omnis latitudo proportionum earundem qualitatum est possibilis causari. Haec propositio difficulter posset improbari; immo forte ita est ut praetendit, quia inspecta tanta multitudine astrorum quanta est in octava sphaera recipientium lumen solare, deinde illud idem reflectentium ad nos, est possibile quod dicitur, quae astra fixa aliter et aliter modificant radios solares, merito quorum astrorum diversi radii, sic reflexi et modificati, diversas videntur habere influentias et proprietates hic inferius.

[29] Ex hiis duabus propositionibus ponuntur conclusiones.

[30] Prima[38] conclusio: merito astrorum potest materia quam philosophi primam vocant quamlibet disponi ad hanc formam vel illam possibilem generari. Patet, quia materia prima disponitur ad hanc formam vel illam in vigore qualitatum elementalium sic invicem proportionatarum. Sed per secundam[39] propositionem, ab astris hic inferius omnis proportio elementalium qualitatum et latitudo proportionum earundem est possibilis generari; igitur, etc.

[31] Corollarium:[40] nullo asino exsistente, contingit asinum et quodlibet aliud animal de inferiori specie per naturam de caelo generari. Patet, quia non ita bene materia potest disponi ad hanc formam vel illam possibilem generari, quin melius quamlibet possit illa materia disponi merito | astrorum sic influentium; quare, etc. V2 36vb

1 et sic de aliis] etc. V2 4 et … qualitatum *om.* V2 5–6 immo … ut praetendit *om.* V2 7 recipientium] recipientia V2 10 et proprietates *om.* V2 12 conclusio *om.* V2 19 caelo] caetero V1 20 bene] ab astris *add.* V2 21–22 influentium] confluentium V2

planetarum sunt hae: Saturnus summe frigidus et ex consequenti siccus est et sic, quantum de se est, omni naturae inimicus. Juppiter vero temperate calidus et humidus est et sic omni naturae amicus. Plus tamen in eo abundat calor, quam humiditas. Mars autem superflue calidus et siccus naturae colericae et sic quodammodo vitae contrarius; plus tamen calidus, quam siccus. Sol vero temperatissime calidus et siccus, vel ut verius dicam, inter omnes qualitates summe aequalis. Venus vero temperate calida, superflue humida, unde fertur habere dominium super usus venereos. Mercurius autem inclinatur per naturam suam ad eum, cui complectitur ex planetis et signis. Dicitur tamen in se esse frigidus et siccus. Lunae vero natura dicitur temperate frigida et summe humida; unde dicitur esse fons humiditatis, sicut sol fons caloris, licet temperate sit calidus.

37 Adn. mg. V1: Secunda propositio.

38 Adn. mg. V1: Prima conclusio.

39 Cf. supra n. 28.

40 Adn. mg. V1: Corollarium.

[32] Secunda[41] conclusio: nulla creatura rationali exsistente, contingit hominem per naturam generari. Patet, quia non ita bene potest materia disponi ab agente particulari, ut ab homine, quin ita bene vel melius possit merito astrorum disponi; ergo conclusio vera.

[33] Corollarium:[42] quod merito huius constellationis vel illius in quovis puncto mediae regionis contingit hominem generari, contradictio non apparet. Patet, quia in media regione non deficit materia, quae materia quantumcumque potest disponi optime ad hanc formam possibilem generari; igitur, etc.

[34] Secundum[43] corollarium: contingit merito huius constellationis ex semine in locum conceptionis mulieris nunc infuso ranam generari, quod si alia hora hanc dilationem habuisses, aquilam volantem inde habuisses. Patet, quia stat hanc constellationem nunc disponere sic istud semen dimissum in locum conceptionis quod merito huius forma ranae educitur de potentia mate- V1 97va riae | dispositae. Sed quia post hoc erit alia constellatio quam nunc, vel alter aspectus quam nunc, ideo data materia aliter disponeretur quam nunc sit disposita, et per hoc satis possibile est quod dispositio quae post hoc inducetur in materia ab hac constellatione aliter se habente arguet adventum formae aquilae ut dixi.

[35] Tertium[44] corollarium: quod in utero mulieris virginitatem observantis, ex materia in utero per constellationem disposita generetur homo per naturam, possibile est. Patet, quia merito astrorum potest aliqua materia recepta in utero virginis aeque bene et melius disponi ad formam humanam quam si in media regione aëris haec materia exsistet. Sed ubi haec materia in media regione esset, posset fieri quod dictum est; ergo longe melius in utero mulie-

---

5 quod] quo *Federici-Vescovini* 5–6 in quovis puncto] puncto in quovis *a.c.* V2 6 contradictio] contradictionem *a.c.* V1 7 in] materia *add. sed del.* V1 8 potest disponi optime] optime potest disponi V2 11 hanc] hoc V1 | dilationem] non *add.* V2 | habuisses] habuisset V2 14 erit] est V2 16 est *om.* V2 18 aquilae] in materia *add.* V2 19 corollarium *om.* V2 22 aeque bene] melius V1 23 exsistet] stetisset V2 24 esset] aeris staret V2 | longe *om.* V2

---

41 Adn. mg. V1: Secunda conclusio.

42 Adn. mg. V1: Corollarium primum.

43 Adn. mg. V1: Corollarium secundum.—Cf. Blasius Pelacanus de Parma *Questiones disputate super octo libros Physicorum* secunda redactio I q. 7 (ed. Biard) 8: contingit inde aliquando mustelam generari, aliquando ranam, aliquando viperam aliquando avem, et sic modo unum modo aliud, secundum quod constellatio celi plus confluit nutu Dei. Et hoc patet evenire, et evenire specialiter in partibus Tuscie, et in riperia Janue in civitate Savone, dicitur quod nunc est mulier que semper, hora sui partus, paulo ante producit aliquid de predictis, ut ranam vel aliud.

44 Adn. mg. V1: Corollarium tertium. *Manicula.*

ris virginitatem servantem. Et tunc sequitur hoc argumentum non valere: 'ista mulier peperit, ergo non est virgo'.

[36] Et per hunc modum dixit Albertus[45] modum istum fuisse alias reparatum, scilicet per generationem aequivocam; et sic ut inquit iste homo fuit post diluvium generatus. Stetit namque tota terra occulta aquis, et sic de specie humana non remansit individuum de arca Noae. Non asseruit illud fuisse.

[37] Nunc respondeamus ad ista ut inde heretici non insurgant. Et in summa concedo primam[46] propositionem et secundam.[47] Et cum hoc volo concedere quod nullo asino exsistente, asinum generari in media regione est possibile, et similiter quodlibet aliud animal de inferiori specie. Video namque ranas moveri ab aëre ibidem genitas, quod idem concedo de bove ut ait Avicenna, et de asino.

[38] Ideo transeo ad secundam[48] conclusionem, quam dico esse simpliciter falsam. Et quando dicitur quod 'contingit materiam quamlibet disponi merito constellationis', negatur hoc. Tamen, sive hoc concedatur sive non, merito astrorum numquam induceretur anima intellectiva, quae a deo creata est speciali gratia ipsius confluente. Et dico etiam quod si a media regione moveretur deorsum unum animal plasmatum ad instar formae humanae, illud animal esset animal brutum; nec esset homo. Non dubito tamen quod deus posset creare unum hominem in superficie terrae et in quovis puncto mediae regionis.

[39] Et sic dico ad corollaria[49] quae subiunguntur quod in utero mulieris virginitatem observantis non potest merito astrorum materia sic disponi quod

1 servantem] observantem V2 | tunc] non *add. sed exp.* V2 2 ergo] illa *add.* V2 3–6 Et … illud fuisse *om.* V1 4 sic] hoc V2; *deest* V1 5 generatus] generale V2; *deest* V1 11 aëre] media regione aëris V2 13 quando] cum V2 16 gratia ipsius] *ord. inv.* V2 | Et *om.* V2 17 deorsum *om.* V2 | formae humanae] *ord. inv.* V2

45 Albertus Magnus *De causis proprietatum elementorum* I tract. 2 cap. 14 (1980) 87a: scimus enim, quod stellae sunt aequivoca generantia et tamen sunt prima generantia, quia non generant, inquantum sunt aequivoca, sed potius in quantum conveniunt. Cf. Blasius Pelacanus de Parma *Quaestiones disputatae De anima* I q. 8 (1974) 78: Decima conclusio: aliquando fuit ita quod non fuit homo. Patet, quia tempore diluvii universalis, quod fuit currente anno 4487°, stetit tota terra cohoperta aquis quadriginta diebus et ultra. Et, tunc, omnes homines et alia animalia submersa sunt et statim mortui per suffocationem. … Et hoc est in capitulo Alberti Magni in libro *De natura locorum*. Cf. Petrus Pomponazzi *Quaestiones physicorum*, cit. Federici Vescovini (1981) 171 n. 57: Ipse ⟨Avicenna⟩ enim fuit magnus astrologus et longe excellentior fuit in astrologia quam Aristoteles ipse; per rationes astronomicas ibidem probat quod sunt diluvia universalia et per ignem et per aquam. Et Albertus in libro *De natura elementorum* ponit has rationes unde probat aliquando nulla esse animalia, qui tota terra potest cooperiri aquis.

46 Cf. supra n. 27.

47 Cf. supra n. 28.

48 Cf. supra n. 32.

49 Cf. supra n. 33–35.

inde generetur homo, licet hoc possit esse, deo specialiter concurrente, uti alias cum dicitur[50] "Verbum caro factum est et habitavit nobis", hora qua filius dei assumpsit naturam humanam in utero Beatae Virginis. Et sic respondendum V2 37ra est illis[51] qui dicerent | talia posse fieri in lumine naturali. Et hoc pro nunc volo V1 97vb sufficere de secundo articulo huius praesentis quaestionis. |

[40] Transeo ad tertium articulum in quo quaero breves difficultates.

[41] Quarum[52] prima sit haec: an contingit aliquod animal generari, nulla causa particulari concurrente, sed tantummodo ex solo influxu caelorum.

[42] Videtur quod sic. Quod patet de animalibus quae generantur ex putrefactione. Unde pro generatione eorum videtur sola causa universalis concurrere.

[43] Ad oppositum, quia causa universalis ratione suae universalitatis non determinavit materiam magis ad recipiendum unam formam substantialem quam aliam, et quia, ut communiter tenetur,[53] non ponitur pluralitas formarum substantialium in eadem parte materiae. Hinc est quod causa universalis ratione eius universalitatis non produceret unam formam nec plures formas.

[44] Ad difficultatem respondetur iuxta tenorem determinationis quae facta est. Dicatur quod a causa universali non generatur animal absque concursu agentis particularis. Et hanc conclusionem probat ratio adducta prius.[54]

[45] Et ad argumentum in oppositum,[55] cum dicitur de animalibus quae generantur ex putrefactione, dico quod ista animalia habuerunt in materia causam particularem, ut fuit certa proportio qualitatum primarum, vigore quarum materia disposita est sicut, gratia exempli, pro forma ranae, quod non pro forma viperae. Et hoc sufficit.

1 hoc *om.* V2 2 habitavit nobis] etc. V2 4–5 Et … quaestionis *om.* V2 6 articulum *om.* V2 7 contingit] contingat V2 8 caelorum] caelesti V2 13 determinavit] deteminaret V2 14 communiter] p *add. sed del.* V2 17 respondetur *om.* V2 19 adducta prius] *ord. inv.* V2 22 vigore *om.* V2

50 Io. 1, 14.

51 Id est, Avicenna: cf. supra n. 6

52 Adn. mg. V1: Prima difficultas. *Manicula.*

53 Cf., e.g., Marsilius de Inghen *Quaestiones super libros De generatione* I q. 6 (1567) 150a: Quinta conclusio et ultima istius articuli est, non sunt ponendae plures formae substantiales totales in eodem individuo; Albertus de Saxonia (attrib.) *Quaestiones in Aristotelis Physicam* I q. 19 (1999) II 253: Et notandum est circa istud quod, quamvis in eadem portione materiae non possunt esse simul plures formae, tamen possibile est quod in una portione materiae, in una eius parte, habeat talem formam et in alia aliam.

54 Cf. supra n. 5.

55 Cf. supra n. 42.

[46] Si vero dicas quod Philosophus septimo *Metaphysicae*[56] dicit aliquam esse generationem aequivocam, ut est generatio quae fit sola universali causa concurrente pro generatione ranae vel alterius animalis imperfecti—

[47] Hic respondetur quod non dicitur generatio aequivoca ista de causa, sed dicitur generatio aequivoca quoniam generans et generatum non sunt eiusdem speciei, ut saepe contingit, in qua tamen generatione agens particulare occultum vel ignotum concurrebat.

[48] Secunda[57] difficultas: an per 'causam universalem' intelligatur solum corpus solis, vel indifferenter totum caelum cum suis astris, taliter lumine et influentia et motu influentibus.

[49] Respondendum quod per 'causam universalem' intelligatur totum caelum etc., non circumscribendo a proposito influentiam quae assidue fit a Prima causa; quia si illa non influeret, nec caelum influeret. Scribitur[58] namque secundo *Metaphysicae*[59] quod in causis invicem essentialiter subordinatis, Prima cessante non agunt aliae. Per 'causam' ergo 'universalem' intelligitur | ut dixi, ita V1 98ra
quod non est astrum in caelo quin hic inferius influat.

[50] Tertia[60] difficultas: utrum sol efficacius influat ⟨per eius motum⟩ quam si quiesceret.

[51] Dico hic quod ad plures partes influit per eius motum quam si quiesceret. Quia per eius motum modo est orientalis, modo occidentalis, nunc septen-

4 ista] hac V2   6–7 occultum vel *om.* V2   9 solis] solare V2 | et *om.* V2   13 namque] enim V2   14 essentialiter] renaliter (*sic*) V1 | Prima] causa *add.* V2   15 cessante] aliae *add. sed del.* V1 | non agunt aliae] aliae non agunt V2   16 astrum] aliquod astrorum V2

56 Aristoteles *Metaph.* VII,7 1030a27–31: Generationes uero alie dicuntur factiones. Omnes autem factiones sunt aut ab arte aut a potestate aut a mente. Harum autem quedam fiunt et a casu et a fortuna, similiter ut in factis a natura; quedam enim et illic eadem et ex spermate fiunt et sine spermate.

57 Adn. mg. V1: Secunda difficultas.

58 Adn. mg. V1: *Manicula.*

59 Aristoteles *Metaph.* IX,8 1050b18–23: ergo actione sunt omnia; nec eorum que ex necessitate sunt, et ea prima; nam si ea non essent, nichil esset motus qui est sempiternus; nec si quid motum sempiternum, non est secundum potentiam motum, sed unde quoque (huius autem nichil materiam prohibet existere), propter quod semper agit sol et astra et totum celum. Cf. Blasius Pelacanus de Parma *Questiones disputate super octo libros Physicorum* VII q. 2 (2017) 21: Et hic noto quod causis invicem essentialiter subordinatis, prima non agente alie non agunt, licet non econtrario. Potest enim agere prima causa aliis non agentibus.

60 Adn. mg. V1: Tertia difficultas.

trionalis, et nunc australis. Et sic per eius motum pluribus partibus applicabitur quam si quiesceret.

[52] Dico secundo quod si quiesceret, gratia exempli, in nostro meridiano, maiorem caliditatem induceret quam inducat per eius motum. Patet, quia per eius quietem radii eius magis figerentur in obiecto corpore. Et hoc patet experimento consimili, quia tu saepe ducis manum tuam per flammam ignis sine laesione, quae manus, si quiesceret in flamma per duplum tempus, laederetur ab igne per caliditatem inductam.

[53] Et per hoc dico tertio quod ubi sol quiesceret in meridiano nostro, hemisphaerium nostrum inhabitabile cito redderetur. Patet per exemplum[61] datum.

[54] Quarta[62] difficultas: utrum sol maiorem caliditatem inducat inferius aestate quam vere vel autumno.

[55] Et videtur quod in vere vel autumno, quia in vere vel autumno velocius movetur quam in aestate, quia vere vel autumno | describit aequinoctialem, in aestate vero circulum Cancri in temporibus aequalibus. Et quia aequinoctialis est maior circulo Cancri, sequitur quod velocius movetur vere vel autumno quam etc., et per consequens plus calefacit quia motus ponitur calefactivus.

V2 37rb

[56] Ad oppositum est experientia.

[57] Respondetur quod, ut experientia docet, plus calefacit in aestate.

[58] Et ad argumentum in oppositum, dico quod ista ratio destruit se ipsam, quia postquam velocius movetur vere vel autumno, minorem moram contrahit iuxta hunc passum vel illud quam si tardius moveretur, ut patet inspecto exemplo[63] dum tu dicis 'manum per flammam ignis' etc.

[59] Sed iuxta hoc dubitabis: in casu in quo sol de caetero moveretur velocitate infinita, sic quod sicut in anno facit unam revolutionem de motu proprio aut in die de motu diurno, ita in quolibet instanti faceret unam revolutionem, et sic de caetero moveatur, iuxta hoc vertatur in dubium an[64] plus calefaciat hemis-

1 partibus] huiusmodi *add.* V2 | applicabitur] applicatur V2 6 sine] absque V2 7 duplum] subduplum V2 9 meridiano] meridie supra orizonta V2 12 caliditatem inducat] *ord. inv.* V2 13 autumno] auptomno *var.* V2 14 in *om.* V2 18 calefactivus] de genere calefacientium V2 20 ut] est *add.* V2 21 in oppositum *om.* V2 22 autumno] etc. V2 22–23 contrahit] trahit V2 24 ignis *om.* V2

61 Cf. supra n. 52.
62 Adn. mg. V1: Quarta difficultas. *Manicula.*
63 Cf. supra n. 52.
64 Cf. supra n. 54–58.

phaerium nostrum an minus. Et iuxta positam determinationem,[65] apparet quod non debet calefacere postquam non contrahit moram ad hoc passum vel illud.

[60] Respondetur quod in hoc casu maiorem | caliditatem induceret quam nunc inducat; et tantam induceret quantam si quiesceret. Quia postquam sic moveretur ut tu ponis in casu, semper esset applicatio ipsius ad hemisphaerium nostrum et ad oppositam partem, ut patet intelligenti. V1 98rb

[61] Ultima difficultas: an astra agunt in hunc modum iuxta ultimum sui conatum.

[62] Respondetur quod sic: semper luminosum praesentatum medio sic agit.

[63] Dico secundo quod potest fortius agere, quia potest melius applicari passo quam nunc, ratione passi et astrorum, quia forte solus est eclipsatus.

[64] Et hoc de tertio articulo.

[65] Ad quartum articulum patet solutio rationum. Et hoc de quarto articulo.

---

1 positam determinationem] *ord. inv.* V2 2 debet] debeat V2 | vel] ad *add.* V2 5 et tantam *rep. sed del.* V1 6 moveretur *om.* V2 | tu *om.* V2 | casu] sem(?) *add. sed del.* V1 8–13 ultima ... eclipsatus *om.* V1 14 articulo *om.* V2 15 quartum articulum] rationes V2 | rationum] per ea quae dicta sunt V2 | et ... articulo *om.* V2

---

65 Cf. supra n. 58.

4

Liber II q. 8: Utrum inter agentia particularia, supposita semper generali influentia superiorum, possit qualitas una vel plures formam substantialem producere

V1 = Vat. lat. 2159 ff. 98rb–100ra; V2 = Vat. lat. 3012 ff. 37rb–38va

[1] Octavo quaeritur utrum inter agentia particularia, supposita semper generali influentia superiorum, possit qualitas una vel plures formam substantialem producere.[1]

[2] Et arguitur primo affirmative.

[3] Primum[2] argumentum: radius solaris occurrens speculo concavo aut vasi vitreo aqua pleno producit formam substantialem ignis in combustibili; igitur, etc.

---

1 Octavo quaeritur] *ord. inv.* V2 | inter … particularia] auctoritate Aristotelis in hoc secundo capitulo de causis et modis causarum *add. sed del.* V2 2 influentia] astrorum *add.* V2 | qualitas] formam sub *add. sed del.* V1 4 affirmative] auctoritate Aristotelis in hoc secundo capitulo de causis et modis causarum sed ratione et sit *add.* V2

---

1 Cf. Gualterus Burlaeus *Tractatus primus: de quattuor conclusionibus circa formas accidentales* (1996) 168: Prima quod qualitas in virtute propria potest producere formam substantialem vel in virtute propria esse principium totale productivum forme substantialis. Cf. Blasius Pelacanus de Parma *Questiones disputate super octo libros Physicorum* prima redactio II q. 4 (ed. Biard): "Utrum qualitas in vigore proprio possit producere formam substantialem". Ed. Joël Biard.

2 Adn. mg. V1: Primum argumentum. *Manicula.*—Cf. Blasius Pelacanus de Parma *Questiones disputate super octo libros Physicorum* prima redactio II q. 4 (ed. Biard) 1: Retenta prima forma questionis arguitur primo sic affirmative: exposito speculo concavo radiis solaribus vel radiis Martis, generatur ignis, ergo sequitur quod radius Solis, qui est una simplex qualitas, est generator forme substantialis. Confirmatur alio experimento, quia cristallus exagona exposito eiusdem radiis idem similiter operatur. Et confirmatur quia vas vitreum ut urinale plenum aqua expositum radiis Solis idem facit. Cf. Blasius Pelacanus de Parma *Quaestiones de optica* q. 10 (1961) 214: Tertium argumentum sumptum ab experientia: cum exposueris speculum concavum vel urinale plenum aqua radiis solaribus, sol suis radiis iuxta speculum vel urinale generat ignem. Cf. Robertus Grosseteste *De natura locorum* (1912) 70–71: Quid etiam

[4] Secundo[3] sic: ferrum supercalefactum applicatum straminibus, sola caliditate informante illud, generat ignem; igitur, etc. Consequentia patet et antecedens similiter. Et dico notanter quod hoc facit ferrum caliditate informante ipsum et non ratione sui, quia de ratione ipsius ferri esset generare ferrum, quia "unumquodque naturaliter est sibi similis generativum" etc.[4] Nec valet ubi dicas quod in ferro sic supercalefacto est ignis; sed hoc dici non potest. Probo, quia forma substantialis ignis non potest informare materiam sub tanta densitate et gravitate ut est ferrum, et etiam ubi ferrum esset uniformiter supercalefactum. Tunc qua ratione in una eius parte esset forma substantialis ignis, et per totum sui esset ignis. Et si sic esset, dicendum quod hoc non esset ferrum, eo quod forma ferri et ignis non compatiuntur se in eadem materia.

[5] Tertium[5] argumentum: qualitas potest, inter agentia particularia se sola, sola influentia superioris causae concurrente, corrumpere formam | substan- V2 37va
tialem; igitur qualitas potest producere formam substantialem. Antecedens patet, quia aqua enim sua frigiditate potest corrumpere formam substantialem ignis, ut constat. Sed consequentia patet. Probo, quia non est difficilius introducere novam formam quam corrumpere praeexsistentem. Quod patet, | V1 98va
quia hora qua corrumpitur forma substantialis, illa forma resistat corrumpenti quantum est possibile; sed hora qua introducitur nova forma a generante nihil videtur resistere generanti. Probo, namque materia non resistit, quia materia appetit formam.[6] Patet etiam consequentia ex alio, quia si qualitas potest

1 sic *om.* V2 5 sibi similis generativum] generativum sibi similis V2 9 eius parte] parte ipsius V2 11 esset] est V2 | forma] substantialis *add.* V2 13 argumentum] calidit *add. sed del.* V2 19 corrumpitur] nova *add.* V2

possit reflexio, patet in corporibus politis et maxime in concavis, quoniam in libro *De speculis* probatur, quod speculo concavo ad solem posito ignis accenditur in eo, quod, ut praetactum est, omnes radii reflexi a superficie corporis concavi concurrunt ad punctum unum, et in illo puncto et circa est locus combustionis, sicut in libro praedicto declaratur.

3 Adn. mg. V1: Secundum argumentum. *Manicula.*

4 Cf. Aristoteles *De anima* II,4 415a26–30 Naturalissimum enim operum uiuentibus est, quecunque perfecta et non orbata aut generationem spontaneam habent, facere alterum quale ipsum, animal quidem animal, planta autem plantam, quatinus ipso semper et diuino et inmortali participent secundum quod possunt.

5 Adn. mg. V1: Tertium argumentum.—Cf. Blasius Pelacanus de Parma *Questiones disputate super octo libros Physicorum* prima redactio II q. 4 (ed. Biard) 2: Quartum argumentum: qualitas in vigore proprio potest corrumpere formam substantialem, ergo qualitas in vigore proprio potest producere formam substantialem.

6 Cf. Aristoteles *Physica* I,9 192a22–23; *Auctoritates Aristotelis* 142 (32): Materia appetit formam sicut femina virum et turpe bonum.

corrumpere formam substantialem, et non datur materia sine forma,[7] oportet ut tunc nova forma inducatur, et non videtur a quo nisi a corrumpente formam priorem; quare, etc.

[6] Quarto[8] sic: qualitas una vel plures potest in totam alterationem praecedentem adventum novae formae substantialis in materia; ergo qualitas una vel plures potest producere novam formam substantialem. Consequentia videtur plana. Unde materia sufficienter disposita pro adventu formae necessario inducitur illa forma. Et antecedens patet. Est namque evidens quod summa caliditas in certa materia potest producere summam caliditatem et summam raritatem, et cum hoc potest producere summam siccitatem, quibus inductis in materia inducetur forma substantialis ignis.

[7] Tu[9] vides etiam quod motus est generativus ignis. Unde fricando lapidem cum fugillo[10] generatur ignis, et non nisi ratione motus.

[8] Idem contingit de plumbo posito in extremitate sagittae quae diffusa est ab arcu, illud namque plumbum liquefit et aliquando inflammatur.[11]

---

1 sine] absque V2 4 sic *om.* V2 11 materia] data *add.* V2 13 cum] lapide aut lapidem cum *add.* V2 | fugillo] fuxilo V1 figulo V2 14 diffusa] condifusa V2

---

7 Cf. Aristoteles *Physica* I,7 190b19–23: unumquodque quod dicitur secundum substantiam, quod fiat ex subiecto et forma; componitur enim musicus homo ex homine et musico quodammodo; resolves enim rationes in rationes illorum. Manifestum igitur est quod fiant utique que fiunt ex his. Cf. Blasius Pelacanus de Parma *Quaestiones in Aristotelis De anima libri* I q. 8 (1974) 57: Septimum argumentum: scriptum est primo *Physicorum* ad textum quod non datur materia sine forma, nee forma sine materia.

8 Adn. mg. V1: Quartum argumentum.—Cf. Blasius Pelacanus de Parma *Questiones disputate super octo libros Physicorum* prima redactio II q. 4 (ed. Biard) 2: Quintum argumentum: qualitas in vigore proprio potest in totam alterationem precedentem formam ignis in materia, ergo qualitas in vigore proprio potest in formam substantialem ignis.

9 Adn. mg. V1: *Manicula.*—Cf. Blasius Pelacanus de Parma *Questiones disputate super octo libros Physicorum* prima redactio II q. 4 (ed. Biard) 1: Et iterum confirmatur quia fricatio ferri cum lapide generat ignem et per consequens producit formam substantialem igno, igitur qualitas in vigore proprio est productiva forme substantialis.

10 De fugillo, cf. DuCange *Glossarium mediae et infimae Latinitatis* (1845) 623b: Fugillus, Ferrum, quo extrahitur ignis de petra.

11 Cf. Aristoteles *De caelo* II,7 289a17–20: Natus est enim motus ignire et ligna et lapides et ferrum; rationabilius igitur propinquius igni, propinquius autem aer; velut et in latis sagittis; hee ipse enim igniuntur sic ut liquescant plumbee; et quoniam igniuntur iste, necesse et eum qui in circuitu ipsarum aerem hoc ipsum pati.

[9] Patet[12] etiam dato ovo gallinae: si tu ponas ovum sub stercore equino, aut sub axilla tua, generabitur inde pullus.[13] Sive etiam ponas illud ovum in furno aliqualiter calefacto, conservando illum locum in caliditate per aliquos dies, inde generabitur pullus.[14] Et istud experimentum satis cogit ut quod qualitas inter agentia particularia sufficiat pro forma substantiali producenda.

[10] Ad[15] oppositum arguitur primo. Namque posset asinus generare hominem. Probatur, quia clarum est quod omnis qualitas, quaecumque sit illa, in perfectione essentiali plus distat a substantia quam asinus ab homine in perfectione essentiali. Semper qualitas una vel plures est substantiae productiva; igitur a fortiori poterit formica generare asinum, et asinus hominem. Confirmatur, quia plus distat quaevis qualitas, ut summa caliditas, a substantia, quam

---

2 axilla] asella *var.* 9 Semper] Sed V2; sed *add. sed del.* V1

---

12 Cf. Blasius Pelacanus de Parma *Questiones disputate super octo libros Physicorum* prima redactio II q. 4 (ed. Biard) 2: Confirmatur, et sit hoc septimum argumentum, quia contingit pullum generari ex dato ovo, et non videtur quod iste pullus generatur a forma substantiali galline sepe ipsum ovum cohoperientis, nec ex gallo, quia ubi tu ponas datum ovum sub ascellis tuis vel sub limo, generabitur sepe pullus. Patet igitur quod certus gradus caloris est causa generationis pulli.

13 Cf. Aristoteles *De historia animalium* VI,2 559b1–7: Aperiuntur quidem igitur incubantibus avibus, non solum sed et spontanea in terra sicut in Egypto, sepelientibus in fimum; et in Syracusa autem potator quidam supponens sub sestorio in terra tanto bibit tempore, ut aiunt, continue, donec exciperentur ova. Iam autem et posita in vasis calidis maturata sunt et exierunt spontanea. Cf. Paracelsus *Summi philosophi ac medici* I (1573) 10: Exemplum videtis in ovis, in quibus est humiditas mucilaginosa, quae quolibet perpetuo calore putrefit et animatur in pullum vivum, non solum calore gallinae, sed quolibet tali calore. Nam in tali gradu ignis possunt ova in vitro et cineribus maturari, et viva avicula fieri; etiam quilibet homo sub sua axilla ovum maturare potest, et pullum procreare tam bene ac gallina.

14 Cf. Petrus de Crescentius *Ruralia commoda* X (1471) f. 75^{v} (s.f.); *ibidem* IX cap. 86 (c. 1490–1495) f. 131vb: Fertur tamen quod in quibusdam mundi partibus inveniuntur homines qui furnos adeo temperate calefaciant, quod eorum calor aequalis est calore gallinae cubantis, atque in uno furno seu clibano ponunt multitudinem pennarum parvarum, et mille gallinacea ova; post viginti dies nascantur ac extra prosiliunt. Cit. Conradus Gesneri *Historiae animalium* (1555) 429: Fertur in quadam regione inveniri homines, qui furnos ita temperate calefaciant, ut eorum calor par sit calori gallinae incubantis, et in furno seu clibano ponere quam plurimas plumas, et mille gallinacea ova, quae post viginti dies nascantur ac erumpant, Crescentiensis.

15 Adn. mg. V1: *Manicula.*—Cf. Blasius Pelacanus de Parma *Questiones disputate super octo libros Physicorum* prima redactio II q. 4 (ed. Biard) 3: Secundum argumentum: si aliqua qualitas est productiva forme substantialis, tunc asinus posset generare hominem. Tenet consequentia, quia longe plus distat omnis qualitas in perfectione essentiali a substantia, quecumque sit illa, quam asinus distat ab homine. Ergo si qualitas poterit producere formam substantialem, a fortiori poterit asinus producere hominem.

V1 98vb remissa caliditas a summa caliditate. Sed | remissa caliditas non potest producere summam caliditatem, igitur a fortiori summa caliditas non est substantiae productiva.

[11] Item, principium est physicum[16] quod “non contingit agens agere ultra gradum proprium.” Et per hoc dicitur quarto huius[17] quod “in omni motu oportet potentiam motoris excedere potentiam rei motae.”

[12] In qua quaestione erunt quattuor articuli: primo problema, secundo determinatio quaestionis, tertio difficultates, quarto ad rationes in contrarium.

[13] Quantum ad primum sit problema: propter[18] quid est quod latitudo frontis in homine est signum timiditatis, parvitas vero eiusdem significat stoliditatem. Et[19] debetis intelligere problema ut comparetis frontem et per consequens caput ad alia membra, sic quod latitudo frontis in homine parvo est signum V2 37vb timiditatis, quod non esset verum in homine magno. | Parvitas vero frontis in homine magno est signum stoliditatis, quod non esset verum in homine parvo.

[14] Respondetur ad problema, et dico quod causa primae partis stat in hoc quia, fronte existente magno, caput est magnum et per consequens cerebrum. Sed cerebrum magnum in parvo corpore reddit corpus illud frigidius quam oportet et humidius, et per consequens in tali corpore est paucitas spiritus

---

5 dicitur] in *add.* V2 7–8 primo ... in contrarium] etc. V2 9 Quantum ... primum *om.* V2 | sit] hoc *add.* V2 13 timiditatis] et pusillanimitatis *add.* V2 14 est signum stoliditatis] significat etc. V2 15 Respondetur ad problema] Ad problema respondetur V2

---

16 Cf., e.g., Nicolaus Oresme *Quaestiones super De caelo* I q. 20 (1965) 333: nihil potest agere ultra gradum proprium. Cf. Blasius Pelacanus de Parma *Questiones circa tractatum proportionum magistri Thome Braduardini* q. 10 (2005) 149: non contingit aliquod agens agere ultra gradum proprium sue activitatis.

17 Aristoteles *Physica* IV,8 216a11–16. Cf. Nicolaus Oresme *Quaestiones super De generatione et corruptione* I q. 11 (1996) 88: in omni actione et motu agens seu movens debet excedere potentiam rei mote, et hoc est quod dicitur quod omnis motus a proportione maioris inequalitatis fit, et habetur ab Aristotele quarto *Physicorum* et septimo et multis locis. Cf. Blasius Pelacanus de Parma *Questiones disputate super octo libros Physicorum* VII q. 9 (2017) 106: unde, ut scriptum est IV° *Physicorum* et in hoc VII°, in omni motu oportet potentiam motoris excedere potentiam rei mote. Cf. Blasius Pelacanus de Parma *Questiones circa tractatum proportionum magistri Thome Braduardini* q. 8 (2005) 125: Queritur ... utrum in omni motu potentia motoris debeat excedere potentiam rei mote vel equari ei.

18 Cf. Razes *Almansor* II cap. 31 (1893) II 167–168: Qui parvam habet frontem, stolidus est. Qui vero magnam habet frontem, est piger. Cuius frons vehementer est rugosa, inverecundus est. Cf. Bartholomeus Coclitis *Chyromantie ac physionomie anastasis*, comm. in pseudo-Arist. *Physiognomonica* II q. 4 (1523) f. 24rb: Dixit Rasys: Qui parvam habet frontem stolidus est. Qui vero magnam habet frontem est piger.

19 Adn. mg. V1: *Manicula.*

et sanguinis. Modo spiritus et sanguis sunt immediatum instrumentum audaciae, et privatio sanguinis et spiritus per consequens est causa timiditatis. Et sic habetur prima pars. Et per hoc, ut tu plurimum videbis, quod tales sunt multum pallidi in facie ratione paucitatis spiritus et sanguinis.

[15] Ad secundam partem respondeo quod parvitas frontis in magno corpore est signum stoliditatis ob hoc quia, fronte existente parva, caput est parvum, et per consequens cerebrum parvum, et per consequens ventriculi qui sunt instrumenta virtutum cognoscitivarum erunt parvi; et si sic, species diffusae ab obiectis exterioribus in eis se confundent propter indistinctionem eorum, et per hoc non fient bene iudicia de obiectis. Modo quando homo non bene iudicat sed male, semper vel ut plurimum, ille dicitur fatuus. Et volo tibi dicere quod habens caput sic parvum | (V1 99ra) habebit ut plurimum faciem bene coloratam et frontem rugosam, cuius causa est quia habens frontem parvam habebit caput parvum, et per consequens cerebrum parvum, respectu magni corporis semper intelligendo. Ergo cerebrum non poterit contraoperare cordi sua frigiditate, et per consequens in isto cor generabit, respectu cerebri sui, magnam copiam spirituum qui, expansi per cerebrum, calefacient illud; et sic cutis faciei reddetur multum colorata. Et dico bene quod frons erit rugosa; haec namque caliditas operatur. Et hoc de primo articulo.

[16] Quantum ad secundum ponam conclusiones determinantes quaestionem pro una parte, et hoc non obstante per aliquas conclusiones poteris defendere partem oppositam.

[17] Pro[20] conclusionibus noto quod tripliciter potest dici aliquod agens posse hunc effectum vel illum producere. Uno[21] modo formaliter, ut quando producens et productum sunt eiusdem speciei; et sic dicimus quod asinus generat asinum. Secundo[22] modo virtualiter, ut quando agens includit in se

---

3 tu plurimum videbis] plurimum V2 9 eis] eos V2 11 sed male *om.* V2 | ille *om.* V2 15 contraoperare] contraoperari V2 | cordi] caliditati cordis V2 | sua frigiditate] *scrip. ante* non poterit V2 16 cerebri sui] *ord. inv.* V2 19 articulo *om.* V2 20 secundum] articulum *add.* V2 26 modo *om.* V2

---

20 Adn. mg. V1: Nota. *Manicula.*—Cf. Blasius Pelacanus de Parma *Questiones disputate super octo libros Physicorum* prima redactio II q. 4 (ed. Biard) 4: Quantum ad primum articulum, primo noto quod multipliciter potest dici agens posse hunc effectum vel illum producere. Cf. infra n. 28.

21 Cf. Blasius Pelacanus de Parma *Questiones disputate super octo libros Physicorum* prima redactio II q. 4 (ed. Biard) 4: Uno modo formaliter, et hoc est quando producens et productum sunt eiusdem speciei, sive quando idem terminus specificus predicatur formaliter de producente et producto.

22 Cf. Blasius Pelacanus de Parma *Questiones disputate super octo libros Physicorum* prima redactio II q. 4 (ed. Biard) 4: Secundo modo potest aliquod agens producere hunc effec-

perfectionem aequalem effectui producto vel producibili, aut maiorem, sicut dicimus quod sol virtualiter producit ignem, non formaliter, quia sol non est ignis, sed virtualiter.[23] Sol namque est maioris perfectionis quam sit ignis. Producit etiam animam vitalem in plantis, non quod sol sit planta, sed bene est altioris gradus in perfectione quam sit planta, capiendo solem pro aggregato ex Intelligentia et orbe.[24] Alioquin sol, Intelligentia non concurrente, non generaret plantam nec formaliter nec virtualiter, quia omne animatum est perfectius inanimato.[25] Tertio[26] modo contingit aliquod agens producere aliud consequenter, et hoc est quando agens producit aliquid formaliter vel virtualiter, ad cuius productionem sequitur talis vel talis effectus. Unde sicut dicit Commentator nono *Metaphysicae*,[27] "dans formam dat consequenter sequentia ad formam."

[18] Et[28] tunc sit prima conclusio: inter agentia particularia, nulla qualitas est formaliter substantiae productiva. Patet, quia in tali productione formali, generans et generatum sunt eiusdem speciei.

---

6 sol *om.* V2 10 unde sicut] et sic V2

---

tum vel illum virtualiter, et hoc est quando agens continet vel includit in se virtualiter perfectionem equalem vel maiorem cum effectu producto vel producendo.

23 Cf. Blasius Pelacanus de Parma *Questiones super perspectiva communi* secunda pars, q. 5 (2009) 288: Secunda conclusio et in parte responsalis: per reflexionem radiorum tamquam per causam agentem et principalem non potest ignis generari. Ista patet ex conclusione precedente. Tertia conclusio: per reflexionem radiorum solarium tamquam per causam instrumentalem potest ignis generari.

24 Cf. Blasius Pelacanus de Parma *Questiones circa libros Metaurorum Aristotelis* I q. 1 (ed. Biard) 3: Dicit etiam Philosophus II° Physicorum "homo generat hominem et Sol." Per Solem intelligatur aggregatum ex intelligentia et orbe.

25 Cf. Nicolaus Oresme *Questiones super Physicam* VII q. 6 (2013) 753: Et patet etiam ex communi usu loquendi, quia dicimus quod animatum ⟨est⟩ nobilius inanimato et perfectius. Cf. Blasius Pelacanus de Parma *Questiones disputate super octo libros Physicorum* prima redactio II q. 5 (ed. Biard) 9: omne animatum est perfectius quocumque inanimato.

26 Cf. Blasius Pelacanus de Parma *Questiones disputate super octo libros Physicorum* prima redactio II q. 4 (ed. Biard) 4: Tertio modo dicimus aliquando agens producere hunc effectum vel illum consecutive, et hoc est quando agens producit aliquid formaliter vel virtualiter ad cuius productionem semper consequitur tale effectus.

27 Cf. immo Averroës *In Phys.* VIII com. 32 (1562) 370va: generans enim est illud quod dat corpori simplici generato formam suam, et omnia accidentia contingentia; *Auctoritates Aristotelis* 159 (235) Generans dat generato formam et omnia accidentia convenientia ad formam. Cf. Beda (attrib.) *Sententiae philosophicae* (PL 90) col. 986: Dans formam, dat omnia consequentia formam (VIII *Phys.*). Cf. Blasius Pelacanus de Parma *Questiones super perspectiva communi* tertia pars, q. 3 (2009) 329: ut scribitur 9° *Methaphysice*, "dans formam dat accidentiam consequentiam ad ipsam."

28 Adn. mg. V1: Prima conclusio.—Cf. Blasius Pelacanus de Parma *Questiones disputate super octo libros Physicorum* prima redactio II q. 4 (ed. Biard) 4: Prima conclusio: nulla qualitas in vigore proprio est formaliter productiva substantie.

[19] Alia conclusio:[29] inter agentia particularia, nulla qualitas est substantiae | virtualiter | productiva. Patet,[30] quia producere aliquid virtualiter est illud esse altioris gradus in perfectione quam productum. V2 38ra V1 99rb

[20] Tertia[31] conclusio: quaelibet qualitas de primis, inter agentia particularia, se sola est effectus sibi similis productiva. Patet inductive: potest namque summa caliditas inducere summam caliditatem, et remissa remissam, et hoc se sola, non excludendo per ly 'se sola' generalem influxum superioris causae, quo circumscripto, ignis applicatus straminibus non ageret nec aliqua caliditas calefaceret. Et dico 'quaelibet qualitas de primis', quia qualitates secundae non videntur esse activae. Albedo enim non est generativa albedinis, et sic de aliis.

[21] Quarta[32] conclusio: qualitas una vel plures potest consecutive inter agentia particularia formam substantialem producere. Probatur, quia summa caliditas potest in A materia producere summam caliditatem et summam raritatem cum summa siccitate, quibus dispositionibus inductis consecutive generatur forma substantialis ignis consequenter. Confirmatur, quia una qualitas vel plures potest in totam alterationem per quam materia sic sufficienter est disposita pro forma substantiali recipienda; ergo summa caliditas etc., quae conclusio est responsalis.

[22] Cum hoc tamen volo quod absque concursu universalis causae, forma substantialis non induceretur in materia data, qualitercumque disposita esset vel disponi posset. Immo dico quod sine concursu causae universalis non posset materia sufficienter disponi per aliquam qualitatem vel plures, nec etiam per aliud agens particulare.

---

1 Alia] Secunda V2 3 altioris gradus] *ord. inv.* V2 6 inducere] producere V2 7 se² *om.* V2 9 Et *om.* V2 15 consequenter] gratia exempli V1 16 sic sufficienter est] redditur sufficienter V2 17 recipienda] recipiente V2 | etc.] potest etc. *add.* V2 21 sine] absque V2 21–22 posset] potest V2 22 materia] A *add.* V2 | qualitatem] unam *add.* V2

---

29 Adn. mg. V1: Secunda conclusio.—Cf. Blasius Pelacanus de Parma *Questiones disputate super octo libros Physicorum* prima redactio II q. 4 (ed. Biard) 4: Secunda conclusio: nulla qualitas in vigore proprio est virtualiter productiva substantie.

30 Cf. Blasius Pelacanus de Parma *Questiones disputate super octo libros Physicorum* prima redactio II q. 4 (ed. Biard) 4; cf. supra n. 17.

31 Adn. mg. V1: Tertia conclusio.—Cf. Blasius Pelacanus de Parma *Questiones disputate super octo libros Physicorum* prima redactio II q. 4 (ed. Biard) 5: Tertia conclusio: quelibet qualitas formaliter vel virtualiter potest producere qualitatem sibi similem in vigore proprio.

32 Adn. mg. V1: Quarta conclusio. *Manicula.*—Cf. Blasius Pelacanus de Parma *Questiones disputate super octo libros Physicorum* prima redactio II q. 4 (ed. Biard) 5: Quarta conclusio: qualitas una vel plures in vigore proprio consecutive potest producere formam substantialem.

[23] Non[33] obstantibus predictis conclusionibus, sit ista conclusio quod 'inter agentia particularia qualitas una vel plures sit sicut principale agens' non potest improbari. Declaratur conclusio, quia in motu locali apparet quod ubi certa portio materiae fuerit informata summa gravitate sine forma substantiali, si ista portio materiae fuerit in concavo caeli, suae naturae dimissa,[34] illa movebitur versus centrum. Et non potest dici quod iste motus fiat a forma substantiali, ex casu. Confirmatur, quia ubi A sit una certa portio ignis, et ad imaginationem in materia istius ignis inducatur summa gravitas, in hoc casu iste ignis exsistens in concavo caeli, dimissus suae naturae, movebitur deorsum. Et clarum est quod iste motus non fit a forma substantiali ignis, quia forma
V1 99va substantialis ignis inclinaret magis ad oppositum. Fit igitur | a gravitate. Ita verisimile est de qualitatibus alterativis quae sunt caliditas, frigiditas, etc. Unde non apparet nobis quod hora generationis ignis requiratur aliud quam quod materia sufficienter disponatur per caliditatem, levitatem, et raritatem, et hoc potest fieri per unam qualitatem vel plures.

[24] Secunda conclusio: A terra supra caelum ignis, dimissa suae naturae, non ratione qua terra movebitur deorsum. Patet, quia nihil movetur deorsum nisi ratione qua grave. Supponatur igitur A terra esse privata omni gravitate.

[25] Tertia conclusio: stat A terra iuxta centrum exsistente moveri usque ad concavum orbis lune, et hoc ab intrinseco, nullo moto extrinseco concurrente. Patet ubi A terra sit summa levitate informata.

[26] Quarta conclusio: nullus ignis ratione qua ignis calefacit. Patet, quia aliquis ignis congelat aquam, ut ignis summe frigidus.

[27] Nunc subiungo conclusiones per quas potes aliter iudicare de proposita materia.

[28] Et[35] sit prima conclusio: inter agentia particularia nulla qualitas potest
V2 38rb se sola formam substantialem | producere. Probatur conclusio, quia si posset, aut igitur posset formam substantialem producere formaliter aut virtualiter aut

1 sit] posita V2 2 qualitas] *mg.* V2 3 ubi] si V2 4 sine] absque V2 5 suae naturae dimissa] dimissa suae naturae V2 6 centrum] mundi *add.* V2 7 ex casu *om.* V2 10 substantiali] ipsius *add.* V2 11 magis] potius V2 12 est *om.* V2 16–23 Secunda … summe frigidus (n. 26) *om.* V1 24 quas] tu *add.* V2 28–225.1 producere … exsecutive] formaliter aut virtualiter aut exsecutive producere V2

33 Adn. mg. V1: Alia conclusio.

34 Cf. Gualterus Burlaeus *De somno*, q. 17, "Utrum cessante primo proiciente possit proiectum moveri" (cod. Vaticano, Vat. lat. 2151) f. 105rb: Nunc autem unumquodque dimissum suae naturae reddit ad naturam suam, sicut aqua calefacta dimissa suae naturae reddit ad frigiditatem.

35 Adn. mg. V1: Prima conclusio. *Manicula.*

executive.[36] Patet quod non formaliter nec virtualiter, ex quid nominis istorum terminorum. Sed quod non exsecutive probatur, quia si posset exsecutive etc., hoc ideo esset quia qualitas una vel plures posset in totam alterationem necessitantem pro forma substantiali inducenda. Sed hoc non, quia da enim quod qualitas una vel plures disponant sufficienter A materiam. Nec ex hoc sequitur quod debeat introduci forma substantialis in materia, quia verteretur in dubium a quo inducetur ista forma? Non a qualitativa dispositione, quia nec formaliter nec virtualiter nec exsecutive, quia tu potes imaginari unam portionem materiae habere summam gravitatem cum summa frigiditate et siccitate et aliis requisitis, et tamen non educetur inde aliqua forma substantialis. Quia aut producens est agens liberum quod non vult producere, aut si est agens mere naturale est impeditum; quare, etc.

[29] Secunda[37] conclusio, quod 'inter agentia particularia qualitas una vel plures possit in totam alterationem necessitantem pro forma substantiali producenda' non potest demonstrari. Patet, quia hoc videtur posse potenter improbari; ergo, etc. Patet antecedens, quia cum idem dicatur esse disponens materiam et inducens formam substantialem, tunc qualitas una vel plures videretur posse introducere formam substantialem.

[30] Tertia[38] conclusio: inter agentia particularia non videtur quod aliqua qualitas possit se sola aliquam formam substantialem corrumpere. Probatur, quia cum ad corruptionem unius formae substantialis sequeretur introductio alterius formae substantialis, tunc ubi qualitas una vel plures inter | agentia V1 99[vb]
particularia posset corrumpere formam substantialem, posset a fortiori introducere unam substantialem. Et hanc conclusionem probavit unum[39] argumentum ad partes quaestionis.

---

1 Patet *om.* V2 | nec] execut *add. sed del.* V2 4 inducenda] producenda V2 5 Nec ex hoc] Modo ex hoc non V1 7 inducetur] induceretur V2 | forma] substantialis *add.* V2 8 nec[2]] quod *add.* V2 14 necessitantem] necessitatem V2 15 potest] potem V2 21 cum *om.* V2 24 unam] novam formam V2 24–25 argumentum] factum in et arguendo *add.* V2 25 quaestionis] ubi dictum est quod facilius est generare quam corrumpere *add.* V2

---

36 Cf. supra n. 17.

37 Adn. mg. V1: Secunda conclusio.—Cf. Blasius Pelacanus de Parma *Questiones disputate super octo libros Physicorum* prima redactio II q. 4 (ed. Biard) 7: Quinta conclusio: nulla qualitas se sola potest in totam dispositionem necessitantem ad formam substantialem inducendam.

38 Adn. mg. V1: Tertia conclusio. *Manicula.*—Cf. Blasius Pelacanus de Parma *Questiones disputate super octo libros Physicorum* prima redactio II q. 4 (ed. Biard) 7: Sexta conclusio: nulla qualitas se sola potest formam substantialem corrumpere.

39 Cf. supra n. 5.

[31] Quarta[40] conclusio, quod 'nulla qualitas inter agentia particularia se sola possit producere qualitatem incompossibilem formae substantiali' est probabile. Patet, quia ubi staret oppositum, diceres quod summa caliditas potest producere summam caliditatem, quae summa caliditas est incompossibilis formae substantiali aquae. Ubi ergo summa caliditas introduceret caliditatem incompossibilem etc., tunc forma substantialis aquae corrumperetur, quod corrumpens videretur summa caliditas. Sed hoc videtur esse contra conclusionem[41] praecedentem.

[32] Ex[42] quo sequitur corollarie quod aut summa caliditas non potest summam caliditatem producere in materia informata forma aquae, aut quod summa caliditas potest stare cum forma substantiali ipsius aquae. Sed quae pars sit eligibilior, per te elige.

[33] Et hoc de secundo articulo.

[34] Ad tertium articulum, prima difficultas[43] sit ista: cum ita sit de facto quod in aqua inducitur una caliditas incompossibilis formae substantiali ipsius aquae, quaeratur a quo inducitur illa caliditas, postquam non inducitur a summa caliditate.

[35] Respondetur[44] quod illa caliditas incompossibilis etc. producitur ab igne et non a caliditate ipsius, nisi mediante ea tanquam mediante instrumento. Et hoc de isto articulo.

---

3 diceres] tu *praem.* V2 | potest] possit V2 4 caliditatem] in aqua *add.* V2 5 aquae] ipsius V2 | ergo] vero V2 6 etc. *om.* V2 7 videretur] esse *add.* V2 14 Ad] Quantum *praem.* V2 | prima difficultas sit ista] sit prima difficultas V2 | cum] quod V2 16 illa] qua *add. sed del.* V1 16–17 non … caliditate] caliditas non est productiva caliditatis V2

---

40 Adn. mg. V1: Quarta conclusio.—Cf. Blasius Pelacanus de Parma *Questiones disputate super octo libros Physicorum* prima redactio II q. 4 (ed. Biard) 7: Septima conclusio: nulla qualitas potest se sola producere caliditatem incompossibilem forme substantiali aque.

41 Cf. supra n. 30.

42 Cf. Blasius Pelacanus de Parma *Questiones disputate super octo libros Physicorum* prima redactio II q. 4 (ed. Biard) 7: Ex hiis sequitur tale corelarium quod aut summa caliditas non potest se sola summam caliditatem producere, aut quod summa caliditas potest stare cum forma substantiali ipsius aque. Istud sequitur ex dictis. Sed que pars sit eligibilior, istud per te ipsum eligas.

43 Adn. mg. V1: Difficultas.—Cf. Blasius Pelacanus de Parma *Questiones disputate super octo libros Physicorum* prima redactio II q. 4 (ed. Biard) 12: Secunda difficultas: a quo inducatur caliditas incompossibilis forme substantiali aque?

44 Cf. Blasius Pelacanus de Parma *Questiones disputate super octo libros Physicorum* prima redactio II q. 4 (ed. Biard) 12: Ad quam difficultatem respondetur, et dicitur quod caliditas incompossibilis etc. inducitur a generante, ut ab igne, gratia exempli, et non a caliditate ignis.

[36] Sequitur quartus articulus, et quia probabilius est tenere quod inter agentia particularia nulla qualitas possit formam substantialem producere, ideo ad rationes | probantes oppositum respondetur. V2 38va

[37] Ad[45] primam, cum dicitur 'radius solaris occurrens speculo concavo etc. generat ignem', negatur. Et si quaeras a quo generatur ignis, respondetur quod a sole virtualiter mediante radio.

[38] Ad[46] secundum, 'ferrum calefactum' etc., respondetur quod in ferro sic ignito est forma substantialis ignis, et sic illa mediante caliditate | producit ignem. Et quando dicis 'in eandem parte materiae non sunt duae formae substantiales', verum est eiusdem perfectionis; et sic est hic, quia forma ferri est maioris perfectionis quam forma ignis. Est enim quodlibet mixtum omni elemento perfectius. Nec inconvenit quod forma substantialis ignis conservetur certo tempore in materia tantae densitatis quantae est terra. V1 100ra

[39] Aliter potest dici quod ignis generatur in straminibus a corporibus supercaelestibus caliditate ferri concurrente. Sed an inter particularia agentia caliditas ferri sit principale agens, dicendum quod non, postquam pro nunc tu vis sustentare istam partem. Sed non inconvenit formam substantialem ferri mediante caliditate informante ferrum producere hoc.

[40] Ad[47] aliud argumentum, 'caliditas potest corrumpere formam substantialem', illud est negandum, ut una[48] conclusio praetendit.

[41] Et sic, si tu inspexeris determinationem quaestionis, ad rationes in oppositum patefiet solutio.

[42] Et hoc de quaestione.

1 Sequitur … articulus] Transeo ad quartum V2 9 ignem] in straminibus *add.* V2 | dicis] tu *praem.* V2 10 quia] quoniam V2 11 forma *om.* V2 14 in] ex V2 19 aliud argumentum] tertium V2 20 ut] secundum quod V2 21–22 oppositum] tibi *add.* V2

45 Adn. mg. V1: Ad primam.—Cf. supra n. 3. Cf. Blasius Pelacanus de Parma *Questiones disputate super octo libros Physicorum* prima redactio II q. 4 (ed. Biard) 13: Ad primam, cum dicitur "speculo concavo exposito radiis etc., generatur ignis, ergo Sol mediantibus suis radiis generat etc.," negatur argumentum. Ymmo dico quod Sol mediantibus suis radiis istud operatur.

46 Adn. mg. V1: Ad secundum.—Cf. supra n. 4.

47 Adn. mg. V1: Ad aliud.

48 Cf. supra n. 30.

5

Liber IV q. 2: Utrum entia naturalia distantia ab eorum locis naturalibus moveantur ad illa impedimentis subtractis

V1 = Vat. lat. 2159 ff. 114ra–117rb; V2 = Vat. lat. 3012 ff. 49ra–51ra

[1] Secundo quaeritur[1] utrum entia naturalia distantia ab eorum locis naturalibus moveantur ad illa impedimentis subtractis.[2]

[2] Arguitur[3] quod non. Primum argumentum: data una terra in media regione, privata gravitate, illa terra remoto omni impedimento non descenderet versus locum naturalem; igitur, etc. Antecedens patet, quia alioquin probaretur quod velocius moveretur terra sine gravitate quam cum gravitate retenta eadem figura et identitate medii. Quia ubi haec terra sine gravitate descendat, V2 49rb descendat igitur velocitate ut quattuor. B terra summae gravis, ut tibi placet, | et sit velocitate ut decem descendat. Et clarum est quod, si remittatur gravitas ad subduplum, remittetur velocitas; et remittatur, gratia exempli, per duo; et iterum remittatur ad subduplum. Et habeo quod remittetur motus quamlibet servata certa latitudine gravitatis; quare, etc.

[3] Secundum argumentum:[4] aqua in vase, quod vas dicitur clepsydra, non movetur deorsum, foramine superiori clauso; et non est ibi aliquod impedi-

1 Secundo quaeritur *om.* V2 4 descenderet] descendit V2 5 locum] eius *add.* V2 6 terra] cum gravitate quam *add. sed del.* V1 | quam cum gravitate] *mg.* V1 7 haec] a V2 9 descendat] *mg.* V1; *om.* V2 10 subduplum] eius *add.* V2 13 Secundum argumentum] Secundo sic V2 | clepsydra] eclipsedra *var.* V1; eclipsidra *var.* V2 14 clauso] obstructo V2

1 Adn. mg. V1: Secundo.

2 Cf. Ioannes Buridanus *Quaestiones super Physicorum* IV q. 9, “Utrum in motibus gravium et levium ad sua loca naturalia tota successio proveniat ex resistentia medii” (2016) 271–291; cf. Nicolaus de Argentina *Summa* liber II tract. 3 sect. 5 q. 4, “Utrum motus gravium et levium sit naturalis” (2009) 72–73. Resp. Aristoteles *Physica* II,1 192b20–3; *Auctoritates Aristotelis* 144 (50): Natura est principium et causa motus et quietis in quo est primo per se et non per accidens.

3 Adn. mg. V1: Primum argumentum.

4 Adn. mg. V1: Secundum argumentum.—Cf. Simplicius *In quatuor libros De celo Aristotelis* II com. 55 (1540) f. 84rb: Sed et clepsedra repleta aqua si obstruat quis os superius ut non ingrediatur, non effluit per foramina aqua ut non maneat. Cf. Albertus de Saxonia (attrib.) *Expositio in Aristotelis Physicam* IV cap. 1 (1999) 189: Secundo hoc idem patet in clepsydris, ubi, foramine

 | DOI:10.1163/9789004734968_014

mentum, cum illud vas habeat plures meatus inferius magnos et notabiles. Nec potest dici quod hoc sit ratione vacui, quia ponimus vacuum non esse.[5] In casu tamen in quo foramen superius aperiatur, aqua movebitur deorsum; et tamen illa aqua nihil acquisivit de novo quod prius non haberet. Assignanda igitur est causa positiva quare ipsa prius non moveretur.

[4] Tertio:[6] grando vel lapis nunc primo genitus in media regione non movetur deorsum, remoto omni impedimento. | V1 114rb Quia si sic, daretur primum instans esse motus, cuius oppositum dicetur in sexto huius.[7]

[5] Quarto:[8] ubi tota aqua esset in medio aëris in interstitio, tunc terra deberet naturaliter moveri sursum. Tenet argumentum, quia in hoc quarto[9] dicit Aristoteles quod "terra est in aqua tamquam in loco eius naturali." Igitur si per imaginationem de facto tota aqua esset sursum detenta, terra moveretur sursum, quod tamen est falsum.

[6] Quinto:[10] tunc non deberet bolus terrae moveri deorsum sub omni aqua. Tenet argumentum, quia "locatum in eius loco naturali non dicitur grave nec leve," sicut vult Archimedes in tractatu *De inundantibus in liquido*.[11] Et tamen videmus quod parvus bolus terrae descendit sub omni aqua.

---

5 quare] quia V2 7 omni] eius V2 8 esse] sui V2 10 Tenet argumentum *om.* V2 16 Archimedes] Alchimides V1; Alqimides V2

---

superiore obstructo, non potest inferius, quamvis sint ibi foramina non obstructa, exire aqua, propter hoc quod non potest aer intrare. Cf. Blasius Pelacanus de Parma *Questiones disputate super octo libros Physicorum* secunda redactio IV q. 4 (2012) 321: Decimo, posita aqua in eclipsedra, obstructo foramine superiori, aqua descendet ad locum eius naturalem, igitur in eclipsedra dabitur vacuum.

5 Cf. Blasius Pelacanus de Parma *Questiones disputate super octo libros Physicorum* secunda redactio IV q. 4 (2012) 335: Ad decimum argumentum, "posita aqua in eclipsedra" etc., dico quod ista aqua est detenta non a vacuo quia vacuum non est, sed detenta a forma substantiali ipsius, prohibente ne descendat ut non detur vacuum.

6 Adn. mg. V1: Tertio. *Manicula.*

7 Cf. Aristoteles *Physica* VI,5 236a7–27.

8 Adn. mg. V1: Quarto.

9 Aristoteles *Physica* IV,5 212b20–1.

10 Adn. mg. V1: Quinto.

11 Archimedes *De iis quae in humido vehuntur* I, Theorema III, Propositio III (1972) 362: Solidarum magnitudinum quae aequalis molis, et aequalis ponderis cum humido, dimisse in humidum demergentur ita ut superficiem humidi non excedant nihil, et non adhuc referentur ad inferius. Cf. ps.-Archimedes *De insidentibus in humidum*, Petitio 1 (1952) 42: Nullum corpus in se ipso grave esse; ut aqua in aqua, oleum in oleo, aer in aere, non est alicuius gravitatis. Cf. Blasius Pelacanus de Parma *De ponderibus* (1952) 272: (Suppositio) 1. Nullum elementum in eius propria regione ponderat. Hoc dicit Archimedes in tractatu *De insidentibus in liquido*.

[7] Sexto:[12] tunc contingeret quod spatium non posset pertransiri nisi in tempore infinito. Probatur consequentia. Et praesuppono duo. Primum est quod terra non est debite collocata nisi dum medium eius fuerit medium mundi. Praesuppono secundo quod "velocitas insequitur proportionem proportionum".[13] Hiis praemissis, patet intentum. Pro[14] cuius deductione imaginor sphaeram terrae esse perforatam per centrum ex utraque parte. Deinde sit in hemisphaerio nostro corpus columnare de simplici terra, quod sit A, C, cuius medium sit B. Nunc permittatur hoc corpus columnare suae naturae moveri per foramen. Et[15] volo probare quod movebitur aeterno tempore. Quia movebitur quousque fuerit debite collocatum, et per consequens quousque B punctus fuerit medium mundi. Patet hoc ex primo supposito. Sed antequam sit debite collocatum, movebitur secundum unam latitudinem proportionum compositam ex proportionibus infinitis. Quia secunda est subdupla ad primam; tertia subdupla ad secundam; et sic in infinitum. Sed per secundam suppositionem, "proportio velocitatum insequitur proportionem proportionum" etc. Igitur ubi

---

4 mundi] mondi *var.* V2 7 hemisphaerio] imispherio *var.* V1; emisperio *var.* V2 | columnare] columpnare *var.* 10 punctus] ponctus *var.* V2 11 sit] fuerit V2 13 Quia] quare V1

---

12 Adn. mg. V1: Sexto. *Manicula.*

13 Cf. Blasius Pelacanus de Parma *Questiones circa tractatum proportionum magistri Thome Braduardini*, redactio prima, q. 7 (2008) 56: Ultima conclusio: velocitas insequitur proportionem proportionum agentium ad eorum passa equo modo applicatorum. Istam conclusionem vult Magister *De proportionibus* et omnes alii proportioniste, una cum Commentatore in pluribus partibus naturalis philosophie.

14 Cf. Richardus Swineshead *Calculator* tr. XI (1966) 175: Per has manifeste concluditur quod simplex habens unam partem ultra centrum et aliam citra ex se ipso ad centrum non poterit devenire, ita quod medium eius fiat medium mundi. Sicut est simplex columnare figurae. Quia, si sic, sit a ista terra cuius medium sit c, et sit d medium mundi. Et sit aliqua pars ultra centrum, et vocetur continue tota pars que erit ultra centrum b…

15 Cf. Albertus de Saxonia *Quaestiones in Aristotelis De caelo* II q. 26 (2008) 426–427: Iuxta istam conclusionem dubitatur: posito quod aliquod grave uniforme in gravitate haberet medium suae gravitatis extra medium mundi, omni impedimento extrinseco amoto, nihil sibi addendo nec aliquid ab eo removendo, an sic moveretur quod tandem medium suae gravitatis esset medium mundi? Dico quod non, quia, priusquam medium suae gravitatis fieret, medium mundi descenderet aliquali tarditate et dupla ad istam, et quadrupla ad istam, et sic in infinitum; et ideo, si perpetue maneret, perpetue descenderet, et nunquam medium suae gravitatis fieret medium mundi. Cf. Richardus Swineshead *Calculator* tr. XI (1966) 176: Consequentia ista nota est. Et consequentia proxima etiam patet. Quia, si super omnem partem istius distantiae moveretur c punctus equaliter sicut movebitur ad principium istius partis, ex quo secunda pars proportionalis est in duplo minor prima, et motus in secunda esset minor quam subduplus ad motum in prima, maius tempus requireretur ad pertransitionem secundae partis proportionalis quam ad pertransitionem primae. Et, per idem argumentum, maius tempus requireretur ad pertransitionem tertiae partis proportionalis quam secundae, et sic in infinitum.

ista terra incipiat moveri secundum proportionem duplam inter proportionem duplam et proportionem aequalitatis sunt infinitae maioris inaequalitatis secundum quas et ⟨secundum⟩ quamlibet earum movetur deorsum. Sequitur quod nunc movetur aliqualiter tarde; deinde in duplo tardius; deinde in quadruplo; et sic in infinitum. Unde tu habes imaginari quod, cum aliquid istius terrae pertransit ad ultra centrum, movebitur | V1 115ra tardius quam prius movebatur. Et cum maior pars pertransiverit, adhuc in duplo tardius. Et sic patet argumentum—sicut si moveretur A mobile et adaequate per horam, transiret primam partem proportionalem certa velocitate, deinde | V2 49va transiret secundam partem proportionalem velocitate subdupla ad primam. Tunc tantum de tempore requiritur ad pertranseundum secundam partem proportionalem quam requirebatur ad pertranseundum primam. Et sic constat quomodo mobile in tempore infinito supra magnitudinem datam movebitur, et numquam venit ad finem. Et sic in proposito. Et hoc sequitur per hoc quod tu fateris entia naturalia naturaliter moveri ad eius loca naturalia.

[8] Septimo arguitur[16] de minimo formali ignis. Unde pono quod A sit corpus orbiculare, sicut minimum formale ignis cuius centrum sit centrum mundi. Nunc remoto omni impedimento, iste ignis non movebitur sursum ad locum proprium. Quod patet, quia aut secundum eius unam semidiametrum tantum; et hoc non, quia non est maior ratio quod moveatur secundum diametrum unam quam secundum quamlibet. Nec secundum quamlibet potest moveri; quia tunc fieret divisio quamlibet, et sic daretur minus minimo. Stabit ergo ignis extra locum eius naturalem non-detentum.

[9] Octavo:[17] contingit terram in sua naturalitate exsistente recedere a centro mundi, et moveri sursum secundum naturam; igitur, etc. Probo antecedens. Et imaginor quod sit unum corpus orbiculare de simplici terra habens concavum et convexum, cuius centrum sit A. Centrum vero mundi sit B. Et sit circumferentia illius corporis C, D. Et sit ita quod D non distet a B centro mundi, ita quod centrum mundi cadat inter concavum et convexum dati corporis. Nunc constat quod in isto casu ista terra, dimissa suae naturae, movebitur quousque

2 infinitae] proportiones *add.* V2 12 pertranseundum primam] primam partem pertranseundum V2 13 movebitur *om.* V2 15 eius loca] loca eorum V2 16 Septimo] Alio V2 20 moveatur] magis *add.* V2 26 habens *om.* V2 27 convexum] convesum *var.* V1 30 ista] quod illa V2 | quousque] usque ad V2

16 Adn. mg. V1: Septimo. *Manicula.*—Cf. Richardus Swineshead *Calculator* tr. XI (1966) 177: Detur mixtum cuius una pars minor tamen quam medietas sit ultra centrum mundi, in qua parte intendatur ignis. Tunc ille ignis appetit moveri versus sursum; sed illo igne ascendente totum necessario descendet. Ergo ille ignis iuvabit ad descensum illius mixti.

17 Adn. mg. V1: Octavo.

eius centrum, ut punctus A sit idem cum centro mundi quod est B. Sic enim appetit terra collocari ut quod centrum eius fiat centrum mundi. Sed hoc non potest fieri nisi terra continue post hoc per tempus plus et plus removeatur a centro mundi a quo nunc non distat. Et pro illo describe circulum in margine cum suis litteris, isto modo:

FIGURA 1
in ima col. fol. 115ra V1

V1 115rb [10] Nono:[18] tunc parva portio terrae in aërem exsistens, demissa suae naturae, | per descensum eius ad locum inferiorem aquae levaret totam aquam universi de loco eius naturali. Sic non posset unum corpus naturaliter locari nisi aliud removeretur a suo loco. Et probatur consequentia: et sit A una portio terrae unius librae vel duarum ad modum figurae sphaericae. Et velis imaginari quod tota aqua universi sit in vase pyramidali inferius acuto et basis eius sit superius. Nunc incidat ista terra deorsum in aqua. Et quia ponis eam naturaliter locari sub omni aqua, ipsa movebitur deorsum ad fundum vasis; et per consequens faciet distare totam aquam a fundo vasis, elevando illam. Et[19] nisi concedatur hoc esse verum, sequitur quod a proportione minoris inaequalitatis fiat actio, et enim longe magis maioris gravitatis est tota aqua universi quam sit praedicta portio terrae; igitur, etc.

[11] Decimo:[20] sequitur quod potentia ut octo et potentia ut novem per aequale spatium moverent idem mobile eodem tempore. Consequens est fal-

---

5 isto] illud V2 6 exsistens] exsistentem V2 8 aliud] naturaliter *add.* V2 13 fundum] fondum *var.* V2 14 distare] stare V1 | vasis] et *add.* V2 15 fiat] fiet V2 | actio] accio *var.* V1; acio *var.* V2 16 enim] *mg.* V1; *om.* V2 17 portio] proportio V1

---

18 Adn. mg. V1: Nono.—Cf. Blasius Pelacanus de Parma *Questiones circa tractatum proportionum magistri Thome Braduardini* q. 8 (2005) 126: Tertio, per ymaginationem ponatur aqua maris in vase piramidalis figure cuius acuties sit deorsum versus terram, latitudo versus celum. Dico quod si ceperis frustum plumbi vel terre cuius corpulentia adequaretur concavo inferiori huius vasis piramidalis, et ponatur ad hanc aquam maris, istud grave movebitur deorsum usque ad fondum et faciet totam aquam maris ascendere, eam removendo a suo loco naturali. Et sic patet quod actio fit a proportione minoris inequalitatis.

19 Adn. mg. V1: *Manicula.*

20 Adn. mg. V1: Decimo. *Manicula.*

sum, quia tunc velocitas non insequeretur proportiones potentiarum moventium ad resistentias. Probatur consequentia. Et gratia argumenti, ponatur quod potentia motiva ut duo moveat mobile ut unum per spatium pedale certo tempore. Quo dato, sequitur quod potentia motiva | ut tria movebit idem mobile V2 49vb per spatium pedale cum dimidio per idem tempus. Tenet consequentia, quia eadem est proportio quae prius. Igitur potentia motiva ut novem movebit illud mobile per spacium tripedale. Tenet argumentum, quia sicut se habet potentia motiva ut duo ad movendum mobile per spatium pedale, et potentia ut tria ad movendum illud per spatium pedale cum dimidio, ita se habet potentia motiva ut novem ad movendum illud per spatium tripedale. Unde qualis est proportio novem ad tria, talis est triplum ad unum. Et ideo cum potentia motiva ut tria moveat hoc mobile per spatium pedale cum dimidio, potentia motiva ut novem movebit illud per spatium tripedale in eodem tempore. Sed quod potentia motiva ut octo idem faciat patet, quia potentia motiva ut duo movebit illud mobile per spatium pedale. Sed cum proportio octupla sit tripla duplae, sequitur quod potentia motiva ut octo movebit illud mobile per triplum spatium quam potentia motiva ut duo. Ergo movebit per spatium tripedale. | Et hoc fuit V1 115va deducendum; igitur, etc.

[12] Undecimo:[21] tunc idem mobile aeque cito pertransiret duplum et subduplum, et sic aeque cito transiret totum spatium sicut partem eius. Probatur[22] consequentia, pro cuius probatione sit A, B unum spatium uniformiter difforme in resistendo terminatum in extremo intensiori ad gradum ut octo, in remissiori vero ad non-gradum. Et sit B, C spatium sibi aequale in resistentia per totum ut quattuor, ut constat quod eodem tempore vel aequali idem mobile pertransibit A, B et B, C. Patet hoc: postquam ista spatia sunt aequalia, et eadem est proportio potentiae movere ad quodlibet istorum, semper supposito quod latitudo uniformiter difformis correspondeat suo gradui medio. Sit ergo potentia motiva, si volumus, ut novem, quae sit H. Tunc ut constat, eodem tempore vel aequali mobile H pertransibit A, B magnitudinem; et B, C, quia sunt aequales magnitudines et eiusdem resistentiae. Postquam latitudo uniformiter difformis correspondet suo gradui medio, et per consequens ipsius mobilis ad utrumque spatiorum est eadem proportio. Sit ulterius tertia magnitudo C, D aequalis magnitudini A, B, resistentiae per totum ut sex. Et tunc constat quod

---

4 dato] facto V2   10 per *om.* V2   11 triplum] tria V2   13 quod] quam *a.c.* V1   16 motiva *om.* V2   19 tunc *om.* V2 | pertransiret] pertransiret V2   22 gradum ut *om.* V1   23–24 aequale in resistentia] aequalis resistentiae V2   24 ut] quod V2   28 si volumus *om.* V2   33 sex] sextuples(?) V2

---

21 Adn. mg. V1: Undecimum argumentum.

22 Adn. mg. V1: *Manicula.*

in duplo tempore mobile H pertransibit magnitudinem C, D. Quod patet, quia proportio novem ad quattuor componitur ex proportione novem ad sex, et sex ad quattuor, tamquam ex duabus aequalibus. Quia utraque earum est una proportio sesquialtera, et per consequens extremi ad extremum erit dupla ad quamlibet earum. Et per consequens in duplici tempore H mobile pertransibit magnitudinem C, D; igitur in eodem tempore vel aequali pertransibit medietatem eius, quae sit C, F et magnitudinem A, B. Sed cum magnitudo C, F sit aequalis medietati magnitudinis A, B et eius resistentiae, sequitur quod eodem tempore mobile H pertransibit magnitudinem A, B et medietatem ipsius, quod fuit deducendum. Quod autem medietas A, B et C, F adaequantur patet, quia si aliqua tota adaequantur et medietates adaequabuntur. Et quod sint eiusdem resistentiae patet simpliciter, quia C, F est resistentiae ut sex, quia latitudo terminata in extremo intensiori ut octo, et in extremo remissiori ad quattuor, est tota ut sex. Et sic erat medietas B magnitudinis, ut patet ex casu. Et sic patet argumentum.

[13] Duodecimum[23] argumentum: ferrum insequitur magnetem situatum in
V2 50ra V1 115vb media | regione aëris; igitur non | apparet quod corpus naturale naturaliter moveatur ad locum eius naturalem. Nec dici potest quod ille motus sit violentus eo quod talis motus est velocior in fine quam in principio; motus autem violentus est per oppositum.[24]

[14] Item[25] contingit lapidem generari in media regione aëris, quo genito movetur deorsum. Igitur sequitur quod non movetur naturaliter ad suum locum. Tenet argumentum, quia locus naturalis corporis videtur esse locus in quo genitum est, quia non est putandum quod aliquod ens naturale generetur in loco sibi contrario disconvenienti; quare, etc.

[15] In oppositum est Philosophus in hoc quarto[26] ad textum.

[16] Et per hoc aliqua naturaliter moveretur sursum et aliqua deorsum sicut ad loca eorum naturalia.

1 mobile *om.* V2 11 si] s.l. V1; *om.* V2 13 ut] ad V1 | ad] ut V2 20 est per oppositum] opposito modo se habet V2 21 quo] *mg.* V2 27 aliqua *om.* V1 | sicut] sic V2

23 Adn. mg. V1: Duodecimum argumentum.

24 Resp. Aristoteles *De caelo* II,6 288a21; *Auctoritates Aristotelis* 164 (58): Omnis motus naturalis est debilis in principio, in fine fortior, sed e converso motus violentus in principio est fortior, in fine debilior.

25 Adn. mg. V1: Aliud argumentum.

26 Aristoteles *Physica* IV,8 215a25–26: Videmus enim idem grave et corpus velocius ferri propter duas causas aut in differendo per quod, ut per aquam aut terram, aut in differendo quod fertur, si alia sint eadem, propter excellentiam gravitatis aut levitatis.

[17] In hac quaestione erunt quattuor articuli. In primo, problema. In secundo, determinatio quaestionis. In tertio, difficultates. In quarto, ad rationes in oppositum.

[18] Ad primum articulum sit hoc problema:[27] propter quid est quod habentes collum breve sunt audaces et vindicativi?

[19] Respondetur[28] quod huius causa est quia isti habent cerebrum in quo viget virtus irascibilis et concupiscibilis non multum distans a corde in quo est generatio sanguinis et spiritus. Hinc est quod isti alterantur a facili causa ab extrinsicis, et in eis faciliter bullit sanguis. Per oppositum vero iudicabitur de habentibus collum longum et gracile; tales namque sunt timidi.

[20] Et[29] ex hoc habetur causa alterius problematis, in hac forma: propter quid est quod anseres intrantes domum, per ostium quamcumque magnum fuerit in altitudine, inclinant capita?

[21] Ratio huius est quia anseres habent colla eorum valde longa, et per consequens eorum capita multum a corde distant tamquam a fonte caloris, et per

---

2 determinatio quaestionis] de quaesito V2 3 oppositum] contrarium V2 4 Ad primum articulum] Quantum ad primum V2 5 collum] colum *var.* V1 | sunt] quattuor articuli *add. sed del.* V1 | vindicativi] vendicativi *var.* V1 9 bullit] bulit *var.* 14 colla] cap *add. sed del.* V2 | eorum] corporis V2 15 eorum capita] corpus capiti V2

---

27 Cf. Albertus Magnus *De animalibus* I tr. 2 cap. 22 (1916–1921) I 155–156: Et ibidem dicit, quod habens collum gracile et longum est sonorus et stolidus. Qui autem habet collum breve valde, est callidus, defraudator et astutus et dolosus. Qui autem habet collum grossum est stolidus et comestor magnus. Cf. Rogerus Bacon *Secretum secretorum* IV cap. 11 (1920) 170: Capitulum 11: de collo. Qui vero habet collum gracile et longum est sonorus in voce et stolidus. Qui vero habet collum breve valde est callidus, defraudator, astutus, et dolosus. Qui vero habet collum grossum est stolidus et comestor magnus.

28 Cf. Bartholomeus Coclitis *Chyromantie ac physionomie anastasis*, comm. in pseudo-Arist. *Physiognomonica* I c. 39 ("cap. 34" ed.) (1523) f. 18[vb]: Unde Blasius de Parma super quarto *Physicorum* ait: Propter quid est quod habentes collum breve sunt audaces et vindicativi. Causa est quia isti habent cerebrum crassibile et concupiscibile non multum distans a corde in quo est generatio sanguinis et spiritus. Hinc est quod alterantur a facili causa ab extrinseco, et in eis faciliter bullit sanguis, et sunt mobiles sine stabilitate.

29 Adn. mg. V1: *Manicula*.—Cf. Bartholomeus Coclitis *Chyromantie ac physionomie anastasis*, comm. in pseudo-Arist. *Physiognomonica* I c. 39 ("cap. 34" ed.) (1523) f. 18[va]: Ad propositum redeundo. Si collum esset longum et distans a fonte caloris sicut dictum est, per hoc parum de calore et spiritu terminantur ad cerebrum et non sunt capaces ad operationes debitas, quia sunt debiles et indigesti, et per hoc sunt operationis exigue, et etiam sunt timidi sicut nos vidimus quod anseres intrantes domum per hostium quantumcunque magnum inclinant capita quia habent collum valde longum et timent per supra notata; et caput eorum non possunt substinere ideo flectuntur eorum capita. Ideo ratione timiditatis sequitur insipientia propter organum intellectus indispositum.

hoc parum de calore et spiritu transmittitur ad cerebrum, et per hoc timent in introitu etc. Et haec de primo.

[22] Sequitur secundus articulus in quo praemittam evidentias; deinde conclusiones et corollaria. | V1 116ra

[23] Prima[30] evidentia. Quia in titulo quaestionis tangitur de loco naturali, est sciendum quod reperitur locus naturalis et locus proprius, quae loca sunt aliquando idem locus respectu eiusdem corporis, et aliquando distinguuntur. Unde grave deorsum exsistens est in loco proprio, et idem locus est eidem corpori naturalis. Aliquando distinguuntur, quia dum grave detinetur sursum in aëre vel in igne, tunc est in igne tamquam in loco proprio, sed non est in igne tamquam in loco naturali. Non enim sphaera ignis est locus naturalis gravium.

[24] Ideo[31] noto pro secunda evidentia quod locus naturalis ideo dicitur 'naturalis' quia habet quandam virtutem conservativam locati.[32] Locus vero proprius ratione qua proprius non dicitur habere virtutem conservativam locati, sed ex eo dicitur 'proprius' quia est aequalis locato, secundum quod dictum est in praecedenti[33] quaestione.

[25] Et[34] tunc, hiis praemissis, sit prima propositio: quodlibet ens naturale ad intra caelum exsistens, si sit grave, appetit substare omni leviori—et si

1 et *om.* V2 3 Sequitur secundus articulus] Transeo ad secundum articulum V2 7 distinguuntur] distinguitur *a.c.* V2 9 naturalis] et *add.* V2 | distinguuntur] distinguitur *a.c.* V2 19 substare] substenstare *a.c.* V1; sustentare V2

30 Adn. mg. V1: Prima.

31 Adn. mg. V1: Secunda.

32 Cf. Aristoteles *Physica* IV,3 210b34–211a1; *Auctoritates Aristotelis* 150 (125): Locus continet et conservat locatum. Cf. Blasius Pelacanus de Parma *Quaestiones disputatae De anima* II q. 12 (1974) 114: Et de isto loco specialiter verificantur propositiones Aristotelis et aliorum philosophorum dicentium quod locus est principium generationis et quod locus habet virtutem conservativam locati.

33 Blasius Pelacanus de Parma *Questiones disputate super octo libros Physicorum* IV q. 1 (cod. Vat. lat. 2159) f. 113ra: Secunda distinctio: locum proprium adaequari locato contingit dupliciter, quia secundum continentiam et secundum quantitatem. Verumtamen improprie dicitur aliquid aequale alteri secundum continentiam. Tertia distinctio: locum proprium adaequari locato secundum quantitatem contingit dupliciter, ut puta simpliciter et secundum quid. Simpliciter, quando secundum omnem diametrum aequantur; et ita etiam in aliis rebus quae dicuntur aequari. Secundum quid ut continuum aequatur alteri secundum unam diametrum vel plures, sed non secundum omnes.

34 Adn. mg. V1: Prima propositio.

leve, graviori omni superesse. Quae propositio patet iuxta doctrinam Aristotelis quarto[35] *De caelo*.

[26] Secunda[36] propositio: quodlibet ens naturale appetit permanere et in esse conservari. Hoc est Aristoteles primo[37] huius.

[27] Prima[38] conclusio: quodlibet ens naturale appetit naturaliter in loco eius naturali residere. Patet, quia quodlibet | tale appetit conservari, ut dicit (V2 50rb) prima[39] propositio. Sed per quandam evidentiam sive differentiam positam inter locum naturalem et proprium, locus naturalis habet virtutem conservativam locati; quare, etc.

[28] Secunda[40] conclusio et responsalis: quodlibet ens naturale extra locum eius naturalem exsistens appetit moveri ad locum eius naturalem. Patet ex conclusione[41] praecedenti propositione et concurrente.

[29] Tertia[42] conclusio: quodlibet ens naturale ad intra caelum exsistens appetit naturaliter quiescere. Patet, quia quodlibet tale appetit esse in loco naturali eius et non removeri ab illo; igitur appetit quiescere.

[30] Quarta[43] conclusio: nullum corpus naturale ad intra caelum exsistens simpliciter appetit moveri. Patet, quia quodlibet tale naturaliter appetit quiescere; igitur, etc. Tenet argumentum, quia moveri et quiescere sunt opposita. Igitur si quodlibet corpus etc. appetit naturaliter quiescere, sequitur quod nullum tale appetit principaliter moveri, quia alioquin appetitus naturalis fieret in opposita et incompossibilia.

[31] Et iuxta hanc materiam quae est de natura loci sequuntur aliqua corollaria. Primum:[44] terra ceteris elementis naturaliter subest. Et post terra, aqua. Patet ex | prima[45] propositione. (V1 116rb)

---

1 leve] con *add. sed del.* V1 2 caelo] ad testum *add.* V2 3 ens *om.* V2 4 huius] Physicorum V2 6 tale *om.* V2 10 naturale] ad intra caelum e *add. sed del.* V1 15 eius] suo V2 22 hanc] hic V2

---

35 Aristoteles *De caelo* IV *passim*.

36 Adn. mg. V1: Secunda propositio.

37 Aristoteles *Physica* I,9 192a19–22. Cf. Averroës *In Phys.* I com. 81 (1562) 46rb: Omne enim ens, ut dictum est, diligit se permanere. Cf. Blasius Pelacanus de Parma *Questiones disputate super octo libros Physicorum* secunda redactio II q. 7 (ed. Biard) 12: Prima propositio: omne agens naturale appetit permanere.

38 Adn. mg. V1: Prima conclusio.

39 Cf. supra n. 25.

40 Adn. mg. V1: Secunda conclusio.

41 Cf. supra n. 25.

42 Adn. mg. V1: Tertia conclusio.

43 Adn. mg. V1: Quarta conclusio. *Manicula*.

44 Adn. mg. V1: Primum corollarium.

45 Cf. supra n. 25.

[32] Secundum[46] corollarium: terra naturaliter est in aqua, et aqua in aëre; aër in igne, et ignis in caelo naturaliter. Patet, quia elementa sic sunt situata in universo secundum quod plus vel minus participant de qualitatibus motivis.

[33] Tertium[47] corollarium: omnis terra optime disposita aquis circumdata esse debet per naturam. Patet: postquam est gravior omni aqua, sequitur quod ipsa maxime debeat substare, et cum ipsa sit rotunda, sequitur quod ubi sit optime disposita, ipsa est concentrica mundo secundum centrum magnitudinis eius. Et si sic, sequitur quod gratia dei specialiter circumscripta aqua, circumdat eam secundum omnem partem. Quare autem dicam in conclusione 'terra optime disposita' patet intelligenti, eo quod cum magna pars ipsius sit arida et discooperta aquis, quae pars continue alteratur a sole, movetur a frigiditate noctis superveniente, hinc est quod ipsa non est optime disposita.

[34] Aliud corollarium:[48] terra sic se habens est altior omni aqua. Patet, quia ubi aqua esset altior quam sit terra, tunc aqua flueret ad aridam terram. Et sic tota terra esset cooperta aquis.

[35] Quintum corollarium:[49] facile est levare totam aquam mundi de loco in quo est. Patet hoc per infusionem cuiuscumque gravioris. Et hoc unum argumentum a principio patefaciebat.

[36] Aliud corollarium:[50] quod terra moveatur ad aquam qualitercumque se habentem tamquam ad locum eius naturalem non est asserendum. Patet, quia ubi aqua esset in sphaera ignis supra omnem aërem, terra non moveretur ad ipsam. Et haec de secundo articulo.

[37] Ad tertium articulum sit prima difficultas:[51] an orbis lunae sit locus naturalis ignis.

---

2 et *om.* V2 7 substare] substentare V2 | rotunda] rotonda *var.* V2 8 optime] maxime V2 12 discooperta] discoperta *var.* 13 est *om.* V2 | ipsa *om.* V2 15 Aliud] Quartum V2 16 aqua] terra V2 17 cooperta] cohoperta *var.* V2 19 hoc] est *add. sed del.* V1 19–20 argumentum *om.* V2 21 Aliud] Sextum V2 25 Ad tertium articulum] Quantum ad tertium V2

---

46 Adn. mg. V1: Secundum corollarium.

47 Adn. mg. V1: Tertium corollarium.

48 Adn. mg. V1: Quartum corollarium.

49 Adn. mg. V1: Quintum corollarium. *Manicula.*

50 Adn. mg. V1: Sextum corollarium.

51 Adn. mg. V1: Prima difficultas.—Cf. Marsilius de Inghen (attrib.; pseudo-Ioannes Duns Scotus) *Quaestiones in VIII libros Physicorum Aristotelis* IV q. 6 (1891) 56–59: "Utrum ignis sit in concavo orbis Lunae tanquam in suo loco naturali."

[38] Arguitur[52] quod non, quia sphaera lunae propter lunam habet infrigidare; et per consequens non potest conservare sed potius corrumpere ipsum ignem. Igitur non est locus naturalis ignis, quia locus[53] naturalis conservat.

[39] Oppositum est Aristoteles in quarto huius;[54] igitur, etc.

[40] Respondetur quod sphaera lunae est locus naturalis ignis.

[41] Et cum dicitur[55] quod 'luna est frigida', negatur. Quia ratione sui motus et luminis, quod lumen est lumen solare, ipsa est calefactiva et conservativa ignis.

[42] Et cum scribant philosophi[56] in libris iudiciorum quod luna est frigida—

[43] Respondetur quod saepe loquuntur iuxta comparationem non expressam in eorum sermonibus; ut saepe dicunt | medici[57] quod caelum est frigidum, V2 50va cum tamen sit calidum, cum sit vivens. Ita luna dicitur frigida in comparatione ad solem vel ad habentem caliditatem adhaerentem et corrumpentem. | V1 116va

[44] Aliter potest dici, concedendo quod luna sit frigifactiva, et quod luna est corruptiva ignis. Et sic ordinavit natura secundaria intentione ratione mixtorum ut per frigiditatem lunae ipsa in eius motu condensaret partem ignis ei subiacentem, per cuius condensationem ille ignis condensaret, ut inde veniret ad locum in quo mixta habent generari. Et pariter est dicendum de quolibet loco proprio respectu sui locati, quod est evidens visis elementorum proprietatibus quorum unum continet aliud.

---

3 non est] *mg.* V1; non sit V2 | conservat] etc. *add.* V2 4 etc. *om.* V2 5 ignis] ipsius V2 9 in libris iudiciorum] iudiciorum libris V1 9–10 est frigida *om.* V2 12 in eorum] illorum V2 14 ad[2]] maiorem *add.* V2 | adhaerentem] advenientem V2 | corrumpentem] corruptivam V2 15 frigifactiva] frigidativa V2 17 lunae] lunam V2 | condensaret] condempsaret *var.* V1 17–18 partem ignis ei subiacentem] parte ignis ei subiacente V2

---

52 Cf. Marsilius de Inghen (attrib.; pseudo-Ioannes Duns Scotus) *Quaestiones in VIII libros Physicorum Aristotelis* IV q. 6 (1891) 56: Arguitur quod non, quia cuiuslibet corporis naturalis locus naturalis debet esse conservativus ipsius, scilicet locati; sed orbis Lunae non est conservativus ignis; igitur concavum eius non est locus naturalis ignis. Maior patet in quarto huius; et minor probatur, quia Luna est frigida et humida virtualiter.

53 Cf. supra n. 24.

54 Aristoteles *Physica* IV,3 210b34–211a1; cf. supra n. 24.

55 Cf. supra n. 27.

56 Cf., e.g., Rogerus Bacon *Secretum secretorum* I cap. 7 (1920) 17: De Luna est sciendum quod Luna est frigida et humida virtute et effectu in istis rebus inferioribus, quia facit humorem et frigus in eis.

57 Cf., e.g., Sadan (Shadhan b. Bahr) *Excerpta de secretis Albumazar* (1998) 297: Dixit Sadan: 'Audivi Albumazar dicentem quod omnis vita viventium post deum est Sol et Luna.' Cf. Petrus de Abano *Conciliator Differentiarum, quæ inter Philosophos et Medicos Versantur* diff. 9 (1496) 13va: Unde Albumasar in Sadan ait: 'solem et lunam etiam existere vitam omnium viventium post deum.'

[45] Secunda[58] difficultas: an locus in quo generatur corpus naturale sit ei locus naturalis.

[46] Videtur quod sic, per hoc quod scribitur in quarto huius,[59] "locus naturalis est principium generationis."

[47] Oppositum arguitur, quia aqua generat fulgur, quod secundum plures[60] est lapis genitus per adustionem. Haec omnia generantur in interstitio aëris, et tamen media regio non est locus naturalis praedictorum. Quod patet, quia praedicta post eorum generationem quiescerent ibidem et non descenderent.

[48] Respondetur negative. Dico tamen quod ubi media regio haberet virtutem generativam praedictorum merito sui quod illa esset locus naturalis praedictorum. Sed licet ibidem generentur praedicta, et multa alia animata et inanimata, ut pisces, ranae, viperae, etc.,[61] hoc est per concursum astrorum, ut patet.

[49] Et[62] si non vis tenere quod dictum est, distingue de loco naturali, scilicet quod quidam est locus naturalis habens virtutem productivam rei naturalis in esse, et quidam conservativam et non productivam, et quidam econverso,

---

1 corpus naturale] *mg.* V1 3 locus] igne *add. sed del.* V1 5 generat] generanda V1; generando *a.c. mg.* V1; generando V2 6 Haec] et cum V2 11 et *om.* V2 12 viperae] vespae V2 | est] *mg.* V2 14 non *om.* V2 15 productivam] et *add. sed del.* V1

---

58 Adn. mg. V1: Secunda difficultas.

59 Cf. immo Porphyrius *Isagoge* 1.25–2.5 (1966) 6: Dicitur autem et aliter rursus genus quod est uniuscuiusque generationis principium vel ab eo qui genuit vel a loco in quo quis genitus est… etenim patria principium est uniuscuiusque generationis quemadmodum pater.

60 Cf. Robertus Grosseteste (attrib.) *Summa philosophia* tract. 16 cap. 10 (1912) 638: Iam autem supra diximus vapores in sublimibus vario modo qualificari et quandoque substantiam sulphuris in nubibus, ubi causatur fulgur et tonitruum inveniri et quandoque lapides vel animalia ibidem generari; cf. Thimo Judaei *In Aristotelem de Meteoris* II q. 11 (1518) 177a: Quantum ad primum notandum quod in nube quandoque includitur exhalatio … tunc quandoque contingit quod exhalatio inflammata dispergitur in latus; et tunc vocatur fulguratio … et aliquando movetur per viam strictam, et tunc vocatur fulmen; cf. Nicolaus Oresme *Questiones in Meteorologica* II q. 1 (2021) 221: in media regione aeris generantur corruscationes et fulgura, que sunt impressiones ignite. Cf. Blasius Pelacanus de Parma *Questiones circa libros Metaurorum Aristotelis* I q. 9 (ed. Panzica) 7: modo cum media regio aeris sit frigida et exalatio calida, exalatio non transibit per ipsam, sed fugiet ab ipsa, dico hic quod illa exalatio ex qua fiunt comete est materia subtilis et rara multum habens de caliditate, et ratione sue subtilitatis est multum penetrativa, hunc est quod ipsa potest petransire. Et verum est quod sepe, dum est grossa et terrestris multum, ipsa nequit transire, sed propellitur in deorsum, et tunc fiunt fulgura et tonitrua, ut postea videbitur.

61 Cf. Blasius Pelacanus de Parma *Quaestio de intensione et remissione formarum* (1994) 487: Item, in locis terre profundis generantur multa animalia et similiter in locis aque, ut pisces, rane, etc. Et clarum est quod huiusmodi animata non generantur sine concursu magni caloris ipsius continentis.

62 Adn. mg. V1: *Manicula.*

et quidam productivam et conservativam, sicut est videre. Quia in aqua maris generantur pisces etc., et ibidem conservantur qui non generarentur in cuspidibus montium, et deterius conservarentur. Et per hanc distinctionem patet quod sit dicendum de difficultatibus.

[50] Tertia[63] difficultas: an aqua sit omni terra altior.

[51] Videtur quod sic, quia remoto impedimento, videmus aquam ascendere super omnem terram scaturriendo.

[52] Secundo, omnis aqua plus removetur a centro mundi quam terra; immo, ut sic dicam, aqua removetur a centro mundi per semidiametrum terrae.

[53] Oppositum, quia magna portio terrae est arida.

[54] Respondetur, praemittendo hanc propositionem: omnis liquor, sive sit vinum sive aqua sive oleum etc., fluit naturaliter ad partem terrae decliviorem. Patet secundo *Caeli*.[64] | V1 116vb

[55] Prima[65] conclusio: omnis terra arida est omni aqua mundi altior. Patet quia aliter aqua flueret ad ipsam terram.

[56] Sed[66] contra: potest adduci ratio Magistri *De sphaera*[67] quod "ubi navis recedat a portu maris, dimittens aliquod signum in littore, et recedat tantum illa navis quod oculus exsistens iuxta pedem mali non possit videre signum dimissum in littore. Sed si ille idem oculus ascendat ad summitatem mali, videbit illud signum." Et non est nisi quod, quando ille oculus erat iuxta pedem mali, aqua mediabat inter oculum et obiectum visibile et abscindebat radios. Et per

---

6 aquam] terram V2 7 scaturriendo] scaturiendo *var.* 11 praemittendo] praemitto V2 15 ad *rep.* V2 18 mali] in parte inferiori eius *add.* V2 19 dimissum *rep.* V2 | ille *om.* V2 | ascendat] descendat(!) V2 20 Et] hoc *add.* V2 21 abscindebat] ascindebat *var.* V2

---

63 Adn. mg. V1: Tertia difficultas.—Cf. Blasius Pelacanus de Parma *Questiones circa tractatum proportionum magistri Thome Braduardini*, redactio prima, q. 12 (2005) 228: omnis terra arida est omni aquam altior.

64 Aristoteles *De caelo* II,4 287b5–7: Sed et quod aque superficies talis, manifestum suppositionem sumentibus quia nata est semper fluere aqua in magis concavum; concavius autem quod centro propinquius. Cf. infra IV q. 3 n. 3.

65 Adn. mg. V1: Prima conclusio.

66 Adn. mg. V1: Contra.

67 Ioannes de Sacrobosco *Sphaera* I (1949) 83: *Quod aqua sit rotunda.* Quod autem aqua habeat tumorem et accedat ad rotunditatem sic patet. Ponatur signum in littore maris, et exeat navis a portu, et in tantum elongetur quod oculus existentis iuxta pedem mali non possit videre signum. Stante vero navi oculus eiusdem existentis in summitate mali bene videbit signum illud. Sed oculus existentis iuxta pedem mali melius deberet videre signum quam qui est in summitate: sicut patet per lineas ductas ab utroque ad signum: et nulla alia huius rei causa est quam tumor aquae. Excludantur enim omnia alia impedimenta, sicut nebulae et vapores ascendentes.

V2 50vb consequens illa aqua videtur altior quam sit obiectum dimissum | in terra arida. Cuius signum est quia quando dictus oculus erat in summitate mali, aqua non poterat abscindere radios protractos ab obiecto ad oculum propter magnam elevationem oculi. Et idem contingeret si illud obiectum elevaretur de loco in quo est secundum lineam rectam. Et tunc oculus exsistens iuxta pedem mali inciperet videre illud. Tunc haberetur de directo in quam aqua est altior terra arida in qua erat obiectum prius.

[57] Respondetur quod licet argumentum sit bene coloratum, tamen secundum rei veritatem non probat propositum. Sed bene probat aquam esse rotundam, et non aliud. Quia sic arguerem quod terra exsistente bene rotunda et sphaerica esset una pars altior alia, eo quod in strata bene recta, gratia exempli, a Papia ad Mediolanum non possum videre socium praecedentem me per duo miliaria. Et hoc non est nisi quia inter me et ipsum est tumor terrae radios abscindens visuales; tamen illa terra non est altior quam sit terra subiecta pedibus meis; quare, etc.

[58] Quarta[68] difficultas: an ex proprietate locorum naturalium sit quod aliqua in uno loco generentur quae in alio non valeant generari.

[59] Arguitur quod sic, quia ut dicebatur in isto quarto,[69] "locus est sicut principium generationis." Et cum distincta loca habeant distinctas influentias, non mirum est si aliqua generantur in uno loco quae alibi non generantur.

[60] Oppositum, quia locus naturalis non est nisi ultima superficies corporis ambientis,[70] ut aëris vel aqua, etc. Sed omnis aër est eiusdem rationis cum V1 117ra alio aëre, et aqua cum alia aqua. Hic est quod illa loca non | habebunt diversas influentias; igitur, etc.

[61] Respondetur quod ex proprietate locorum est quod aliqua generentur in uno loco quae in alio non generantur. Et ut dictum[71] illud sit verum, capiendum est hic locum secundum rationem eius formalem, et non solum per ultimam

2 non] erat in summitate mali *add. sed del.* V2 3 abscindere] ascindere *var.* V2 11 altior] maior V2 12 a] per quam inter V2 | ad *om.* V1 13 ipsum] non *add. sed del.* V2 17 in uno loco generentur] generentur in loco V2 | valeant] valerent V2 18 isto *om.* V2 24 igitur] quare V2 26 quae] et V2 27 est *om.* V2 | et] s.l. V2

68 Adn. mg. V1: Quarta difficultas.

69 Cf. Aristoteles *Physica* IV,2 209b1–2. Cf. Porphyrius *Isagoge* 1.25–2.5 (1966) 6. Cf. supra n. 46.

70 Cf. Aristoteles *Physica* IV,4 212a20–21: Quare continentis terminus inmobilis primum, hoc est locus; *Auctoritates Aristotelis* 149 (120): Locus est terminus corporis continentis immobilis primum, id est locus est ultima superficies corporis continentis.

71 Cf. supra n. 48.

superficie aëris ambientis, quia ut sic locus non est de genere activorum. Quod patet, quia quantitas non ponitur esse de genere activorum; et locus pro superficie est quantitas;[72] etc. Quid autem intelligatur per locum formaliter sumptum dictum est in praecedenti[73] quaestione.

[62] Et quod conclusio proposita sit vera patet in multis experientiis. Unde[74] est ad hoc videre quod in quibusdam locis est aqua flumalis in qua, si ponatur lignum, transmittatur in ferrum. Et hoc non faciunt omnes aquae.

[63] Et[75] sunt aliqua loca in quibus est pirus producens pira suo tempore, quae cum matura fuerunt, et ad terram cadant, ex eis statim generantur aves. Si autem in aquam ceciderint, convertuntur in pisces—quod non faciunt omnia loca.

---

1 superficie] Et non solum per ultimam superficie *add. seu rep.* V2 5 in *om.* V2

---

72 Resp. Aristoteles *Cat.* 6 4b23–5: Est autem discreta quantitas ut numerus et oratio, continuum vero linea, superficies, corpus; amplius autem praeter haec tempus et locus.

73 Blasius Pelacanus de Parma *Questiones disputate super octo libros Physicorum* IV q. 1 (cod. Vat. lat. 2159) f. 114[rb]: Quinta conclusio: locus secundum eius rationem formalem est penitus immobilis. Pro quo noto unum dictum Commentatoris quod locus potest considerari dupliciter: materialiter et formaliter. Locus materialiter accipitur pro corpore vel superficie eius ambiente locatum sine alio respectu. Et sic locus est mobilis. Sed formaliter accipitur pro corpore vel superficie in ordine ad quattuor puncta universi, quae puncta sunt oriens, occidens, medium caeli, et alvus terrae; et cum hac relatione locus est omnino immobilis. Et sic intelligo quod quocumque loco signato in hemisphaerio tuo, ille sic se habet ad ista quattuor puncta, quod nullus alius locus proprius sic se habet ad ista. Et iuxta hoc patet conclusio. Ratio in oppositum probat rationem; ideo, etc.

74 Cf. Patricius (attrib.) *Versus sancti Patricii episcopi de mirabilibus Hibernie* (1955) 58: Terraque nostra tenet stagnum quod continet istam / Vim qua ligna solent lapides mox esse sub undis / Post tamen annorum, ceu dicunt, tempora septem. Cf. Giraldus Cambrensis, *Topographia Hiberniae* d. 2 c. 7 (ed. J. Dimock V 86): In Britannia maiore, non procul a coenobio Wiburnensi, est nemus quoddam fructicosum. Cuius cum ligna in aquam forte vel terram eo in loco deciderint, post annum reperies in lapidem conversa; *Encyclopaedia Britannica* vol. 15 p. 849 (13th ed.) 'Knaresborough': Opposite the castle is the Dropping Well, the waters of which are impregnated with lime and have petrifying power. Cf. Ioannes Duns Scotus *Quaestiones super libros Metaphysicorum* VII q. 20 n. 41 (OPh IV) 390: Ad tertium: negatur maior, quia manus arida continuatur carni vivae, vel ramus arbori, sicut ligno lapis, secundum aliquos, de fonte convertente lignum in lapidem. Cf. Guillelmus de Ockham *Tractatus de corpore Christi* c. 19 (OTh X 129–130).

75 Cf. Petrus Damianus *De divina omnipotentia*, cap. XI "Quod naturae conditor naturae sit etiam immutator" (1853) col. 614: Unde et terra illa ⟨*i.e., Thilon Indiae insula*⟩ occiduis partibus hanc consecuta est dignitatem, ut ex arborum ramis volucres prodeant, et ad pomorum similitudinem animati atque pennati fructus erumpant? Sicut enim referunt, qui se vidisse testantur, paulatim incidit pendulum quid ex ramo suspendi, deinde in imaginem volucris speciemque formari; postremo quantulumcunque plumescens, hiatu rostri sese ab arbore dividit sicque novus aeris habitator ante pene discit volare, quam vivere. Cf. Michael Maier *Tractatus de volucri arborea, absque patre et matre* (Frankfurt, 1619).

[64] Et sunt aliqua loca in quibus palea producit unicam spicam frumenti, et alia sunt loca in quibus bifurcatur versus medium et ultra, et producit duas vel tres spicas.

[65] Sunt[76] etiam alia loca in quibus homines nunquam moriuntur. Et cum vita fuerit eis taediosa, statim in confinibus sunt loca ad quae, cum accedant, moriuntur.

[66] Item[77] sunt loca in quibus homines producuntur et non augentur ad quantitatem cubitui, contra quos pugnant grue eos vincentes.

[67] Sunt[78] etiam alia loca in quibus homines sunt caudati quantitate unius pedis vel duorum, ut in aliquibus partibus Angliae.

V2 51ra [68] Et etiam multa alia de proprietatibus | locorum possent narrari; tamen sufficiant haec in praesenti. Et haec volo sufficere pro nunc de tertio articulo,
V1 117rb etc. |

---

1 Et *om.* V2 | frumenti] furmenti *var.* V1 2 sunt] supra V2 | bifurcatur] bifulcatur *var.* V1 8 quantitatem] unius *add.* V2 9 sunt *om.* V2 | quantitate *om.* V2 11 Et etiam *om.* V2 12 haec *om.* V2 12–13 volo … etc.] de tertio V2

---

76 Cf. Plinius Secundus *Historia Naturalis* IV c. 89 (2002) 176: Pone eos montes ultraque Aquilonem gens felix, si credimus, quos Hyperboreos appellavere … Mors non nisi satietate vitae epulatis delibutoque senio luxu e quadam rupe in mare salientibus. Cf. Albertus de Saxonia (attrib.) *Quaestiones in Aristotelis Physicam* I q. 21 (1999) 273: Quinto. Auctoritate Lincolniensis in tractatu suo *De reflexionibus radiorum*, ubi narrat quod circa polum sunt quidam montes super quibus sit ita uniformis et regularis reflexio radiorum solis quod est ibi tanta temperies quod homines ibi existentes vivunt ita diu quod tandem habentes taedium de vita saltant ad quendam fluvium fluentem sub illis montibus et submerguntur; quod non esset, si quodlibet ens appeteret se permanere. Cf. Robertus Grosseteste *De natura locorum* (1912) 68–69: Similiter expertum est, ut dicunt auctores ut Marcianus in geometria sua et Plinius in naturalibus suis et Solinus de memorabilibus mundi et multi alii describentes regiones mundi, quod apud Hyperboraeos montes iuxta polum sunt homines, qui a montibus illis dicuntur Hyperboraei. Et habent aerem saluberrimum et temperatissimum, quoniam in tantum vivunt, quod fastidiunt vitam et gratis se praecipitant in mare a rupibus altis et moriuntur.

77 Cf. Aristoteles *Historia animalium* VIII,12 597a4–9: … velut grues faciunt. Permutant enim ex Scythicis campis ad paludes superiores Egypti unde Nilus fluit; est autem locus iste circa quem Pigmei habitant; non enim est hec fabula, sed est secundum veritatem genus parvum, ut dicitur, et ipsi et equi, in cavernosis habitantes autem sunt per vitam. Cf. Anonymus *Sentencia probleumatum Aristotilis* (Brugge, Stedelijke Openbare Bibliotheek 481, f. 36rb): Sunt enim animalia ut homines formata, parva ad longitudinem cubiti extensa; unde dicuntur homines cubitales, cum quibus ciconie et grues pugnant; equitant, habitant in cavernis, nec sunt homines, sed hominibus assimilantur.

78 Cf. Jacobus de Vitriaco *Historia Occidentalis* cap. 7 (1972) 92: anglicos potatores et caudatos affirmantes… Cf. Richardus Divisensis *De rebus gestis Ricardi Primi, regis Angliae* (1838) 20: Tota injuriarum de rege Anglorum et caudatis suis ultio quaeritur. Graeculi et Siculi omnes hunc regem sequentes Anglos et caudatos nominabant. Cf. George Neilson *Caudatus Anglicus: A Mediæval Slander* (1895).

[69] Quantum ad quartum in quo specialiter respondetur ad primum[79] argumentum, in quo facta fuit mentio de una terra privata omni gravitate, etc. Ad hoc potest responderi quod illa terra privata non descendet propter ipsam non habere illud quod requiritur pro descensu eius.

[70] Aliter potest dici quod probabiliter, ut probabiliter dicatur, quod illa terra descendat.

[71] Et quando arguitur quod non, quia tunc contingeret quod terra habens gravitatem tardius descenderet quam terra absque gravitatem—

[72] Respondetur quod, quidquid sit de consequentia, negandum est consequens.

[73] Et ad probationem dico in summa quod terra habens gravitatem, gratia exempli, movebitur in decuplo velocius alia.

[74] Et quando dicitur quod 'ubi remitteretur ista gravitas ad decuplum, ista terra moveretur in decuplo tardius quam nunc moveatur'—

[75] Respondetur negando illud, eo quod velocitas istius terrae non insequitur proportionem gravitatis ad gravitatem eiusdem; sed insequitur proportionem gravitatis ad resistentiam.

[76] Ad alias patet solutio per ea quae dicta sunt.

3 privata *om.* V2 4 requiritur] requisitum est V2 5 quod] quod probabiliter *add.* V2
18 alias] rationes *add.* V2 | ea *om.* V1

79 Cf. supra n. 2.

6

Liber IV q. 3: Utrum corpora naturalia appetant moveri ad eorum loca naturalia per lineas breviores

V1 = Vat. lat. 2159 ff. 117rb–119rb; V2 = Vat. lat. 3012 ff. 51ra–52rb

[1] Tertio quaeritur utrum corpora naturalia appetant moveri ad eorum loca naturalia per lineas breviores.[1]

[2] Arguitur[2] negative. Primum argumentum: grave in aequilibri pendens deorsum movetur usque ad locum eius naturalem per lineam curvam, eo quod brachium aequilibris descendens describit quandam quartam circuli.

[3] Secundum[3] argumentum: aqua et quilibet alius liquor naturaliter movetur ad locum terrae decliviorem, et in hoc descensu non describitur linea recta sed curva, cum terra sit rotunda seu ad rotunditatem accedat.

[4] Patet etiam quod fulgura, similiter nix et grando, haec omnia moventur deorsum per lineas curvas.

[5] Tertio[4] sic: si corpora naturalia moverentur per lineas rectas etc., hoc ideo esset quia talis incessus per rectam esset facilius quam per curvam, et per consequens velocius descenderet per rectam quam per curvam. Cuius falsi-

---

1 Tertio quaeritur] *om.* V2 1–2 appetant … naturalia] ab eorum locis naturalibus distantia remoto impedimento moveantur ad illa V1 3 negative] quod non V2 | pendens] pendes V2 5 quartam] unius *add.* V2 6 Secundum argumentum] Secundo sic V2

---

1 Cf. Witelo *Witelonis perspectivae liber quintus* prop. 5 (1983) 195: Natura agit in omnibus secundum lineas breviores. Cf. Albertus de Saxonia *Quaestiones in Aristotelis De caelo* III–IV q. 11, "Utrum gravia appetant descendere per lineas brevissimas" (2008) 512–516. Cf. Iohannes Dumbleton *Summa logica et philosophiae naturalis* cap. 14 q. 3 "Utrum grave appetat descendere per lineas rectas brevissimas vel alias obliquas" (cod. Vatican Pal. lat. 1056) ff. 95^{ra-vb}. Cf. Blasius Pelacanus de Parma *Questiones disputate super octo libros Physicorum* VII q. 5 (2017) 47: Et per hoc scriptum est IV° *Physicorum* quod omne corpus naturale, cum fuerit extra eius locum naturalem, appetit moveri ad ipsum per lineam brevissimam que recta est.

2 Adn. mg. V1: Primum argumentum.—Cf. Blasius Pelacanus de Parma *Questiones disputate super octo libros Physicorum* secunda redactio I q. 8 (ed. Biard) 4: Sciendum ulterius quod brachium equilibris in eius descensu describit quartam circuli.

3 Adn. mg. V1: Secundum argumentum.—Cf. supra IV q. 2 n. 54.

4 Adn. mg. V1: Tertium argumentum.—Cf. Blasius Pelacanus de Parma *Questiones disputate super octo libros Physicorum* secunda redactio I q. 8 (ed. Biard) 4: sciendum quod omne corpus naturale extra locum eius existens inclinatur moveri ad locum eius secundum lineam rectam, et velocius movetur per illam quam si per curvam moveret.

 | DOI:10.1163/9789004734968_015

tas probatur, quia si capiatur vesica plena aëre et ponatur in aequilibri, in alio vero | brachio ponatur aliud uter vacuum eiusdem gravitatis cum priori aëre V1 117va
circumscripto, tunc ut patet uter plenum movetur deorsum. Si vero removetur ab aequilibri, tunc uter plenum tardius descendit quam vacuum; igitur videtur quod uter plenum aëre habeat minorem resistentiam in spatio recto aut in descensu per lineam rectam quam per curvam. Et per consequens linea recta videtur maioris resistentiae quam curva in eodem media.

[6] Quarto[5] sic, et ostendo quod idem grave movetur velocius per curvam quam si per rectam moveretur, servata identitate medii et gradu. Arguitur: sit A unum grave quod positum in media regione aëris ac divisum suae naturae moveatur velocitate ut octo. Deinde ponatur in aequilibri; et quaeratur de ipso: aut hoc grave movetur aeque velociter sicut prius dum movebatur per rectam; et sic habetur propositum. Si vero dicatur quod moveatur tardius per curvam, tunc quaeratur in qua proportione tardius. Si dicatur quod in quadruplo tardius, tunc elongetur brachium illud in quo est pondus datum ad quadruplum. Et tunc illud pondus datum movebatur in quadruplo velocius quam prius, ut quarta conclusio Iordani[6] | *De ponderibus* vult. Et experientia idem docet. Igi- V2 51rb
tur ubi plus elongaretur quam ad quadruplum, adhuc movebitur velocius quam prius. Et per consequens movebitur velocius quam dum movebatur per rectam.

[7] In oppositum erat Philosophus in hoc quarto[7] et eius Commentator.[8] Similiter pro ista parte est omnis experientia.

[8] In hac quaestione erunt quattuor articuli. Primo problema. Secundo de quaesito. Tertio difficultates. Quarto ad rationes in oppositum.

1 capiatur] capivatur(?) V2 | vesica] inter V2 | plena] aquae *add. sed del.* V1 2 ponatur] aliud aër *add. sed del.* V1 | uter] uterque V2 3 vero] ponitur *add. sed del.* V1 5 in² *om.* V2 8 velocius *om.* V2 15 datum] et elongaretur *add.* V2 18 plus] magis V2 19 per²] lineam *add.* V2 22 Primo] In primo erit V2 | Secundo] In secundo V2 23 Tertio] In tertio V2 | Quarto] In quarto V2

5 Adn. mg. V1: Quarto.

6 Cf. Iordanus Nemorarius *Liber Jordani de ponderibus* prop. 4 (1952) 156. Cf. Blasius Pelacanus de Parma *Questiones disputate super octo libros Physicorum* VII q. 5 (2017) 47: per unam propositionem Iordani *De ponderibus*, grave appensum brachio movebitur in quadruplo velocius quam prius movebatur. Cf. melius Nicolaus Oresme *De proportionibus proportionum* (1966) 282: Quarta conclusio. Quavis velocitate demonstrata et qua placet proportione proposita, si proportio ignota a qua venit velocitas sit commensurabilis proportioni proposite investigare, quod si fuerit commensurabilis fiet nota.

7 Aristoteles *Physica* IV,1 208b11–2: fertur enim unumquodque in sui locum non prohibitum.

8 Averroës *In Phys.* IV com. 4 (1562) 122vb: ... quaerunt per istum motum ubi conveniens sibi, quae convenientia non est nisi per potentiam naturalem exsistentem in illo ubi.

[9] Ad primum articulum sit problema. Propter quod est quod interfectus per vulnus illatum sibi a Platone vel ab alio, quod si prima die vel secunda usque in tertiam diem supervenerit cadaveri homo qui intulit sibi vulnus illo, tunc sanguis emanabit exterius; et ob hoc multi executores iustitiae saepe venerunt in notitiam de illo qui interfecit, cum prius lateret.

V1 117vb [10] Ad[9] problema respondetur | quod ratio problematis potest esse quod hora qua interficiebatur fuerunt impressae species a vulnerante in vulnerato repraesentativae obiecti sub ratione mali, et adeo fuerunt sic species impressae quod adhuc conservatae sunt in interfecto ut in parte vitali ipsius. Sed sive ibi sit pars vitalis sive non, sunt tamen adhuc conservatae species, quae species vigorantur hora qua secunda vel tertia die praesentatur obiectum a quo dictae species fuerunt causatae. Et in hoc non inconvenit spiritum et sanguinem agitari, et consequenter fluere ad partem laesam tamquam ad partem debiliorem, sicut contingit faciem pallidam rubefieri hora qua quis verecundatur per delictum.

[11] Et[10] iuxta illam materiam reperitur aliud problema, propter quid est si pulsetur tympanum de corio lupini, et cum hoc pulsentur multa alia tympana quae sunt de corio ovium, illa non audiuntur, sed solum auditur tympanum de corio lupi.

---

1 Ad primum articulum sit] Quantum ad V2 3 supervenerit] homo *add. sed del.* V1 4 saepe *om.* V2 6 quod[2]] quia V2 8 repraesentativae] repraesentatione V1 | ratione] specie V2 9 conservatae] conservata V1 12 sanguinem] commoveri et *add.* V2 16 problema] ut *add.* V2 17 tympanum] timpanum *var.*; lupinum seu *add.* V2 18 ovium] aliorum animalium V2 | solum] solummodo V2

---

9 Cf. Aegidius Romanus *Quodlibeta* XXV q. 2, "Utrum ad praesentiam occisoris plaga occisi effundant sanguinem?" (1646) 341a: Dixerunt autem alii quod hoc fiat ex spiritibus existentibus vel remanentibus in corpore occisoris. Nam cum aliquis occiditur ab aliquo cum quodam rancore et torvis oculis occisus respicit occidentem et e converso, propter quod aliqui vapores vel aliqui spiritus exeunt ab occiso in occisorem et e converso. Dicunt autem isti quod quando supervenit occisor, quia naturale est quod pars redeat ad suum totum, aliqui vapores et spiritus ex corpore occisi ad occisoris redeunt cuius prius fuerunt et e converso. Et quia ex reditu talium vaporum sit quaedam commotio in corpore occisi; consequens est quod inde emanet sanguis. Cf. Aegidius Romanus *In tertium librum Sententiarum quaestiones* d. 33 q. 6 art. 1 (1632) 643b.

10 Cf. Albertus Magnus *De animalibus* XXII tr. 2 cap. 1 (1916–1921) II 1411: Timpana de coriis luporum pulsata insurdant audientes. Cf. Albertus Magnus *De animalibus* XXII tr. 2 cap. 1 (1916–1921) II 1407: Corium etiam eius ⟨leonis, sc.⟩ cum corio lupi positum, depilat ipsum et similiter alia coria. Cf. Albertus Magnus *De animalibus* XXII tr. 2 cap. 1 (1916–1921) II 1411: Odium autem quod est inter lupum et oves in omnibus membris est radicatum, ita quod cordae de visceribus ovium factae cum cordis de visceribus lupi factis permixtae non sonant. Cf. Nicolas da Foligno *Epistola*, cit. Lynn Thorndike, *Science and Thought in the Fif-*

[12] Et ratio problematis assignatur, quia sicut ovis in praesentia lupi multum timet lupum et remanet ovis in praesentia lupi sicut se in mortua, sic in corio lupino et in corio ovis relinquuntur quaedam species sive proprietates contrariae. Et haec de primo.

[13] Pro secundo articulo erunt primo evidentiae, deinde conclusiones et corollaria. Prima[11] evidentia sit haec: hoc verbum 'appeto' positum in titulo quaestionis, cum ponitur in aliqua locutione, facit terminum sequentem appellare formam et rationem.[12] Quare non sequitur 'equum Iohannis appeto, igitur appeto

---

1 sicut] sicuti V2 4 contrariae] etc. *add.* V2 5 Pro secundo articulo] Quantum ad secundum articulum V2 6 evidentia] etc. *add.* V2

---

teenth Century (1967) 313: Item dicit Albertus, 22° *De animalibus*, capitulo de lupo, cordas quae fiunt ex visceribus lupi et ovis nequaquam facere armoniam, quoniam odium eorum in vita est radicatum in omni membro quoad formam mixtionis. Cf. Bero Magni de Ludosia *Disputata super libros De anima* I q. 6 n. 115–116: Respondent aliqui quod complexiones lupi et ovis sint inimicae. Unde dicunt quod complexio lupi est valde contraria complexioni ovis. Et hoc probant per signum, quia pellis lupina supraposita pelli ovis corrodit eam. Secundo, quia cordae factae ex visceribus ovis et lupi tractae super eodem instrumento musicali numquam consonant.

11 Adn. mg. V1: Prima evidentia.—Cf. Blasius Pelacanus de Parma *Questiones super tractatus logice magistri Petri Hispani* De predicamentis III, 6 (2001) 275: Et iterum dicunt quod equum regis Francie tibi promitto, et equum qui cras generabitur tibi promitto. Hoc statim probant: si dedero equum regis Francie non amplius sum tibi obligatus de promisso. Vel igitur solvitur promissum, vel non. Si solvitur tibi promissum in dando tibi equum regis, evidenter sequitur quod equum regis tibi promisi. Et si non do tibi promissum, evidenter sequitur quod adhuc teneor tibi et sum obligatus, quod est contra casum. Et sic arguitur de equo qui cras generabitur, et propter istud argumentum conceditur quod dum equum tibi promitto, quemlibet equum ibi promitto. Omnes iste tamen sunt negande 'promitto tibi equum regis', 'promitto tibi Brunellum album' etc.

12 Cf. Ioannes Buridanus *Summulae de suppositionibus* 4.5.3 (1998) 83–84: Magna est differentia quantum ad appellationes inter verba significantia actus animae cognoscitivae, sicut sunt ista verba: 'cognoscere', 'intellegere', 'significare', 'supponere', 'promittere' et huiusmodi, et inter alia verba… Termini autem respectu verborum significantium huiusmodi actus animae, si sequantur illa verba et construantur cum eis tamquam terminantia transitum eorum, appellant rationes secundum quas significant ea quae significant. Si vero illi termini praecedant illa verba, non sic appellant illas rationes. Cf. Ioannes Buridanus *Summulae de practica sophismatum* cap. 4 soph. 14 (2004) 94: Dicendum est tamen quod bene differt ponere 'equum' a parte ante et a parte post, quia praedicta verba, propter appellationem rationis, confundunt quodammodo terminos sequentes, ita quod non possit fieri descensus ad singularia per propositionem disiunctivam. Cf. Guillelmus Heytesbury *De sensu composito et diviso* (1494) 2rb: Secundus modus est mediante termino habente vim confundendi, sicut sunt hiusmodi verva requiro, indigeo, praesuppono, incipio, desidero, cupio, volo, teneo, debeo, necessarium, semper, inaeternum, eternaliter, immediate, et sic de aliis.

equum Iohannis'; 'vinum[13] Gradense appeto, ergo appeto vinum Gradense'. Et ita non sequitur 'equum regis Scotiae tibi promitto, igitur promitto tibi equum regis Scotiae'.[14] Verum est quod logica communis[15] oppositum scribit, de quo non curo in praesenti.

[14] Secunda[16] evidentia: appetitus[17] naturalis non est aliud nec alia quam naturalis inclinatio rei non praesentis. Per hoc quod appetitus naturalis non V1 118ra est aliqua entitas superaddita appetenti, sed est ipsamet | res naturalis naturaliter se inclinans taliter ad hoc vel ad illud, ut puta ad hunc locum vel ad illum per rectam vel per curvam. Hoc dico quantum ad propositum spectat de appetitu.

V2 51va [15] Tertia[18] evidentia: cum hic sit sermo de motu naturali tantum, ideo | ausculta quid sit motus naturalis et quid non. Dico[19] quod motus naturalis est

---

1 Gradense] grande V2 | Gradense] grande V2 3 logica] loica *var.* V1; loyca *var.* V2 6 non praesentis] nunc praesentis *a.c.*; nec putetis(!) *in ima pag.* V1; appetentis V2 | Per hoc] Et per hoc tenendum est V2 6–7 non est] sit V1 7 superaddita] rei *add.* V2 | ipsamet] ipsa V2 11 motu] appetitu V2 12 ausculta] asculta *var.* V1 | sit] loc *add. sed del.* V1

---

13 Id est, Grado, Italia. De 'vino grandi' cf. Baverius de Baveriis *Consilia de Baveriis* (1521) f. 38vb: Vitet vinum grande et vinum dulce et grossum similiter turbidum.

14 Cf. Ioannes Buridanus *Summulae de practica sophismatum* cap. 4 soph. 14 (2004) 98–99: Ad aliam, dico quod equum regis debeo, et ad equum regis solvendum me obligavi. Ideo si illum equum emerem vel daretur mihi et solverem eum, ego essem quitus. Sed nec debeo equum regis nec me obligavi ad solvendum equum regis, nec tu potes petere equum regis, nisi sub disiunctione ad alios, quia illum non promisi secundum rationem secundum quam dicitur equus regis, sed secundum indefinitam rationem huius termini 'equus'.

15 Cf., e.g., Albertus de Saxonia *Logica* tr. II, cap. 7 (2010) 328: Quarta regula est: Ista verba habent vim confundendi terminos sequentes se confuse tantum, scilicet *appeto, desidero, promitto, opto, requiro, teneor, debeo* et similia… Propter hoc omnes istae consequentiae non valent 'Debeo tibi denarium, ergo denarium tibi debeo', 'Promitto tibi equum, ergo equum tibi promitto', 'Appeto vinum, ergo vinum appeto'… In omnibus enim istis arguitur a termino stante confuse tantum ad eundem stantem determinate, et tunc consequentia non valet.

16 Adn. mg. V1: Secunda.

17 Cf. Thomas de Aquino *Quaestiones disputatae de veritate* q. 25 a. 1 co.: Appetitus ergo naturalis tendit in ipsam rem appetibilem sine aliqua apprehensione rationis appetibilitatis: nihil enim est aliud appetitus naturalis quam quaedam inclinatio rei, et ordo ad aliquam rem sibi convenientem, sicut lapidem ferri ad locum deorsum.

18 Adn. mg. V1: Tertia.

19 Cf. Marsilius de Inghen (attrib.; pseudo-Ioannes Duns Scotus) *Quaestiones in VIII libros Physicorum Aristotelis* V q. 7 (1891) 238^{a-b}: ideo dicitur III *Ethicorum* quod naturale est quod fit a principio intrinseco, passo conferente vim agenti.

"qui fit a principio intrinseco passo conferente vim", secundum Aristotelem tertio *Ethicorum*,[20] et per oppositum habes quod sit motus[21] violentus.

[16] Nunc ponuntur conclusiones, habito hoc fundamento. Quia ubi possibile foret grave vel leve moveri subito ad locum eius naturalem, sic subito moveretur quod non temporaliter. Et hoc debet esse evidens, quia omne carens perfectione ad quam inclinatur, appetit illam subito obtinere patet.

[17] Prima[22] conclusio: motus gravis in aequilibri est naturalis, patet quia talis motus fit a principio intrinseco absque concursu extrinseci moventis; ergo talis est naturalis. Confirmatur, quia per illum motum grave acquirit locum eius naturalem; igitur talis motus est naturalis. Tenet consequentia, quia "omnis motus est de ratione termini ad quem."[23]

[18] Secunda[24] conclusio: grave extra eius locum exsistens per curvam appetit descendere. Patet ex praecedenti, quia postquam grave in aequilibri descendit per curvam naturaliter, et in descendendo describit lineam curvam; igitur ipsam appetit describere et per illam descendere. Quod autem grave in aequilibri descendat per curvam patet, quia grave brachio affixum movetur secundum quod brachium movetur; sed brachium in tali motu describit unam quartam circuli cuius circumferentia est arcus.

[19] Tertia[25] conclusio: grave extra locum eius naturalem exsistens non appetit descendere per rectam. Et ista est conclusio responsalis. Probatur conclusio, quia si appeteret descendere per rectam, ergo naturalis eius inclinatio

---

1 vim *om.* V2; *spat. vac. 6 littera* V2 | secundum] igitur *a.c.* s.l. V1 4 sic] sicut V2 5–6 perfectione] ante *add. sed del.* V1 9 acquirit] appetit V2 | eius] suum V2 11 ratione] genere V2 14 descendendo] descendit *add. sed del.* V1 16 grave] *mg.* V1 | affixum] aequilibris *add.* V2 21 eius *om.* V2

---

20 Aristoteles *Ethica ad Nicomachum* III,2 1110b15–6: Videtur utique violentum esse, cuius extrinsecus principium, nichil conferente vim passo.

21 Cf. Marsilius de Inghen *Quaestiones super libros De generatione* I q. 24 (1567) 209a: Tertio modo ⟨'violentus'⟩ dicitur proprie, quod fit a principio activo extrinsece non conferente vim passo. Et sic describitur tertio *Ethicorum*.

22 Adn. mg. V1: Prima conclusio.—Cp. Blasius Pelacanus de Parma *De ponderibus* (1952) 238–240: Ex hoc sequuntur tria corollaria: 1. Primum est: motus gravis in equilibri non est simpliciter naturalis. ... Tertium corollarium: Talis motus gravis in equilibri est medius inter motum naturalem et violentum. Cf. Leonardo da Vinci *Il codice atlantico* (2000) 1702: Il terzo [corollario] e che 'l moto grave nella bilancia e mezzo tra 'l naturale e violento.

23 Cf. Averroës *In Phys.* VI com. 48 (1562) 237rb: sed essentia duorum motuum qui ponuntur contrarii secundum contrarietatem eorum ex quo sunt, et secundum contrarietatem eorum ad quod sunt. Cf. Albertus de Saxonia (attrib.) *Quaestiones in Aristotelis Physicam* III q. 6 (1999) 505: secundum Commentatorem motus est de essentia termini ad quem.

24 Adn. mg. V1: Secunda conclusio.

25 Adn. mg. V1: Tertia conclusio.

esset ad rectam. Modo "quod naturale est non aliter assuescit", ut scribitur secundo *Ethicorum*.[26] Ergo postquam appetitus esset iste moveri per rectam, numquam moveretur per curvam—quod est falsum et contra conclusionem praecedentem.[27] Unde quod 'grave appetat descendere per rectam' idem est quod grave appetat moveri per lineam secundum rationem secundum quam V1 118rb illa linea est | recta. Et per hoc talis appetitus videtur esse limitatus ad hoc. Per hoc dixi in prima[28] evidentia quod hoc verbum 'appeto' facit terminum sequentem se appellare suam rationem. Et ideo in summa grave non appetit moveri per rectam nec per curvam, licet per rectam appetat moveri. Similiter per curvam appetat moveri, ex[29] quo sequitur quod grave non appetit moveri per rectam per quam appetit moveri. Patet hoc inspecta prima evidentia.

[20] Quarta[30] conclusio: ceteris paribus, velocius movetur grave per rectam ad eius locum naturalem quam si per curvam moveretur. Patet de se.

[21] Quinta[31] conclusio: ceteris paribus, grave potens indifferenter moveri per rectam et per curvam, sic per rectam movetur quod non per curvam. Patet, quia velocius movetur per rectam quam si per curvam moveretur, ceteris paribus; igitur per rectam in minore tempore attingit ad locum naturalem, ad quem naturaliter inclinatur, et in quo habet conservari quam si per curvam moveretur. Antecedens est conclusio praecedens.[32] Et consequentia tenet in vigore conclusionis praemissae.

[22] Sexta[33] conclusio: ceteris paribus, ubi elective moverentur gravia linea recta in eius motu praeferretur curvae. Patet ex dictis. Postquam per illam velocius moverentur. Sed ausculta diligenter de causa quod cum cetera paria supV2 51vb ponantur praeterquam quod linearum una ponitur recta alia curva, | et idem

1 assuescit] asuesit *var.* V1; asuescit *var.* V2 2 esset *rep. a.c.* V1 | iste *om.* V2 | per *rep.* V2 8 se *om.* V2 11 prima] principali V2 13 paribus] g *add.* V2 15 paribus] potest *add. sed del.* V1 | indifferenter *rep.* V2 18 igitur] quia V2 | locum] eius *add.* V2 | naturalem] in *add. sed del.* V1 21 conclusionis] propositionis V2 22 elective] ellective *var.* V2 23 eius *om.* V2

26 Aristoteles *Ethica ad Nicomachum* II,1 1103a23: neque aliud aliquod eorum que aliter innatta sunt, aliter utique assuescat. Cf. Blasius Pelacanus de Parma *Quaestiones in Aristotelis De anima libri* III q. 9 (1974) 140: modo quod naturale est non aliter assuescit *Ethicorum* secundo.

27 Cf. supra n. 18.

28 Cf. supra n. 13.

29 Adn. mg. V1: *Manicula*.

30 Adn. mg. V1: Quarta conclusio.

31 Adn. mg. V1: Quinta conclusio.

32 Cf. supra n. 20.

33 Adn. mg. V1: Sexta conclusio.

velocius movetur per rectam quam per curvam. Causa est "meliora applicatio mobilis ad rectam quam ad curvam". Et idem scribit Dormenton in octava parte suae *Physicae naturalis*.[34]

[23] Et[35] ex hoc sequitur alia conclusio: ceteris paribus, maior est proportio mobilis ad rectam quam ad curvam. Patet quia postquam velocius movetur per rectam quam per curvam, et proportio[36] velocitatum insequitur proportionem proportionum, etc. Patet propositum. Ex hoc sequitur quod linea curva, ceteris paribus, plus resistit quam recta. Quia non potest utique aeque bene fieri applicatio.

[24] Si autem quaeris: quae res est haec applicatio?

[25] Respondetur quod huiusmodi applicatio non est tertia entitas mobili et magnitudini superaddita, sed est mobile et magnitudo sic se habentes.

[26] Cui namque dubium est quod ignis distans a stuppa per miliare non aget in stuppam quam si applicaretur per contactum inflammaret | eam. Et V1 118va ideo dicet Philosophus in septimo huius[37] quod "omnis actio fit per contactum." Et declarabitur hoc in motu locali et in motu augmentationis, et similiter

1 movetur] per curvam quam *add. sed del.* V1 | quam] quam per curvam *mg.* V1 | Causa] quia V2 2 Dormenton] Dormeton V2 3 *naturalis om.* V1 6 per] rectam *add. sed exp.* V2 7 propositum] et *add.* V2 8 resistit] mobili *add.* V2 | utique] curvae V2 13 stuppa] stupa *var.* 14 stuppam] in *add.* V2; in *add. sed del.* V1 | quam] qua V2 | inflammaret] inflamaret *var.* | eam] *del.* V1

34 Iohannes Dumbleton *Summa logica et philosophiae naturalis* cap. 14 q. 3 "Utrum grave appetat descendere per lineas rectas brevissimas vel alias obliquas" (cod. Vatican Pal. lat. 1056) f. 95va: Ita intelligendum est in proposito quod forma gravis nec alia tendit appetitu suum determinate ut moveat per lineam rectam nec obliquam, sed ex applicatione unius medii vel alterius accedit sibi moveri. Cf. Blasius Pelacanus de Parma *Questiones disputate super octo libros Physicorum* VII q. 5 (2017) 47: melior applicatio agentis ad passum est occasio maioris accelerationis in effectu.

35 Adn. mg. V1: Alia conclusio.

36 Cf. Blasius Pelacanus de Parma *Questiones circa tractatum proportionum magistri Thome Braduardini* q. 10 (2005) 149: iuxta hoc sit prima conclusio: proportio velocitatum in motibus id est effectuum productorum vel producibilium insequitur proportionem suarum causarum. Cf. Blasius Pelacanus de Parma *Questiones disputate super octo libros Physicorum* VII q. 4 (2017) 38: omnis velocitas sequitur proportionem et proportio velocitatum sequitur proportionem proportionum.

37 Aristoteles *Physica* VII,2 245a15–b18. Cf. Aristoteles *De generatione animalium* (trans. Michaeli Scoti) II,1 734a3–4: quoniam nihil potest mouere nisi secundum contactum. Cf. Aristoteles *De gen. et cor.* I,6 322b22–24: Neque enim facere hec et pati possunt principaliter que non possibile ad inuicem tangere. Cf. Averroës *Libri Aristotelis De sensu et sensato* (1949) 35: Omnis enim actio et passio fit per contactum aut per medium aut sine medio. Cf. Blasius Pelacanus de Parma *Quaestiones de anima* II q. 25 (2019) 4: ut scribitur, *VII Physicorum* et *I De generatione*, omnis actio fit per contactum.

alterationis. Et[38] non vult aliud dicere Philosophus nisi quod, ad hoc quod agens agat et passum patiatur, requiritur optima applicatio agentis ad passum, sic quod "inter agens et passum non sit aliquod medium." Et hoc scribit ipse ad litteram in illo septimo,[39] ut videbitur.

[27] Nunc[40] sequuntur corollaria. Et primum est: quibusdam veris coassumptis, quanto linea est curvior, tanto plus resistit et ei mobile deterius applicatur. Hoc est evidens de se. Quia quanto linea est curvior, tanto plus recedit a linea recta per quam gravia naturaliter et velocius moventur ad eorum loca naturalia.

[28] Secundum[41] corollarium: eiusdem circuli maior portio plus curvatur minore, quam veritatem dico patere ex secunda conclusione primae *Dictionis* Almagesti.[42]

[29] Hiis[43] verbis sive veritatibus sic praemissis sit primum corollarium: quanto grave in aequilibri plus descendit, tanto tardius et tardius movetur. Patet, quia quanto plus descendit, tanto maiorem arcum describit eiusdem circuli. Sed per secundam veritatem de praemissis, maior portio circuli eiusdem plus curvatur minore. Ideo per primam veritatem, tardius et tardius movetur, cuius oppositum contingit de gravi descendente per lineam rectam, cuius motus semper est velocior et velocior.

---

5 veris] *spat.vac. 6 litt.* V1 8 naturaliter et *om.* V2 11 secunda] quinta V2 | Dictionis] additionis V2; dicciones *a.c. mg.* V1 13 sive veritatibus] seu veris V2; suae veritatibus *a.c.* V1 | sic *om.* V2 | primum] tertium V2 15 arcum] archum *var.* V2

---

38 Adn. mg. V1: *Manicula.*

39 Aristoteles *Physica* VII,2 245a21–23: Si quidem igitur sensibiles quidem sunt passiones, per hec autem alteratio fit, hoc igitur manifestum est quod patiens et passio simul, et horum nullum est medium.

40 Adn. mg. V1: Primum corollarium.—Cf. Blasius Pelacanus de Parma *Questiones disputate super octo libros Physicorum* secunda redactio I q. 8 (ed. Biard) 4: Propter quod dicendum est secundo (secumdo *ed.*) quod quanto aliqua linea est curvior, tanto per illam grave tardius descendit, et quanto minus curva, tanto velocius movetur.

41 Cf. Blasius Pelacanus de Parma *Questiones super perspectiva communi* III q. 3-B (2009) 349: Respondetur quod huius non est alia causa nisi quia eiusdem circuli maior portio plus curvatur minore, ob hoc minor portio rectior est, et per hoc portio maioris circuli. Vult enim quinta conclusio prime dictionis Almagesti quod arcus non sunt proportionales suis chordis. Esta enim proportio arcuum maior quam proportio chordarum.

42 Ptolemaeus *Almagestum* dictio I cap. 9 (1515) 6v: erit proportio chordae longioris ad chordam breviorem minor proportione arcus chordae longioris ad arcum chordae brevioris. Cf. Regiomontanus *Epitome Almagesti* (2021) I prop. 12. Arcuum inequalium in semicirculo maioris ad minorem est proportio maior quam corde maioris ad cordam minoris.

43 Adn. mg. V1: Primum corollarium.

[30] Secundum corollarium:[44] quanto grave longiori brachio appensum extiterit, tanto plus gravitabit et deorsum velocius movebitur. Patet, quia quanto brachium est longius, tanto minus oblique descendit appensus; quare, etc. Et hic ausculta quod in corollariis dicitur quod deorsum velocius movebitur. Hic tangitur comparatio huius ponderis appensi in brachio longiori ad ipsum brachium ubi in brachio longiori permaneret. Alioquin corollarium per hac parte falsum esset, propter corollarium[45] quod praecessit.

[31] Tertium corollarium:[46] in infinitum potest gravitare granum milii in aequilibri. Patet, quia elongando brachium quantumlibet a centro motus aequilibris, semper granum milii minus et minus oblique descendet.

[32] Quartum corollarium:[47] longe plus potest gravitare | grave in aequilibri V1 118vb
quam ubi in recta linea extra locum eius quiesceret. Patet, quia non tantum potest gravitare in aequilibri | quin magis possit gravitare per maiorem remo- V2 52ra
tionem ipsius a centro aequilibris; sed in recta non potest hoc fieri; quare, etc.

[33] Quintum corollarium:[48] non contingit grave ita velociter moveri in aequilibri, ceteris paribus, sicut ubi per rectam lineam moveretur. Patet, quia non contingit, ceteris paribus, lineam rectam sic resistere descendenti per ipsam quin magis linea curva quantumcumque ad rectitudinem accedat per hoc. Quia non contingit ita bonam fieri applicationem ipsius mobilis ad spatium curvum sicut si per rectam descenderet.

[34] Ex hiis ulterius potest haberi causa huius, scilicet propter quid pondus supra caput tuum minus ponderat quam si illud sustentares in brachio erecto, cum velocius tunc moveatur quam per curvam.

---

1 Secundum] Quartum V2 | grave] cuius *add. sed del.* V1 5 huius] *mg.* V2 8 Tertium] Quintum V2 | milii] milei *var.* V1 9 brachium] granum milii V2 10 granum milii *om.* V2 11 Quartum corollarium] Quare sequitur corollarium sextum V2 12 quiesceret] naturalem descenderet V2 15 Quintum] Unum V2 22 erecto *om.* V1

---

44 Adn. mg. V1: Secundum corollarium.

45 Cf. supra n. 29.

46 Adn. mg. V1: Tertium corollarium.—Cf. Blasius Pelacanus de Parma *De ponderibus* (1952) 264: Et sequitur unum corollarium: Quod stat granum milii in brachio equilibris positum sua gravitate sursum impellere pondus in millecuplo gravius. Patet, potest enim granum milii ad duplum eius extendi, et erit quadruplum et sic quantumlibet. Quo facto applicetur brachio equilibris equalium brachiorum in continuum et directum. Alıud autem pondus quantumcumque vis pendeat in altero brachiorum sic quod medium ponderis iacentis quantumlibet volueris plus distet a centro motus altero pondere. Et tunc patet propositum.

47 Adn. mg. V1: Quartum corollarium.

48 Adn. mg. V1: Quintum corollarium.

[35] Sextum corollarium:[49] pondus oblongum aeque distans horizonti cum brachio cui affixum est, plus ponderat quam si in termino brachii ad circumvolubile verteretur. Patet, quia totum oblongum continuatum cum brachio et quaelibet eius pars describit quartam maiorem quam ubi in circumvolubili tangeret. Et per consequens, plus ponderat secundum illum situm quam secundum istum. Et hoc de secundo articulo.

[36] Ad tertium articulum sit prima difficultas:[50] utrum gravius alio in aequilibri sit gravius illo absque aequilibri.

[37] Quod sic arguitur, quia ubi sic esset scientia de ponderibus videretur plena erroribus per hoc quod faceret nos errare, quod non est credendum cum adinventa sit ut per eam comprehendamus proportiones corporum in gravitate et levitate.

[38] Ad oppositum arguitur, quia uter plenum aëre positum in uno brachio aequilibris et aliud uter vacuum, sic se habent in aequilibri quod plenum plus gravitat reliquo vacuo aut quam si ipsummet vacuum foret. | V1 119ra Ubi vero ponatur in sphaera aëris cum alio, tunc reliquum velocius movetur; plenum vero tardius cuius oppositum erat in aequilibri.

[39] Respondetur quod gravius alio in aequilibri est gravius illo extra aequilibrem, semper servata identitate medii in raritate et densitate, servatis etiam aliis quae concurrunt ad velocitatem motus, ut figura etc.

[40] Et tunc ad argumentum[51] in oppositum, dico quod uter plenum gravius est alio. Et ad argumentum in oppositum, 'quia in aëre vacuum uter velocius movetur alio absque aequilibri', concedatur. Sed hoc ideo est quia non servatur identitas figurae. Unde uter plenum habet figuram latam quae multum est inepta ad dividendum medium. Uter vero vacuum haberet eius latera insimilis complicata, propter quod abilius est ad dividendum medium quam si esset plenum.

---

1 Sextum] Octavum V2 | oblongum] oblungum *var.* | horizonti] orizonti *var.* 2 si *om.* V1 4 maiorem] *mg.* V1 5 tangeret] ponderet V2 | situm] scitum V1 6 Et hoc de secundo articulo *om.* V2 7 Ad] Transeo *praem.* V2 | articulum] et *add.* V2 8 illo] eodem V2 9 Quod ... esset] Et apparet quod non quia V2 | videretur] esse *add.* V2 10 plena] de *add. sed del.* V2 16 alio] reliquo V2 18 illo] eodem V2 21 plenum] aëre *add.* V2 24 plenum] aëre *add.* V2 25 ad] a V1 26 complicata] implicata V2 26–27 esset plenum] plenum foret V2

---

49 Adn. mg. V1: Sextum corollarium.

50 Adn. mg. V1: Difficultas ⟨⟨prima⟩⟩.

51 Cf. supra n. 38.

[41] Secunda[52] difficultas: utrum homo possit levare maius pondus mediante aequilibri quam sine aequilibri, vel econtra. Et loquor de aequilibri brachiorum aequalium, quia ubi esset sermo de aequilibri brachiorum inaequalium, contingeret hominem debilem levare sursum castrum domini,[53] ut videlicet dicere quod contingit hominem debilem in aequilibri brachiorum inaequalium esse maioris gravitatis quam castrum domini, ut bene scis per aliqua problema[54] in quaestionibus primi huius explicata, quia per longationem unius brachiorum. Sit igitur sermo de aequilibri aequalium brachiorum.

[42] Et ponatur ista conclusio: homo maius pondus non potest levare aequilibri | quam absque aequilibri. Declaratur quia sine aequilibri potest homo V2 52rb levare lapidem maioris gravitatis quam ille homo sit, ut patet de plano. Tu vides enim quod unus homo levat et fert duos homines insimul sine aequilibri; tamen idem homo non potest levare mediante aequilibri pondus nisi fuerit minoris gravitatis quam ipse sit. Unde si homo fuerit in gravitatem decem ponderum, et fuerit A lapis decem ponderum, si iste homo ponatur in uno brachio, et lapis in alio, non fiet motus, ut scis; quia a proportione aequalitatis non fit motus, et a fortiori non poterit levare gravius ipso.

[43] Et haec sufficiant de tertio articulo.

[44] Ad quartum articulum, ad rationes in oppositum patet solutio. Et hoc de tota quaestione etc. | V1 119rb

---

1 pondus] q *add. sed del.* V1 6 quam] totum *add.* V2 7 huius explicata] complicata V2 | longationem] ellungationem V2 9 non *om.* V1 11 ut patet de plano] *om.* V2 12 insimul sine] et fert absque V2 13 fuerit] ipse sit V2 15 brachio] aequilibris *add.* V2 16 alio] reliquo V2 | ut … motus] *om.* V2 18 sufficiant *om.* V2 19 Ad] Quantum *praem.* V2 19–20 Et … etc. *om.* V2

---

52 Adn. mg. V1: Difficultas secunda.

53 Cf. Blasius Pelacanus de Parma *Questiones disputate super octo libros Physicorum* secunda redactio I q. 8 (ed. Biard) 5: Verum est quod diversitas est ex parte gravitatis situalis, propter quod in operibus necesse est reperire libram feri positi in brachio equilibris posse substentare vel levare castrum domini positum in alio brachio. Patet hoc quia clarum est quod castrum domini non est infinite gravitatis, et sic gravitas istius castri ad libram feri est determinata proportio.

54 Blasius Pelacanus de Parma *Questiones disputate super octo libros Physicorum* secunda redactio I q. 8 (ed. Biard) 4–5.

7

Liber IV q. 7: Utrum rarefactio sit possibilis

V1 = Vat. lat. 2159 ff. 127vb–130rb; V2 = Vat. lat. 3012 ff. 57va–59rb

[1] Septimo quaeritur utrum rarefactio sit possibilis.[1]

[2] Arguitur[2] negative. Condensatio non est possibilis, igitur nec rarefactio. Consequentia patet, quia ubi aliquid rarefieret et nihil condensaretur, statim corpora suis dimensionibus se penetrarent,[3] ut patet intelligenti. Et antecedens patet, quia ubi posset aliquid condensari, illud inter alia elementa erit ignis. Sed probo quod non, quia cum ignis sit uniformis per totum; ita poterit condensari versus convexum sicut versus concavum eius. Condensetur igitur sicut possi- V2 57vb bile incipiendo a convexo eius, et sit condensans aequaliter, | applicatum toti convexo et cuilibet eius parti; et tunc in hoc sequitur quod totum convexum ignis incipit distare orbiculariter a concavo caeli, igitur datur vacuum.

[3] Secundum argumentum:[4] si rarefactio esset possibilis, tunc esset concedendum aliquid esse rarefactum et non esse rarefactum. Consequens falsum,

---

1 Septimo quaeritur] *ord. inv.* V2 4 Et] Sed V2 5 elementa *om.* V2 6 cum *om.* V2 7 sicut[2]] si est V2 11 Secundum argumentum] Secundo arguitur V2 12 esse *om.* V2

---

1 Cf. Guillelmus de Ockham *Quaestiones in libros Physicorum Aristotelis* q. 92 "Utrum rarefactio vel condensatio sit possibilis" (1984) 645–646; Ioannes Buridanus *Quaestiones super Physicorum* IV q. 11 "Utrum rarefactio et condensatio sint possibiles vel utrum possibile sit aliquid rarefieri vel condensari" (2016) 299–305; Albertus de Saxonia (attrib.) *Quaestiones in Aristotelis Physicam* IV q. 13, "Utrum condensatio et rarefactio sint possibiles" (1999) 708–718. Resp. Aristoteles *Physica* IV,9 216b21–4: Sunt autem quidam qui per rarum et per densum opinamur manifestum esse quod sit vacuum. Si quidem enim non est rarum et densum, neque coire et inculcari possibile est.

2 Adn. mg. V1: Primum argumentum.

3 Resp. Aristoteles *Physica* IV,5 211b24–6: Neque enim simul augmentari necesse est locum, neque puncti esse locum, neque duo corpora in eodem loco, neque spatium aliquod esse corporeum; cf. Albertus de Saxonia (attrib.) *Expositio in Aristotelis Physicam* I cap. 7 (1999) 185: Hic Aristoteles solvit tertiam rationem quae erat: sequitur quod corpora se penetrarent. Ipse iterum negat consequentiam; sed hoc bene sequitur, si locus poneretur spatium seu dimensio separata contenta inter latera corporis continentis, sed locus non est ponendus talis dimensio, sicut prius probatum est, et adhuc, posito quod locus esset corpus continens, non sequeretur penetratio corporum, ex eo quod corpus continens ponitur esse contentum.

4 Adn. mg. V1: Secundum.

 | DOI:10.1163/9789004734968_016

quia contradictoria non verificantur de eodem.[5] Probatur argumentum, quia capiatur unum pedale terrae cuius una medietas rarefiat per horam, et reliqua medietas per idem tempus condensetur, sic quod ista duo in eodem tempore aequales latitudines acquirent raritatis et densitatis. In hoc casu patet quod hoc pedale est rarefactum, quia plusquam medietas est rarefacta; igitur hoc totum est rarefactum. Tenet argumentum, quia ut scribitur primo[6] huius, "denominatio fit ab excessu." Et antecedens patet. Probatur quod idem sit non-rarefactum, quia quantum dat de denominatione raritatis acquisita, tantum videtur auferre de denominatione densitatis acquisita, ubi latitudo densitatis adaequetur raritati. | Sed sic est in casu dato; igitur, etc. V1 128ra

[4] Tertium argumentum:[7] si quaestio haberet locum, tunc posset esse possibile quod hoc per horam fieret albius et albius, et per eandem horam continue haberet minus et minus de albedine. Deduco argumentum, quia sit A unum pedale cuius medietas sit alba, et reliqua nigra. Et constat quod in hoc casu A nec est album nec nigrum. Nunc si rarefactio est possibilis, rarefiat pars alba; et cum hoc continue remittatur albedo in parte albam. Et tunc patet propositum. Quia in fine horae est verum dicere quod 'hoc est album', quia plus quam medietas est alba. Et continue est verum dicere quod 'hoc est albius quam prius', quia denominatio sequitur partem.[8] Et quia continue maior et maior pars erit alba, sic proportionaliter continue erit denominatio secundum albedinem maior et maior. Patet aliud, scilicet quod continue illud habebit minus et minus de albedine.

[5] Quartum argumentum:[9] si rarefactio esset possibilis, tunc poterit aliquid rarefieri. Aut igitur uniformiter aut difformiter hoc rarefit. Sed probatur quod non, quia si aliquid potest uniformiter rarefieri, tunc illud in eius rare-

1 de] o *add. sed del.* V2 3 per *om.* V1 4 acquirent] aquirant *var.* V2 | patet] apparet V2 5 hoc *om.* V2 6 huius] *Physicorum* V2 8 videtur] dare de de *add. sed del.* V1 9 densitatis] densitas V2; densitatis *a.c.* V2; dempsitas *var.* V1 11 Tertium argumentum] Tertio V2 12 horam] hora V2 | continue] fieret *add. sed exp.* V2 13 argumentum *om.* V2 15 si] plusquam V1 | pars alba] per albam V2 16 cum *om.* V2 18 prius] fuit *add.* V2 19 sequitur] insequitur V2 | alba] illa V2 20 continue erit *om.* V2 21 scilicet] videlicet V2 23 Quartum argumentum] Quarto V2

5 Cf. Albertus Magnus *In Priora an.* I tr. I c. 8 (1890); *Opera omnia* II (1890) 470A: contradictoria non verificantur de eodem simul.

6 Cf. immo Aristoteles *De anima* II,4 416b23: Quoniam autem a fine omnia appellari iustum est. Cf. Thomas de Aquino *Summa theologiae* IaII q. 25 a. 2 arg. 1: Sed denominatio fit a potiori, ut dicitur in II *De anima*. Cf. Iacobus Zabarella *Commentarium in Meteora* (1602) 1001: denominatio non fit, nisi ab excessu.

7 Adn. mg. V1: Tertium argumentum.

8 Id est, "denominatio sequitur partem potiorem"; cf. supra. n. 3.

9 Adn. mg. V1: Quartum argumentum.

factione acquiret quantitatem infinitam. Et tenet argumentum, quia si aliquid rarefieret uniformiter, igitur omnia duo puncta aequaliter distantia ante rarefactionem per amplius distabunt quam prius per actum rarefactionis. Et quia infinita sunt puncta in corpore quod augeretur, sequitur quod infinitae quantitates acquirentur non communicantes, ut velim dicere quod inter quaecumque puncta aequaliter distantia cadunt aequales quantitates non-communicantes. Et ideo si hoc rarefit uniformiter, quaelibet quantitas istius situata inter duo puncta rarefit et acquirit novam quantitatem, quae quantumcumque parva sit, habeo quod infinitae quantitates acquirentur non-communicantes et aequales inter se. Modo tales videntur facere quantitatem infinitam. Si vero rarefactio sit difformis, ad idem inconveniens deduceretur. Et signabo quantitatem exsistentem inter duo puncta valde propinqua versus extremum quod remissius rarefit. [V1 128rb] Et procedit argumentum. |

[6] Quintum[10] argumentum: et probatur quod nihil possit condensari. Quia videmus notitia experimentali, quod dato aëre in folliculo cuius omnia foramina sunt bene obstructa, ille aër per quamlibet pondus suprapositum folliculo non potest condensari, eo quod latera non fiunt propinquiora invicem quam prius. Et sic dicendum est de igne, et pariformiter de quolibet alio corpore naturali. [V2 58ra] Cum | igitur nihil possit condensari, sequitur applicando proposito quod nihil potest rarefieri.

[7] Sextum argumentum:[11] si quaestio haberet locum, tunc rarefactio et condensatio possent simul intendi in eadem parte materiae. Et sic possibile esset idem simul secundum eandem partem rarefieri et condensari—quod est impossibile, licet multi medicorum[12] non habeant hoc pro inconvenienti. Sed probatur falsitas huius, quia ubi rarefactio et condensatio possent simul intendi etc., tunc contingeret idem mobile moveri motibus contrariis, ut simul et semel ascendere et descendere; et sic idem fieret centro mundi propinquius et pro tunc continue plus a centro distaret, ut patet intelligenti, cum ad rarefactionem

---

2 puncta] poncta *var.* V2 6 aequales] les *add. seu rep.* V2 14 argumentum] sic V2 15 videmus] quod *add.* V2 | quod *om.* V2 | folliculo] foliculo *var.* | omnia *om.* V2 16 obstructa] ostruta *var.* V1 21 Sextum argumentum] Sexto V2 25 intendi *om.* V2 27 mundi *om.* V2 27–28 pro tunc *om.* V2

---

10 Adn. mg. V1: Quintum argumentum.—Cp. Albertus de Saxonia (attrib.) *Quaestiones in Aristotelis Physicam* IV q. 13 (1999) 710: Tertio. Hoc idem patet de vesica repleta aere, undique obstructa, quae si comprimatur, aer in ipsa condensatur, et non nisi per motum localem.

11 Adn. mg. V1: Sextum argumentum.

12 Resp. Avicenna *Canon medicinae* I fen 1 doc. 3 cap. 3 (1564) 19b: substantia carnium ⟨viri⟩ est rarior. Cf. Iacobus Forliviensis *Expositio super primo Canonis Avicenne cum questionibus eiusdem*, q. 37 (q. 33 *Puccinotti*) "Utrum caro mulieris sit rarior quam caro viri" (1495) f. 21va (bis).

sequatur motus localis et ad condensationem sequitur motus oppositus. Sed probatur prima consequentia, videlicet quod 'si rarefactio sit possibilis, quod tunc rarefactio et condensatio possent simul intendi' etc. Quia si aliquid rarefit, cum rarefactio insequatur operationem calidi et condensatio operationem frigidi, igitur si contingit idem simul et secundum eandem partem calefieri et frigefieri, habetur propositum. Sed quod hoc sit possibile patet, quia sit A unum summe calidum et B unum summe frigidum; et applicentur invicem in distantia pedali, et hoc in sphaera aëris; tunc sic A agit in B, igitur agit in medium, igitur medium calefit. Et ex alia parte B agit in A, igitur agit in medium datum, igitur medium frigefit. Nec est dicendum quod medium calefit secundum unam partem et frigefit secundum aliam, quia de ratione cuiuscumque agentis est agere in modum sphaerae.[13] Et sic patet illud medium intendi in caliditate et frigiditate secundum eandem partem. Et per consequens rarefactio et condensatio in eadem parte intenduntur.

[8] Septimum argumentum:[14] si rarefactio esset possibilis, posset igitur aqua in tantum rarefieri quod in ea submergeretur quodlibet grave, et per consequens | navis vacua submergeretur. Et quia latitudo raritatis latitudini densita- V1 128va
tis aequatur, tunc in tantum posset aër condensari. Et sic aër posset in tantum condensari quod per aërem nataret homo et navis onerata; consequenter posset per ipsum deferri.

[9] In oppositum erat Philosophus in hoc quarto capitulo de vacuo.[15] Saepe enim dicebat[16] aërem posse quamlibet rarefieri et condensari.

[10] In hac quaestione erunt quattuor articuli. Primo problema. In secundo de quaesito. Tertio difficultates. Quarto ad rationes in oppositum.

---

3 Quia] Quare V2 15 Septimum argumentum] Septimo V2 16 submergeretur] sumergeretur *var.* V2 18 aequatur] adaequatur V2 | sic] de pari *add.* V2 19 onerata] honerata *var.* 21 Saepe] Saepius V2 23 Primo] In primo pro V2 23–263.1 In secundo … In secundo articulo] *spat.vac. 9 lin.* V2

---

13 Cf. Robertus Grosseteste *De luce* (2011) 79: Rediens igitur ad sermonem meum dico quod lux, multiplicatione sui infinita in omnem partem equaliter facta, materiam undique equaliter in formam sphericam extendit. Cf. Nicolaus Oresme *Questiones super Physicam* VII q. 3 (2013) 731: Tertio, confirmatur auctoritate perspectivorum naturalium, qui dicunt quod agens multiplicat virtutem suam circa se ad modum sphere.

14 Adn. mg. V1: Septimum argumentum.

15 Aristoteles *Physica* IV,9 216b21–4: Sunt autem quidam qui per rarum et per densum opinamur manifestum esse quod sit vacuum. Si quidem enim non est rarum et densum, neque coire et inculcari possibile est.

16 Cf., e.g., Aristoteles *Physica* IV,9 217a32–3: Similiter igitur et cum multus aer sit in minori fit corpore, et ex minori maior: potentia cum sit materia fit utraque.

[11] Pro[17] primo articulo sit problema: propter quod est quod praegnantes desiderantes, gratia exempli, cerasa, nec potentes adimplere quod optant, si contingat eas apponere manum alicui parti sui corporis, ut applicare manum faciei aut alteri parti, in consimili parte habebit foetus cum venerit ad lucem maculas quae erunt similes cerasis. Et sic dico si desiderassent vinum rubeum nec potuissent, hae illud habuisset foetus maculas in modum vini rubei.

[12] Ad problema respondetur quod ubi problema sit universaliter verum, causa huius potest esse quia hora qua praegnans optat quod hae non potest, in eam insurgit fortis aestimatio de absentia obiecti desiderati, de qua aestimatione fit commotio sanguinis et spiritus, de ratione cuius commotionis in parte foetus tamquam in parte melius disposita imprimitur idolum proportionatum obiecto quod desiderabatur, quae pars est illa quae correspondent parti corporis praegnantis percussae manu ipsius. Et haec ratio valet tantum quantum potest valere in proposito. Verumtamen multi[18] dicunt problema non esse universaliter verum. Sed forte isti sunt specialiter negantes communem modum loquendi, qui communis modus loquendi habet aliquid veritatis. Scriptum enim est septimo *Physicorum*[19] "fama quam omnes famam non ex toto perit." Et idem scribitur libro *De somno*, capitulo de divinatione[20] etc. Et hoc
V1 128[vb] de primo. |

1–2 desiderantes] dexiderantes *var.* V1; *deest* V2 2 cerasa] cerexa *var.* V1; *deest* V2 5 desiderassent] dexiderasent *var.* V1; *deest* V2 9 aestimatio] extimatio *var.* V1; *deest* V2 | desiderati] dexiderati *var.* V1; *deest* V2 11 foetus] fectus *var.* V1; *deest* V2

17 Cf. Blasius Pelacanus de Parma *Questiones disputate super octo libros Physicorum* VII q. 3 (2017) 27: Et primo patet in pregnantibus que dum optant quod habere non possunt sicut cerasa aut alia comestibilia aut potabilia et hora extimationis tangant corpora earum manu, convincuntur quod fetus productus ad lucem habebit ad eius personam maculam consimilem ei de quo erat extimatio. Cf. Petrus Pomponazzi *De naturalium effectuum causis sive de incantationibus* (1556) 54: mulier praegnans ex vehementi desiderio ut pote ciceris similitudine ciceris foetum signatum parit. Cf. Marsilio Ficino *De amore* VII, 8 (1956) 252: Praegnantes saepe mulieres vinum, quod avidissime cupiunt, vehementer excogitant. Vehemens cogitatio spiritus movet interiores atque in iis excogitate rei pingit imaginem. Isti sanguinem movent similiter et in molissima fetus materia vini imaginem exprimunt.

18 Non invenimus.

19 Cf. immo Aristoteles *Ethica ad Nicomachum* VII,14 1153b27–8; *Auctoritates Aristotelis* 241 (126): Fama quam omnes famant, omnino non deperditur.

20 Aristoteles *De somno et vigilia* 1 462b14–17. Cf. Averroës *In De sompno et vigilia* 1 (1949) 95: Et res que sunt famose apud omnes aut sunt necessarie secundum totum aut secundum partem: impossibile enim est ut famosum sit falsum secundum totum; cf. *Auctoritates Aristotelis* 204 (101): Famosum secundum rationem non est falsum.

[13] In secundo articulo, condescendo ad determinationem, notando quod si per actum rarefactionis acquiratur aliqua nova quantitas, vel non acquiratur sed solum partes fiant distantiores quam prius, non calculo in praesenti. Et pariformiter dico de condensatione, an per actum eius deperditur nova quantitas, vel partes solum fiant propinquiores quam prius, demittatur in praesenti. Insuper an quodlibet corpus naturale sit | condensabile an ne, non curo in praesenti. Hoc dico pro tanto quia aliqui[21] namque estimant nullum corpus simplex posse condensari, nec per consequens rarefieri, dicentes corpora simplicia non esse porosa, et per consequens partes talium corporum non possunt fieri propinquiores quam sint solum finaliter. Isti[22] volunt quod corpora mixta sint condensabilia propter corpora mixta habere porositates propter quas partes possunt fieri propinquiores. V2 58[rb]

[14] Similiter an possit fieri condensatio absque alteratione praecedente vel non—haec et praedicta per difficultates declarabo. Accedo igitur ad propositum.

[15] Et[23] sit prima conclusio: possibile est materiam informatam forma terrae vel aquae informari post hoc forma ignis. Patet conclusio, quia ut scribitur secundo *De generatione*,[24] "potest quodlibet elementum immediate in alterum transmutari". Et sic terra potest transmutari in ignem et econtra.

---

4 eius] condensationis V2 5 vel] an V2 | solum] solummodo V2 | demittatur] omittatur V2 7 estimant] extimant *var.* 9 talium] huiusmodi V2 11 corpora mixta] ista V2 17 Patet] haec *add.* V2

---

21 Cf. Albertus de Saxonia (attrib.) *Quaestiones in Aristotelis Physicam* IV q. 13 (1999) 714: Tertio est sciendum quod rarefactibilium et condensabilium quaedam sunt mixta et quaedam simplicia elementa ... difficile est salvare ista de corporibus simplicibus non porosis, propter hoc quod aliquibus difficile videtur quomodo talia condensantur et fiunt minora quam prius erant sine penetratione corporum et subingressu unius in reliquum.

22 Cf. Albertus de Saxonia (attrib.) *Quaestiones in Aristotelis Physicam* IV q. 13 (1999) 714: Mixta solent poni porosa habentia poros repletos quibusdam corporibus subtilibus de facili cessibilibus et cedentibus in condensatione et approximatione partium ad invicem talium corporum, et in rarefactione talium quaedam corpora subtilia, sicut aer, intrant tales poros. Propter huiusmodi poros non est difficile salvare rarefactionem et condensationem corporum mixtorum.

23 Adn. mg. V1: Prima conclusio.

24 Aristoteles *De gen. et cor.* II,4 331a13–16: Quoniam quidem igitur omnia innata sunt ad invicem transmutari, manifestum.

[16] Secunda[25] conclusio et responsalis: rarefactio est possibilis, et similiter condensatio. Patet prima pars conclusionis, quia materia quae nunc informatur forma terrae potest informari forma ignis. Aut igitur ista materia remanebit sub tanta densitate sub quanta est. Et hoc non quia tunc daretur ignis ita densus sicut est terra in sua dispositione exsistens; relinquitur igitur quod ista materia non stabit sub tanta densitate, et per consequens rarefiet. Et sic habetur conclusio vera in parte.

[17] Secunda pars patet, quia quam bene potest terra in ignem transmutari ad sensum datum, tam bene potest ignis transmutari in terram. Et in hoc quaeratur: aut materia ignis stabit sub tanta raritate sub quanta est. Et hoc non, quia
V1 129ra tunc daretur terra | sub tanta raritate sub quanta est ignis in sua dispositione exsistens. Relinquitur igitur istam materiam ignis condensari. Et patet propositum.

[18] Tertia[26] conclusio: non ex hoc dicitur aliquid rarefieri quia illud occupet maiorem locum quam prius. Patet conclusio, quia per actum augmentationis repertae in animali, animal occupat maiorem locum quam prius occupabat, et tamen hoc non est per rarefactionem; immo stat continue animal fieri densius et densius, et cum hoc occupare maiorem locum et maiorem.

[19] Quarta[27] conclusio: non ex hoc dicitur aliquid rarefieri quia materia eius fiat in maiori loco quam prius. Patet conclusio, quia ubi de corpore sphaerico fiat tetrahedron quod est corpus quattuor superficierum, materia huius corporis occupabit maiorem locum quam prius absque rarefactione; immo stat ipsum aliqualiter condensari et occupare maiorem locum quam dum erat sphaericum.

---

1 et *om.* V2 | responsalis] secunda *add.* V2 5 sicut] sicuti V2 12 istam *om.* V2 12–13 Et patet propositum *om.* V2 16 occupabat *om.* V2 19 aliquid *om.* V2 20 corpore *om.* V2 21 tetrahedron] tetracedron *var.* | superficierum] istud c *add. sed del.* V1

---

25 Adn. mg. V1: Secunda conclusio.—Cf. Ioannes Buridanus *Quaestiones super Physicorum* IV q. 11 (2016) 301: Et omnino manifestum est pluribus experientiis et concessum quod rarefactio est possibilis et similiter condensatio illo modo.

26 Adn. mg. V1: Tertia conclusio.—Cp. Marsilius de Ingen *Abbreviationes super octo libros Physicorum Aristotelis* IV tr. 2 (1521) f. 19ra: Et per oppositum condensatio dicitur mutatio qua corpus redditur minus quam prius erat sine egressu alicuius extrinseci quod prius erat in eo. Cp. Marsilius de Inghen (attrib.; pseudo-Ioannes Duns Scotus) *Quaestiones in VIII libros Physicorum Aristotelis* VI q. 14 (1891) 113a: condensatio dicitur quando aliquod corpus sit minus sine recessu alicuius partis ab eo.

27 Adn. mg. V1: Quarta conclusio.—Cp. Marsilius de Ingen *Abbreviationes super octo libros Physicorum Aristotelis* IV tr. 2 (1521) f. 19ra: Proprie autem rarefactio est mutatio qua corpus redditur maius quam prius erat sine ingressu extrinseci seu ab extrinseco. Cp. Marsilius de Inghen (attrib.; pseudo-Ioannes Duns Scotus) *Quaestiones in VIII libros Physicorum Aristotelis* VI q. 14 (1891) 113a: rarefactio dicitur quando aliquod corpus sit maius non adveniente aliquo alio ab extrinseco.

[20] Confirmatur ista conclusio: quia stat aliquod corpus continue rarefieri, et in hac rarefactione continue occupabit minorem locum; igitur, etc. Antecedens declaro, quia velis imaginari quod annihilata tota aqua universi, rarefiat aër replens locum quem occupat et quem aqua occupabat; et cum hoc velis imaginari quod ignis aliqualiter rarefiat versus concavum. Tunc in hoc casu aër est rarior quam primo esset, et tamen occupat minorem locum quam prius. Quod declaratur, quia prius aër erat in concavo ignis et in convexo aquae tamquam in loco sibi proprio. Nunc vero est in concavo ignis, quod est minus quam prius; et in convexo terrae, quod etiam est minus quam esset convexum aquae; quare conclusio vera.

[21] Quinta[28] conclusio: non ex hoc dicitur aliquid condensari quia materia eius fit in minori loco quam prius. Patet, quia stat aliquid rarefieri et occupare minorem locum quam prius. Patet per illud quod dicebatur[29] de aëre, quia per actum rarefactionis occupat minorem locum etc. Patet etiam conclusio, quia fiat de sphaerico ut prius tetrahedron, tunc istud | [V2 58va] corpus tetrahedron occupabit maiorem locum quam prius. Et hoc notabiliter condensetur, igitur aliqualiter istud corpus. Et tunc ut constat illud corpus occupat maiorem locum quam dum erat sphaericum, et per transmutationem | [V1 129rb] ipsius sphaerici ad tetrahedron condensabitur illud corpus; igitur, etc.

[22] Sexta[30] conclusio: ex hoc non dicitur aliquid rarefieri quia partes eius extremales magis distent a medio quam prius. Patet conclusio, quia ubi A B sit unum corpus cereum vel de alia materia ducibili et extensibili, tunc ut est evidens potest istud corpus elongari ad duplum et ad quadruplum eius; et in hac extensione extrema eius fiunt distantiora a medio quam prius, et tamen in hoc contingit illud corpus fieri densius et densius. Et hoc patefiet tibi ubi inspicias motum per quem fit candela cerea quae hora qua fit densior et densior per compressionem eius inter duo pondera in cuius compressione extrema fiunt distantiora.

[23] Septima[31] conclusio: ex hoc aliquid dicitur rarefieri, quia figura eius non variata omnium duarum partium puncta remotiora fiunt quam prius, et hoc secundum omnem diametrum materia eius nequaquam addita vel subtracta.

---

3 annihilata] anichilata *var.* | rarefiat] rere fiat V2 5 versus] eius *add.* V2 9 et] etiam *add.* V2 12 quia] cum primo(?) *add.* V2 | aliquid] rare condensari *add. sed del.* V1 15 tetrahedron] vel econtra *add. sed del.* V1 17 Et *om.* V2 | ut constat *om.* V2 22 cereum] terreum V2 23 potest] patet V2 24 extrema] extranei V2 26 candela] candella *var.* V2 | cerea] terrea terrea V2 | fit *rep.* V1

---

28 Adn. mg. V1: Quinta conclusio.

29 Cf. supra n. 20.

30 Adn. mg. V1: Sexta conclusio.

31 Adn. mg. V1: Septima conclusio. *Manicula.*

Ista conclusio videtur esse clara. Dico enim in conclusione ‘secundum omnem diametrum’, quia licet extrema in candela cerea fiant remotiora quam prius, hoc est, solum secundum diametrum longitudinalem, verbi gratia, distarent partes latitudinales plus quam prius, haberetur rarefactio dixi ‘materia nequaquam addita’ etc. quia per additionem materiae ut cerae ad candelam possent partes plus distare absque rarefactione.

[24] Octava[32] conclusio: quamquam condensatio sit possibilis, et cum hoc rarefactio. Non est tamen possibile per naturam quodlibet corpus rarefieri, nec potest per naturam quodlibet corpus condensari. Declaratur, quia si est possibile, ponatur in esse; et habetur ista conclusio, ‘quodlibet corpus naturale naturaliter condensatur, igitur quattuor elementa insimul sumpta cum mixtis occupant minorem locum quam prius’. Igitur datur vacuum, quod non contingit per naturam.[33] Patet alia pars, quia si est possibile quodlibet corpus rarefieri, ponatur in esse; et tunc habetur: ‘quodlibet corpus naturale naturaliter rarefit, igitur quattuor elementa simul cum mixtis occupant maiorem locum quam prius.’ Sed ante rarefactionem occupabant totum concavum caeli, igitur per actum rarefactionis occupant maiorem locum. Sed non contingit hoc nisi latera caeli dilatentur vel cedant, quod non contingit per naturam. Concedendum est tamen ‘quodlibet corpus naturale naturaliter posse rarefieri’. Et concedendum est, ‘quodlibet corpus naturale posse naturaliter condensari’. Hoc patet inductive et ex praecedenti conclusione. |
V1 129va

[25] Sequitur illud corollarium: necesse est si per naturam aliquid rarefiat, aliquid condensari et econtra. Et notanter dico ‘per naturam’, quia ubi octava sphaera rarefieret ab eius convexo usque ad eius concavum, non expediret aliquid condensari; sed hoc per agens naturale fieri non posset. Et hoc de secundo articulo.

[26] Transeo ad tertium articulum. Prima difficultas:[34] dato uno pedali cuius medietas sit rarefacta et reliqua condensata aequaliter, in hoc casu quaeritur an hoc pedale debeat dici rarefactum et non-rarefactum.

---

1 secundum] sed V2 2 cerea] terrea V2 3 distarent] sic aliae *add.* V2 5 etc.] vel subtracta V2 | ut cerae ad candelam *om.* V2 9 potest … corpus] est possibile quodlibet corpus per naturam V2 11 mixtis] mistis *var.* V1 14 habetur] illa conclusio *add.* V2 18 vel cedant *om.* V2 19 Et] similiter *add.* V2 19–20 Et concedendum … naturaliter] *mg.* V1; vel *add. sed del.* V1 21 et *om.* V2 25 fieri *om.* V2 27 Prima difficultas] in quo quaero difficultates quorum prima sit illa V2

---

32 Adn. mg. V1: Octava conclusio.

33 Cf. Aristoteles *Physica* IV,7 216a20–21; *Auctoritates Aristotelis* 150 (130): Non est vacuum in natura.

34 Adn. mg. V1: Prima difficultas.

[27] Respondetur. Et primo dico quod hoc debet dici rarefactum et nonrarefactum ut sit de composito extremo. Patet hoc. Dico secundo quod in dicto casu plus quam eius medietas est rarefacta. Patet, quia data medietate rarefacta, ipsa cum medietate mediantis partis | condensatae est rarefacta iuxta V2 58vb illud; illud est rarefactum cuius plus quam medietas est rarefacta.

[28] Dico tertio quod quocumque puncto dato in parte condensata ultra illud terminatur rarefactio. Patet.

[29] Dico quarto quod quacumque parte data quae est rarefacta est reperire partem maiorem rarefactam in pedali. Dato istud patet intelligenti. Et sic suo modo de condensatione.

[30] Dico quinto quod, dato uno pedali cuius medietas rarefiat et reliqua aequaliter condensetur cum illa, quia secundum aequalem latitudinem in hoc casu hoc pedale rarefit et in fine totum erit rarefactum. Patet, quia in fine pars rarefacta est plus quam medietas totius corporis, et pars condensata est parva valde respectu totius. Igitur iuxta illud quod dicitur,[35] 'illud est rarefactum cuius plus quam medietas' etc., patet totum.

[31] Dico sexto quod dictum pedale habens tantum de raro quantum de denso et praecise tantum est rarum et non est densum.

[32] Dico septimo quod, si esset unum pedale cuius una medietas esset rara et reliqua medietas esset privata omni raritate et densitate, in hoc casu totum pedale esset rarum. Patet quia postquam in medietate est latitudo raritatis et in reliqua medietate nihil est quod obiciat denominationi, sequitur quod latitudo raritatis informans medietatem potest communicari toti, et denominare totum 'rarum'—non tamen sub eodem gradu, sed solum sub gradu subduplo. Quia si latitudo raritatis ut octo informat medietatem, tunc hoc est ita rarum per totum sicut si esset per totum rarum ut quattuor.

[33] Dico octavo quod, dato pedali cuius medietas est rara et reliqua | medie- V1 129vb tas habet aequaliter de raritate et densitate per totum, tunc totum est rarum. Patet, quia medietas habens tantum de raritate quantum de densitate nec est rara nec est densa. Et ista medietas sic se habet acsi esset privata omni gradu raritatis et densitatis; igitur ista medietas non habet in se causam impedientem denominationem alterius partis.

[34] Secunda[36] difficultas: an sit possibile aliquid uniformiter rarefieri.

2 sit] ille sermo V2 | Dico] de V2 3 data *om.* V2 8 Dico] de *add.* V2 13 et] etiam V2 20 esset] esse V1 21 Patet] statim *add.* V2 28 tunc] in hoc casu V2 29 Patet] statim *add.* V2 33 aliquid] nl *add. sed del.* V1

35 Cf. supra n. 27.

36 Adn. mg. V1: Secunda.

[35] Respondetur quod sic. Patet, quia nihil prohibet agens rarefactivum esse applicatum corpori quod debet rarefieri aequaliter et uno modo toti et partibus.

[36] Dico[37] secundo quod stante rarefactione uniformi, partes huius quod sic rarefit difformiter moventur. Patet, quia partes corporis distantes a medio moventur velocius partibus propinquis centro, quia moventur a pluribus motoribus. Moventur enim per impulsionem ab aliis partibus per rarefactionem earum, et etiam moventur suis propriis rarefactionibus. Ex quo patet hoc argumentum valere, 'hoc uniformiter rarefit, igitur difformiter movetur secundum eius partes'. Et econverso valet, 'partes huius corporis uniformiter moventur per solum actum rarefactionis, igitur istud corpus difformiter rarefit.'

[37] Dico tertio quod in rarefactione uniformi per quam pedale augetur ad duplum, quaelibet eius pars sic augetur ad duplum. Et hoc est verum tam de partibus proportionalibus quam de aequalibus. Hoc patet intelligenti.

[38] Tertia[38] difficultas: an sit possibile idem simul rarefieri et condensari secundum eandem partem, ita quod rarefactio et | condensatio sese intendant in eadem parte subiecti. V2 59ra

[39] Respondetur, et dico primo probabiliter quod sic. Quod persuadetur sic: possibile est causas de per se rarefactionis et condensationis intendi in eadem parte subiecti, igitur possibilis est effectus intendi etc. Consequentia videtur tenere. Et antecedens patet, quia possibile namque est caliditatem et frigiditatem intendi in eadem parte subiecti, ut declaravi in arguendo[39] ad partes in proposito de reactione, dum calidum agit in frigidum, et econtra. Tunc

---

16 condensatio] intensio V2

---

37 Adn. mg. V1: *Manicula.*

38 Adn. mg. V1: Tertia.

39 Cf. supra n. 7. Cf. Blasius Pelacanus de Parma *Questiones disputate super octo libros Physicorum* V q. 9, Vat. lat. 2159, f. 153vb (cit. Stefano Caroti, "La 'reactio' in Italia: Jacopo da Forlì" (2006) 16): Unde dicemus reactionem (*corr. ex* rarefactionem) esse possibilem ad hunc sensum, quia dum calidum agit in frigidum repatitur a frigido ... non quod frigidum inducat aliquam qualitatem positivam in calido. Et quod hoc sit verum patet, quia ubi summe calidum ageret in remisse frigidum et in agendo repateretur, ita quod maiorem gradum frigiditatis reciperet in se, tune cum summa caliditate staret aliquis gradus frigiditatis, aut summe calidum non amplius esset summe calidum, quare numquam summe calidum posset assimilare sibi passum, quia dum ⟨remisse frigidum⟩ (*ms.*: summe calidum) incipit ⟨re⟩agere, desinit ⟨summe calidum⟩ esse summe calidum, et si desinit esse summe calidum, cum nullum agens agat ultra gradum proprium, non poterit inducere in passum summam (*del.* frigiditatem) caliditatem.

medium per quod haec distant videtur intendi in caliditate et frigiditate. Modo ut constat de ratione caliditatis est rarefacere, et de ratione frigiditatis condensare.[40]

[40] Confirmatur:[41] pro quo accipiatur unum corpus columnare rectum ut asta deinde adaequate per horam sic incurvetur quod tandem fiat circulus. Tunc volo probare quod a concavo ad convexum huius sit acquisita | certa latitudo densitatis, quoniam totum concavum est brevius quam fuit ante dum lignum erat rectum,[42] et non est subtracta materia in toto nec in parte; igitur facta est condensatio a concavo usque ad convexum. Nisi dicatur quod ista latitudo densitatis est solummodo acquisita in concavo quod distat a convexo per totam spissitudinem ligni. V1 130ra

[41] Sed hoc dici non potest. Primo, quia sic per concavum solum intelligeres superficiem mathematicam[43] absque omni profunditate; modo talis superficies non est nata condensari, nec talis superficies datur nisi apud imaginationem. Patet etiam quia ubi separetur concavum huius circuli usque ad punctum medium inter concavum et convexum, apparet adhuc quod in residuo est certa latitudo densitatis acquisita usque ad convexum, et per consequens tota latitudo densitatis videtur informare totum istud lignum sic incurvatum ab eius concavo usque ad convexum. Verumtamen uniformiter difformiter terminatur in extremo intensiori versus concavum, et in extremo remissiori ad convexum. Eodem modo probatur in dicto casu quod a convexo ad concavum sit latitudo raritatis acquisita, quia totum convexum huius circuli est longius quam foret dum erat rectum, sic quod partes videntur ab invicem plus distare quam primo; igitur est rarefactum. Et sic dico, procedendo a convexo usque ad concavum.

---

4 columnare] columpnare *var.* 10 densitatis] raritatis V2 12 solum] solummodo V2 15 ubi *om.* V2 16 inter] circulum *add. sed exp.* V2 | adhuc] evidenter *add.* V2 17 densitatis] raritatis V2; raritatis *a.c. mg.* V1 | usque *om.* V2 18 densitatis] raritatis V2 19 terminatur] terminata V2 21 in dicto casu *om.* V2 22 raritatis] densitatis V2; densitatis *a.c. mg.* V1 24 ad] eius *add.* V2

---

40 Cf. Blasius Pelacanus de Parma *Questiones circa libros de generatione et corruptione* I q. 1 (ed. Biard): de ratione frigoris est condensari et ingrossari.

41 Cf. Ioannes Buridanus *Quaestiones super Physicorum* IV q. 11 (2016) 302: Et forte quod in violentis incurvationibus lignorum habent locum huiusmodi condensatio et rarefactio. Nam cum arcus quasi est rectificatus et habens superficiem concavam quasi aequalem secundum longitudinem superficiei convexae, tamen quando multum incurvatur, oportet superficiem concavam fieri multo breviorem et superficiem convexam longiorem.

42 Cf. Marsilius de Ingen *Abbreviationes super octo libros Physicorum Aristotelis* IV tr. 2 (1521) f. 19ra: Condensatio est possibilis per solum motum localem. Patet, quia incurvatione virgae, medietas concava condensatur.

43 Cf. Robertus Alyngton *Litteralis sententia super Praedicamenta Aristotelis* (1993) 250: superficies mathematica est quantitas moli permanens, longa et lata, sed non profunda.

Et hoc non potest esse nisi in eadem parte materiae sint raritas et densitas acquisitae. Et sicut latitudo densitatis terminabatur in extremo intensiori ad concavum et remittebatur versus convexum, sic latitudo raritatis terminatur in extremo intensiori ad convexum. Et semper remittitur versus concavum. Et haec est causa propter quam sagitta ab arcu sic impetuose per aërem fertur, quia in hoc, partes ligni arcus quae sunt indebitae invicem situatae redeunt ad situm suum per deperditionem raritatis versus convexum, et deperditionem densitatis versus concavum.

[42] Quarta[44] difficultas: an sit possibile aliquid condensari ad non-quantum.

[43] Respondetur quod sic, saltem per naturam simpliciter immensam; sed alio modo non apparet. Ubi tamen quantitas sit ipsa res quanta, non apparet
V2 59rb quod aliquid bene possit | condensari ad non-quantum, sicut non apparet quod possit linea recta esse linea recta indivisibilis, quia eo ipso quod linea est, vide-
V1 130rb tur habere partem situaliter extra partem,[45] sed ubi extensio | sit rei extensae condistincta.

[44] Tunc ad difficultatem dicendum quod per solam approximationem partium et remotionem quantitatis potest aliquid condensari ad non-quantum. Et in hoc casu concedendum est illud non-quantum esse infinite densum, quia in tali partium approximatione fit continue plus et plus de materia in minori loco et tandem in non-quanto loco.

[45] Quinta difficultas:[46] an penetratio corporum sit condensatio.

[46] Videtur quod sic. Quia in eodem loco fit plus de materia quam prius.

[47] In oppositum patet, quia dum corpora beatorum transiverunt per sphaeras usque ad octavum caelum, hoc fuit absque condensatione; quare, etc.

[48] Respondetur quod penetratio corporum non est condensatio secundum quod probat argumentum.

---

2 sicut] sicuti V2 | densitatis] raritatis V2; raritatis *a.c. mg.* V1 3 raritatis] densitatis V2; densitatis *a.c. mg.* V1 5 arcu] archu *var.* V2 6 invicem situatae] situatae ad invicem V2 8 versus] convexum *add. sed exp.* V2 9 condensari] usque *add.* V2 16 approximationem] appropinquationem V2 18 densum] dempsum *var.* V; dempnsum *var.* V2 21 Quinta difficultas] Quinto V2

---

44 Adn. mg. V1: Quarta.

45 Cf. Aristoteles *Metaph.* V,13 1020a7–8. Cf. Guillelmus de Ockham *Summa logicae* I c. 44 (OPh I 137): quantitas continua permanens nihil aliud est nisi res una habens partem situaliter distantem a parte; Ioannes Buridanus *Expositio et quaestiones in Aristotelis De caelo* I q. 2 (1996) 240: Notandum est quod magnitudo proprie non est nisi extensio qua res habet partem extra partem situaliter et continue.

46 Adn. mg. V1: Quinta difficultas.

[49] Et ad motivum in oppositum, dicendum quod argumentum non valet.

[50] Sexta[47] difficultas: utrum per rarefactionem acquiratur aliquid materiae quae rarefit; et ita per oppositum de condensatione, an in condensatione fiat deperditio alicuius rei.

[51] Respondetur quod ubi velis tenere extensionem esse rei extensae condistinctam per rarefactionem, acquiratur nova quantitas; sed ubi non, non acquiritur.

[52] Septima[48] difficultas et ultima: cum raritas et densitas privative opponantur, quaeritur an raritas habeat rationem privativi, an densitas.

[53] Respondetur primo quod raritas habet rationem privativi, quia per raritatem continue fit minus et minus de materia in loco, et per oppositum per densitatem continue fit plus de materia in eodem loco. Et ideo densitas videtur habere rationem positivi.[49]

[54] Aliter dico, ut conceditur satis probabiliter, quod raritas habet rationem positivi, quia per raritatem acquiritur nova quantitas; per densitatem vero deperditur. Igitur videtur quod raritas habeat rationem positivi; alioquin privativum esset causa productiva positivi, et positivum esset causa privativa positivi, ut patet. Et haec de tertio articulo.

[55] Ad quartum articulum pro solutione rationum patet solutio, inspectis difficultatibus et determinatione quaestionis. Et haec de quarto.

---

2 utrum] an V2 3 an in condensatione fiat] sit V2 9 an[2]] vel V2 10 primo *om.* V2 11 continue *om.* V2 18 ut patet *om.* V2 | articulo *om.* V2 19 pro solutione rationum *om.* V2 20 et … quarto] positis una cum demonstratis quantitatis proposita V2

---

47 Adn. mg. V1: Sexta difficultas.

48 Adn. mg. V1: Septima difficultas.

49 Cf., e.g., Richardus Swineshead *Calculationes* tr. V (1520) 18[va]: Pro istis sustinenda est ultima positio, quae scilicet ponit quod densitas se habet positive.

Liber V, q. 1: Utrum agens naturale, hora qua agit in passum, agat in ipsum secundum se totum et quamlibet eius partem

V1 = Vat. lat. 2159 ff. 134vb–137ra; V2 = Vat. lat. 3012 ff. 62rb–64ra

[1] Circa quintum librum *Physicorum* quaeritur primo utrum agens naturale, hora qua agit in passum, agat in ipsum secundum se totum et quamlibet eius partem.[1]

[2] Arguitur primo quod non.

[3] Primum argumentum: ubi medietas solis aufereretur a toto, reliqua medietas tantum illuminaret quam nunc illuminat intensive et extensive. Quod intensive patet, quoniam cum corpus solis sit uniformiter lucidum, tantum est de luciditate in medietate eius quam a toto; et per consequens intensive poterit medietas inducere tantam latitudinem sicut primo luminis, et per consequens medietas sumpta non concurrebat. Patet etiam quod extensive, quoniam basis pyramidis lucis erit in hoc casu tota diameter solis. Et ante abscisionem non poterat esse maior; quare, etc.

[4] Secundum argumentum: ubi corpus solis sic condensaretur quod per eius condensationem adaequaretur grano milii, adhuc tantum illuminaret quam nunc intensive et extensive. Et tamen in hac condensatione plures partes deperderetur, ut constat; igitur illae partes deperditae non concurrebant ad actionem. Quod autem in hoc casu producat sol de lumine intensive et extensive patet; postquam habet tantum de luce quam primo habuit, nec medium ei repugnat. Igitur ad quamcumque distantiam primo agebat ad tantam nunc

---

1 librum *Physicorum* *om.* V2 2 passum] passuiva V2 | in[2]] se *add.* V2 7 cum] s.l. V2 8 in medietate eius] eius in medietate V2 | a] in V2 9 sicut] s.l. V1; *om.* V2 11 pyramidis] pyramidis *var.* | abscisionem] absisionem *var.* V1 13 condensaretur] condempsaretur *var.* V1 14 adhuc] adhunc *a.c.* V1 15 quam] et *add.* V2 17 actionem] acionem *var.* V2 | sol *om.* V2

---

1 Resp. Aristoteles *Physica* V,1 224a22–26. Cf. Ioannes Buridanus *Quaestiones super Physicorum* V q. 1 "Utrum generatio sit motus" (1509) ff. 83rb–84va; Albertus de Saxonia (attrib.) *Quaestiones in Aristotelis Physicam* V q. 1, "Utrum omnis transmutatio ad substantiam sit successiva" (1999) 760–766; Nicolaus Oresme *Questiones super Physicam* V q. 3, "Utrum forma substantialis divisibiliter acquiratur ita quod generatio sit mutatio successiva" (2013) 582–588; Marsilius de Inghen (attrib.; pseudo-Ioannes Duns Scotus) *Quaestiones in VIII libros Physicorum Aristotelis* V q. 2 "Utrum omnis mutatio ad substantiam sit successiva" (1891) 183^{b}–187^{a}.

 | DOI:10.1163/9789004734968_017

aget, vel ad maiorem, propter illud quod scriptum est in commento quintae decimae propositionis *Libri de causis*[2] in hac forma: "virtus unita fortior est ipsa dispersa."

[5] Tertium argumentum: ubi A pedale summe calidum esset applicatum B, remisse frigido, non maiorem latitudinem caliditatis induceret in B passum quam si medietas ipsius A auferatur a toto A; quoniam semipedale summe calidum applicatum, remisse frigido, induceret tantam latitudinem caliditatis.

[6] Quartum argumentum: ubi A pedale summe calidum per condensationem fieret indivisibile ut punctus, tunc in hoc casu, si quaestio | foret vera pro V1 135ra parte affirmativa, non posset A agere in B passum; quia tunc non ageret in passum secundum aliquam partem, quia nullam haberet partem. Tamen consequens est falsum, quia in infinitum velocius ageret quam primo dum erat extensum. Et maiorem latitudinem induceret, quod patet per hoc, quoniam in hoc casu esset sicut si esset unum pedale uniformiter calidum, et caliditas totius poneretur in medietate eius, tunc illa medietas esset in duplo calidior quam primo. Et si in quarta parte poneretur, esset in quadruplo calidior quam primo. Igitur si caliditas informans pedale uniformiter poneretur in puncto, ille punctus haberet caliditatem infinitam intensive, et per consequens posset inducere | caliditatem infinitam intensive. Et ita in proposito. V2 62va

[7] Quintum argumentum: hora qua passum patitur, non patitur secundum quamlibet eius partem; igitur nec agens hora qua agit. Antecedens patet, quia tunc parvus ignis hic exsistens calefaceret maius spatium in centuplo quam sit tota terra. Et tenet argumentum 'quoniam hic ignis calefacit aliquem aërem, ergo calefacit quamlibet partem aëris'. Ergo totus aër universi patetur secundum quamlibet eius partem; et sic totus aër universi calefit ab isto igne. Modo totus aër universi est centuplus ad ipsam terram. Isto pro nunc supposito quod quodlibet elementum sit decuplum ad elementum immediatum sibi inferius.[3]

1 maiorem] altiorem V2 2 fortior est] est fortior se V2 5 in] D *add. sed del.* V2 8 Quartum argumentum] Quarto V2 11 aliquam] alicam *var.* V1 16 in² *om.* V2 17 puncto] poncto *var.* V2 20 Quintum argumentum] Quinto V2 22 centuplo] centecuplo *var.* 24 patetur] patitur V2 27 quod] quamlibet *add. sed del.* V1

2 Cf. *Liber de Causis* XVI (ed. Pattin) 138; *Auctoritates Aristotelis* 232 (13): Omnis virtus unita fortior est se ipsa dispersa. Cf. Ioannes Peckham *Perspectiva communis* I prop. 18a (16a) (1970) 94: Omnis autem virtus unita (unica *ed.*) fortior est quam divisa. Cf. Blasius Pelacanus de Parma *Questiones super perspectiva communi* prima pars, q. 4 (2009) 101: Modo virtus unita fortior est quam si foret dispersa, 15a propositione *De causis*.

3 Cf. Aristoteles *De gen. et cor.* II,6 333a20–23: Si quidem igitur secundum quantum, necesse idem aliquid esse existens in omnibus comparabilibus quo mensurantur, verbi gratia si ex

[8] Sextum argumentum: si passum pateretur etc., sequitur alterum duorum: aut quod agens ageret ultra gradum proprium; aut quod agens numquam posset sibi assimilare passum. Et patet argumentum, quoniam ubi A sit unum summe calidum, et B unum remisse frigidum, tunc facta applicatione A, agit in B; et per consequens A patitur a B. Tenet ista consequentia, supposita reactione. Sed per te, A patitur a B secundum omnem eius partem, igitur totum A et quaelibet eius pars remittitur; et per consequens A desinit esse summe calidum. Si igitur A, quod a principio applicationis erat summe calidum, non assimilabit sibi B, stante applicatione, habetur primum inconveniens. Si autem assimilabit sibi, habetur aliud inconveniens, ut constat.

[9] Item, omne agens physicum habet materiam, quae non est alicuius activitatis; ergo, etc.

[10] In oppositum est Philosophus a principio[4] huius quinti, distinguens de mobili et de moventi, dicens quod de per accidens est quod agens agat ratione V1 135rb partis; sed de per se et primo est quod agens agat | secundum se et quamlibet eius partem.

[11] Sed non obstantibus auctoritatibus arguitur: quia ubi in media regione esset magna terra, dimissa suae naturae, ipsa movetur deorsum magna velocitate. Et si ab illa magna terra tu abscinderes medietatem eius, tunc eius motus multum retardaretur. Et non potest esse alia causa, nisi quod tota terra et quaelibet eius pars concurrebat ad eius motum; immo quacumque parte data magna vel parva secundum hoc retardabitur motus.

[12] In hac quaestione erunt quattuor articuli. In primo problema. In secundo determinatio quaestionis. In tertio difficultates. In quarto ad rationes in contrarium.

---

5 in *om.* V2 | patitur] agit V2 10 aliud] alliud *var.* V1 11 Item] si *add.* V2 12–13 ergo ... Philosophus *om.* V2 13 de] motu et *add. sed del.* V1 15 agens] s *add. sed del.* V2 17 arguitur *om.* V2 19–20 eius ... retardaretur] tunc retardaretur motus eius multum V2 21 concurrebat] concurrebant V2 | eius] hunc V2 | data] sumpta V2 23–25 In ... contrarium] etc. ut supra V2

---

aque pugillo erunt pugilli aeris decem. Cf. Blasius Pelacanus de Parma *Questiones circa tractatum proportionum magistri Thome Braduardini*, redactio prima, q. 12 (2008) 216: ut communiter dicitur, quodlibet elementum est decuplum ad elementum inferius sibi immediatum. Cf. Blasius Pelacanus de Parma *Questiones circa libros Metaurorum Aristotelis* I q. 9 (2019) n. 29: quodlibet elementum est decuplum ad elementum sibi inferius immediatum.

4 Aristoteles *Physica* V,1 224a22–26: Transmutat autem transmutans omne aliud quidem secundum accidens, ut cum dicimus musicum ambulare, quoniam cui accidit musico esse, hoc ambulat: in quo autem huius aliquid mutare simpliciter dicitur mutari, ut quecumque dicuntur secundum partes.

[13] Quantum[5] ad primum articulum sit problema: propter quid est quod homo morsus a cane rabido, post diem nonam, si ei praesentetur aqua, mirabiliter eam abhorret, iudicans videre in aqua canem rabidum et eius intestina.

[14] Respondetur quod hoc ideo est quia morsus a cane rabido, hora laesionis, fecit sibi magnam estimationem de cane rabido, quae estimatio continue intenditur secundum quod alteratio plus intenditur. Venenatur namque totum corpus eius et generatur in eo venenatus sanguis et spiritus. Ideo propter augmentum laesionis fit continue fortis estimatio de cane rabido, hinc est quod praesentata ei aqua clara sive crystallo a quibus fit reflexio, reflectuntur species ad oculum obiectum repraesentantes, ut puta canem rabidum.

[15] Quare autem fiat hoc magis ab aqua, etc.?

[16] Respondetur, quia alia corpora sunt colorata quae efficacius movent quam species reflexae moveant; aqua vero non proprie est colorata. Similiter etiam nec crystallus. Et ideo hoc apparet sibi in aqua. \Et hoc est verum ubi visio fiat extramittendo. Sed ubi non, dicendum est quod anima mediante corde alterato convenient⟨⟨ter⟩⟩ generat sanguinem et spiritus caninum. Et quia "anima insequitur complexionem corporis",[6] hic est quod etc./

[17] Et haec sufficiant de primo articulo quaestionis, etc. Amen. | | V1 135va V2 62vb

[18] Transeo ad secundum articulum, in quo erunt duo processus. Et primo referam me ad passum. Et videbo an passum, hora qua patitur, patiatur secundum quamlibet eius partem, an ne. Et propter hoc addita fuerunt aliqua argumenta a principio de patiente. In secundo processu quaeram de agente, utrum agens, hora qua agit, agat secundum quamlibet sui partem.

[19] De primo sit prima evidentia: hic terminus 'pars' ponitur in titulo quaestionis. Ideo aliquando capitur 'pars' pro ipsa materia quae est pars compositi;

1–18 articulum ... Amen (n. 17) *om.* V2 5 estimationem] extimacionem *var.* V1 9 crystallo] cristalo *var.* V1 | reflexio] reflecxio *var.* V1 19 secundum articulum] tertium V2 | quo] primo V2 22 utrum] scilicet *add.* V2 24 sit prima evidentia *rep.* V2 | ponitur *om.* V2

---

5 Cf. Blasius Pelacanus de Parma *Questiones disputate super octo libros Physicorum* VII q. 3 (2017) 27: Similiter morsus a cane rabido post nonum diem dicit se videre intestina illius canis cum exponitur sibi aqua. Cp. Petrus de Abano *Conciliator Differentiarum* diff. 179 (1568) 236b: mingunt etiam quandoque pinguedinem, in qua res carnosae mirabiles, acsi essent animalia, et quasi catuli apparent, et alia huiusmodi versuta contingunt symptomata.

6 Cf. pseudo-Aristoteles *Physiognomia* 1 (1562) 132va: Quod et animae sequuntur corpora. Cf. Blasius Pelacanus de Parma *Questiones super Metheorum* (cod. Vat. Chig. O IV 41) 68ra–b; cit. Christophe Grellard, "Une histoire naturelle des religions" (2019) 64 n. 16: secundum quod scribitur iuxta principia physonomie Aristotelis, anima insequitur complexionem corporis. Cf. Blasius Pelacanus de Parma *Quaestiones disputatae De anima* I q. 8 (1974) 58: anima insequitur complexionem corporis.

aliquando capitur 'pars' pro parte composita ex materia et forma. 'Pars' primo modo non est alicuius activitatis; immo materia prima ponitur infinitae passibilitatis. Ideo de tali parte non habet quaestio locum, ut quod 'quaelibet pars agentis, hora qua agit, active ad actionem concurrat'.

[20] Secunda evidentia: duplex est passum. Quoddam est distans ab agente secundum totum et quodlibet sui, sicut approximando calidum frigido. Quoddam est passum indistans ipsi agenti. Et idem dico de agente. Exemplum huius: data aqua calefacta, et remoto calefaciente, ipsa infrigidatur ab agente; quod agens non distat ab ipsa aqua. Infrigidatur namque a sua forma substantiali extensa per totam aquam et quamlibet eius partem.

[21] Et tunc ponuntur conclusiones de patiente. Prima conclusio: aqua calefacta, remoto calefaciente, secundum se totam et eius quamlibet partem, ad frigiditatem alteratur. Patet statim, quoniam aqua calefacta habet in se principium alterandi ipsam ad frigiditatem; quod quidem principium est aequaliter applicatum toti aquae et cuilibet eius parti. Igitur ipsa secundum quodlibet sui patitur a tali principio.

[22] Hic tamen nota quod in conclusione, cum dicitur quod 'ipsa alteratur secundum quamlibet sui partem', hic non capitur 'pars' pro parte formali; quia forma substantialis aquae non infrigidatur, quia "formae non est pati sed agere".[7] Capitur igitur ly 'pars' pro parte materiali. Quia secundum rei veritatem materia prima est illa quae calefit, modo frigefit, modo dealbatur, etc.

[23] Secunda conclusio: non contingit passum ab agente distans pati secundum totum et quodlibet sui. Et hoc a principio temporis mensurantis actionem V2 63ra usque | ad finem. Patet conclusio, quoniam semper agens prius agit in partem propinquam quam remotam.[8] Nec potest agere in partem remotam nisi partes V1 135vb intermediae | fuerint primo informatae vel dispositae secundum dispositionem secundum quam agens agit in passum.

3 locum] quantum ad hoc *add.* V2 6 totum] se V2 7 Et *om.* V2 10 extensa] extensiva V2 11 patiente] calefaciente(!) V2 | conclusio] ergo V2 15 secundum] se *add. sed del.* V1 16 principio] ergo *add. sed del.* V1 19 substantialis] ipsius *add.* V2 | quia[2]] forma *add. sed del.* V1 20 ly] hic V2 26 quam] in partem *add.* V2

7 Cf. Thomas de Aquino *Scriptum super Sententiis* IV d. 12 q. 1 a. 2 qc. 5 co.: operatio formae non est pati, sed agere. Cf. Blasius Pelacanus de Parma *Quaestiones in Aristotelis De anima libri* I, 8 (1987) 61: formae non est pati, sed agere.

8 Cf. Gualterus Burlaeus *Expositio super libros I et II De generatione* (2005) 125: Sed agens prius agit in partem propinquam quam remotam. Cf. Blasius Pelacanus de Parma *Questiones circa tractatum proportionum magistri Thome Braduardini*, redactio prima, q. 11 (2005) 203–204: prius agens agit in partem propinquam quam in remotam.

[24] Tertia conclusio: possibile est hoc passum pati secundum se et quodlibet sui ab agente distanti ab ipso passo. Patet conclusio, quoniam ut tetigi in conclusione praecedente,[9] 'prius agens agit in partem propinquam quam in remotam'. Tamen non obstante, antequam passum assimiletur agenti versus finem actionis, agens agit in totum passum et in eius quamlibet partem. Unde si A esset unum summe calidum, applicatum B remisse calido, licet prius ageret in partem propinquam quam remotam a principio applicationis, tamen stante applicatione est dare tempus in quo A summe calidum agit per totum B, et per quamlibet partem eius. Et sic patet conclusio.

[25] Quarta conclusio: possibile est in certo casu hoc passum sic dispositum pati secundum se totum et quodlibet sui ab hoc agente, et hoc a principio suae actionis huius agentis usque ad finem. Pono casum in terminis: volo quod A sit unum calidum uniformiter ut quattuor, et B sit unum remisse calidum, cui A stetit applicatum per horam. Et nunc A desinat assimilare sibi ipsum B. Deinde volo quod A moveatur ipsum A, et loco eius ponatur C summe calidum uniformiter. Tunc arguo sic: 'ante hoc A agebat per totum B et quamlibet eius partem, et C est potentius ad agendum quam A, et est facta applicatio agentis ad passum; igitur sequitur quod B secundum se totum et quodlibet sui patitur a C.' Et hoc a principio huius agentis usque quo assimilaverit sibi ipsum B. Ex quo sequitur quod in certo casu stat agens non prius agere in partem propinquam quam in remotam. Et loquor semper de passo resistente agenti. Quoniam postea[10] fiet sermo 'an luminosum cui nihil obviat prius agat in partem propinquam quam remotam, mediis sibi non resistentibus'. Et haec sufficiant de passo.

[26] Nunc de agente sit prima conclusio: omne agens habens partes, hora qua agit, agit secundum se totum et quodlibet sui.[11] Probatur conclusio inductive, quia omne agens habens partes aut est motivum localiter, aut alterativum, aut augmentativum. De movente localiter patet conclusio, quoniam magna terra velocius movetur deorsum | quam pars eius. Si per se seorsum exsisteret, V1 136ra igitur in signum huius illa pars efficiebat partem huius motus. Et intelligatur

1 se] totum *add.* V2 2 ab] hoc *add.* V2 | distanti] distantem V2 | ipso] hoc V2 5 in eius quamlibet partem] quodlibet sui V2 7 remotam] in partem distantem V2 16 A] s.l. V2 20 casu] non *add. sed del.* V1 21 in] partem *add.* V2 23 quam] in *add.* V2 28 augmentativum] augmentivum V2 30 illa] ista V2

9 Cf. supra n. 23.

10 Cf. infra n. 36–39.

11 Cf. Aristoteles *Physica* V,1 224a25: in quo autem huius aliquid mutare simpliciter dicitur mutari, ut quecumque dicuntur secundum partes. Cf. Nicolaus Oresme *Questiones super Physicam* V q. 3 (2013) 579: illud quod est primo motum, mutatur secundum quodlibet sui, ut patet primo capitulo.

conclusio servata identitate medii et fine. De agente alterationis patet etiam conclusio, quoniam magnus ignis ad maiorem distantiam agit, et plura combustibilia transmutat in tanto tempore quam parva pars eius, si per se seorsum esset. Et idem patet de augmentate.

[27] Corollarium: quibuscumque corporibus eiusdem speciei datis in maiori corpore est virtus activa. Maior patet, inspecta conclusione proposita cum eius probatione.[12] Non dico tamen quod in maiori corpore sit maior virtus quam in parte eius; quia tanta est caliditas intensive in favilla ignis quanta in toto igne universi, et tanta est gravitas in parvo bolo terrae quanta in tota terra mundi. V2 63rb Verumtamen gravitas | quae est in tota terra universi activior est quam sit gravitas partis eius.

[28] Secundum corollarium: quantitas est de numero activorum. Patet ex praedictis, cum postquam maior terra cum eadem gravitate velocius fertur deorsum quam etc.[13] Verum quia Commentator quarto huius[14] oppositum scribit, dicens quod "quantitas non est de genere activorum." Intelligatur sermo eius in hac forma: 'agens ratione qua est tantae vel tantae quantitatis, non agit'. Et hoc est verum, quia ita scintilla ignis potest inducere summam caliditatem in passum sicut totus ignis mundi—licet non ita cito, quia non tot partibus passi potest applicari scintilla ignis sicut magnus ignis.

[29] Secunda conclusio: in aliqua actione ubi reactio adest, idem agit secundum se totum et quodlibet sui; praecise autem patitur secundum partem. Et haec conclusio sequitur ex praedictis. Bene prima pars patet ex conclusione praemissa;[15] secunda vero per hoc quod 'omne agens prius agit in partem propinquam quam remotam'.[16]

---

4 esset] exsisteret V2 6–7 eius probatione] probatione eiusdem V2 7 Non *om.* V2 8 in] parva *add.* V2 | favilla] favila *var.* 9 gravitas] intensive *add.* V2 | mundi] universi V2 13 cum *om.* V2 17 quia] quoniam V2 18 mundi] universi V2 19 applicari *add.* V1 | sicut] totus ignis *add. sed del.* V2 20 aliqua] omni V2 | actione] reactione *a.c.* V2 22 pars] huius conclusionis *add.* V2 23 vero] patet, patet *add.* V2 24 quam] passi quam in parte V2 | remotam] a passio eius distante *add. mg.* V2

---

12 Cf. supra n. 25–26.

13 Id est, maior terra cum eadem gravitate velocius fertur deorsum quam quam pars eius; cf. supra n. 26.

14 Cf. Averroës *In Phys.* IV com. 15 (1562) 126vb–127ra. Cf. Blasius Pelacanus de Parma *Questiones disputate super octo libros Physicorum* VII q. 9 (2017) 127: scriptum est quarto *Physicorum* a Commentatore, quod quantitas non est de genere activorum.

15 Cf. supra n. 26.

16 Cf. Aristoteles *Physica* VII,2 243a3–7; *Auctoritates Aristoteles* 155 (185): Movens et motum simul sunt et inter ea non est dare medium. Cf. Nicolaus Oresme *Questiones super Physicam* V q. 3 (2013) 584: agens prius disponit partes sibi propinquas quam remotas. Cf. Blasius

[30] Tertia conclusio et ultima: probabile est A et B invicem mutuo agere quorum residua unius est gradualiter maior quam activitas alterius. Pro intellectu conclusionis tamen noto fore existimandum quod omne agens, hora qua agit in passum, densitas cuius excedit densitatem medii fingere eius radios in partem propinquam passi, qui radii occurrentes illi parti flectuntur, et in eorum flexione contingit agens agere ultra gradum proprium, et resistentiam maiorem gradualiter corrumpere quam sit qualitas | informans agens, per quam (V1 136rb) agens agit in passum. Et ideo videmus saepe quod per congregationem radiorum solarium ignis generatur in combustibili, quod combustibile est modo foenum, modo aliud. Et ex hoc patet conclusio. Et ex hac probatione potest inferri omnem actionem esse reflexiva. Et haec sufficiant de secundo articulo.

[31] Ad tertium articulum, prima difficultas: utrum haec sit concedenda, agens naturale aeque cito agit in partem propinquam quam et in remotam ipsius passi.[17]

[32] Et sine argumentis, respondetur quod sic. Tamen ausculta quod ly 'aeque cito' facit tantum sequentem stare confuse tantum; et per hoc istae stant simul: 'quacumque parte passi data in partem propinquam prius agit agens quam in remotam'; et 'nulla est in qua agens primo agat secundum se totam, et cum hoc tamen agens agit in partem propinquam'. Quod patet quia agens, hora qua agit, non agit in partem indivisibilem; quia nulla est talis.[18] Sed quacumque parte data in ea est assignare propinquum et remotum.

[33] Secunda difficultas: an calidum ut quattuor et frigidum ut quattuor invicem applicata agant et reagant.

[34] Ad quod aliqui[19] dixerunt quod non, quia tunc actio fieret a proportione aequalitatis.

---

1 Tertia ... ultima] Ultima conclusio V2 3 existimandum] extimandum V1 4 densitas] dempsitas *var.* V1 | eius radios] radios suos V2 6–7 resistentiam ... corrumpere] corrumpere resistentiam maiorem gradualiter V2 7 quam] qualitatem *add.* V2 10 foenum] modo lignum *add.* V2 | potest] dici et *add.* V2 11 sufficiant *om.* V2 12 Ad] Quantum *praem.* V2 | articulum] sit *add.* V2 13 quam *om.* V2 15 ausculta] asculta *var.* V1 18 in] partem *add.* V2 21 remotum] quare etc. *add.* V2

Pelacanus de Parma *Questiones disputate super octo libros Physicorum* VII q. 8 (2017) 93: alterans prius disponit materiam propinquam quam remotam.

17 Cf. supra n. 23.

18 Cf. Aristoteles *Physica* VI,1 231a24–25: inpossibile est indivisibilibus esse aliquod continuum. Cf. Blasius Pelacanus de Parma *Questiones circa tractatum proportionum magistri Thome Braduardini*, redactio prima, q. 9 (2008) 142: non est ymaginandum quod *a* sic applicatum ipsi *b* agat in partem indivisibilem ipsi *b*, quia nulla est talis.

19 Non invenimus.

[35] Sed absque pluribus argumentis, dicatur quod agunt invicem et reagunt. Quoniam in actione non est comparanda potentia activa unius ad potentiam activam alterius, sed bene ad resistentiam et econtra. Stat ergo quod calidum ut quattuor habeat resistentiam ut duo, et frigidum ut quattuor habeat resistentiam ut duo. Quia in hoc casu utrumque agit in utrumque.

[36] Tertia difficultas: utrum ubi sol fieret per condensationem aequalis grano V2 63va milii, ageret tantum lumen intensive | et ad tantam distantiam ut nunc stantibus rebus ut stant.

[37] Ad quam difficultatem respondetur quod, retento casu, sol ageret tantum lumen intensive etc.,[20] eo quod tunc lucidum est productivum ad extra totam latitudinem sui luminis etc.

[38] Dico secundo quod, quocumque puncto signato in distantia per quam agit sol in hac condensatione, producet lumen intensius quem primo producebat. Et hoc patet pro tanto, quia infinitae partes lucidi factae sunt propinquiores huic puncto et illi; quare, etc.

[39] Dico tertio quod ad tantam distantiam poterit agere postquam non est V1 136va minoris activitatis | quam primo erat. Verumtamen plures partes ipsius terrae relinquerentur obumbratae quam nunc, quoniam luminosum minus illuminaret infra medietatem eius.

[40] Quarta difficultas: utrum remota medietate solis a sole, reliqua medietas sit productiva tanti luminis et ad tantam distantiam sicut totum.

[41] Quod non, per determinationem[21] quaestionis. Quia omne agens, hora qua agit, agit secundum totum et quodlibet sui.

[42] In oppositum, quia ista medietas habet tantum de luce intensive quam habet totum corpus solis, supposita uniformitate lucidi in luce sua; igitur, etc.

[43] Ad hoc est dicendum quod subtracta medietate etc.,[22] reliqua medietas potest tantam latitudinem lucis inducere extra se quam etc.,[23] sed non ad

1 pluribus argumentis] plura V2 3 ad] s *add. sed del.* V1 3–4 calidum] caliditas V2 4–5 et … duo *om. hom.* V1 5 Quia] Quoniam V2 7 intensive] etc. *add.* V2 10 tunc *om.* V2 13 agit] agit *add.* V2 | intensius] intensive V2 14 quia] quoniam V2 | partes] ipsius *add.* V2 17 erat] esset V2 20 sole] toto V2 23 secundum] se *add.* V2 24 In] Ad V2 | quam] quantum V2 25 igitur, etc. *om.* V2 27 se] argumentum illam *add. sed del.* V1 | quam] quantum V2

20 Id est, sol ageret tantum lumen intensive et ad tantam distantiam ut nunc stantibus rebus ut stant; cf. supra n. 36.

21 Cf. supra n. 26.

22 Id est, subtracta medietate solis a sole; cf. supra n. 40.

23 Id est, quam habet totum corpus solis; cf. supra n. 42.

tantam distantiam. Et hoc est simile, quoniam quantam caliditatem potest inducere magnus ignis, tantam potest inducere parva scintilla ignis, sed non ad tantam distantiam.

[44] Aliter posset dici, tenendo oppositum huius quod dictum est. Unde est imaginandum, dato lucido, una medietas agit per aliam. Et per hoc volo dicere quod plures radii lucis diffunduntur a toto ad punctum datum quam a parte lucidi. Modo plures radii et aeque intensi concurrentes ad unum punctum, lumen intensius inducunt quam si pauciores radii essent. Ratio vero adducta[24] de igne diversa est a proposito isto; eligatur tamen pars prima, quia eligibilior est.

[45] Quinta difficultas: an data latitudine uniformiter difformiter, medietas intensior agat in remissiorem eam intendendo.

[46] Ad quam difficultatem respondendum quod aut talis alteratio est disconveniens alterabili, aut conveniens.

[47] Et iuxta hoc dico duo. Primum est quod, dato uno corpore uniformiter difformiter calido caliditate non sibi conveniente, una medietas intensior non intendit aliam. Quod declaro tibi notitia experimentali. Quoniam velis imaginari quod A sit una aqua calefacta ab igne uniformiter difformiter, sicut contingit de facto. Tunc medietas intensior huius aquae, remoto igne, non intendit aliquam medietatem. Ut constat, quoniam sic medietas medietatis intensioris intenderet medietatem, et sic ultra; et per consequens tota ista aqua se intenderet in caliditate, remoto calefaciente.

[48] Idem, tu vides de medio illuminato a sole uniformiter difformiter, quia | V1 136vb
si una medietas intenderet aliam, tunc in medio reperiretur lumen intensius quam in corpore lucido—quod non contingit. | V2 63vb

[49] Dico secundo quod ubi ista 'caliditas sit conveniens' etc.,[25] quod in hoc casu medietas intensior intendet medietatem remissiorem. Et hoc pro tanto patet, quoniam de ratione agentis naturalis est quod, cum sit applicatum passo secundum debitam proportionem, quod agens agat et passum patiatur.

1–3 Et ... distantiam *om. hom.* V2 2 ignis] tantum *add. sed del.* V1 | inducere] ta *add. sed del.* V1 5 imaginandum] quod *add.* V2 6 toto] lucido *add.* V2 12 remissiorem] remissionem V1 13 Ad quam difficultatem *om.* V2 14 aut] est ei *add.* V2 18 sicut] Nunc V2 19 remoto igne *scrip. ante* medietas V2 20 sic] si *a.c.* V1 21 se *rep.* V2 23 sole] p *add. sed del.* V1 | quia] quoniam V2 27 intendet] intendat V2

---

24 Cf. supra n. 43.

25 Id est, si caliditas sit conveniens uniformiter difformiter uno corpore; cf. supra n. 46–47.

[50] Sexta difficultas: an data latitudine uniformiter difformiter terminata ad non-gradum, medietas intensior separata a toto sit maioris activitatis quam tota latitudo.

[51] Et videtur quod non, per determinationem[26] quaestionis.

[52] Et ex alia parte apparet intuenti quod sic.

[53] Ad difficultatem respondendum est quod sic. Et causa est quia illa medietas est intensior quam tota latitudo. Ubi enim tota latitudo sit ut quattuor, stabit quod medietas intensior erit ut sex, ut patet intelligenti. Et hic praesupponitur illud quod latitudo uniformiter difformis correspondet suo gradui medio.

[54] Septima[27] difficultas: an magna terra et parva aequaliter distantia a medio mundi per idem medium aequaliter moveantur ad medium mundi.

[55] Videtur quod sic. Quia secundum aequales proportiones proveniunt aequales velocitates. Modo eadem est proportio magnae terrae ad medium, et parvae terrae ad ipsum, cum utraque earum sit summe gravis.

[56] In oppositum est experientia.

[57] Ad quam difficultatem quidam magnus Proportionista concessit propositum illud propter rationem assignatam.

[58] Sed quia experientia est in oppositum, ideo respondendum est ad rationem, negando quod sit aequalis proportio.

[59] Et cum dicitur[28] 'utraque earum est summe gravis', concedatur. Et non sequitur ex hoc quod istorum mobilium ad medium sic est aequalis proportio, quoniam tota activitas non procedit ex parte gravitatis.

[60] Oportet igitur dicere quod istae gravitates sint aequales, cum tanta materia aequaliter densa.

[61] Hic tamen est dubium, an gravius alio velocius moveatur deorsum per idem medium et cum eadem figura.

[62] Dicendum quod sic. Ex quo infero tibi quod stat terram aequalem grano milii moveri velocius deorsum quam magnam terram. Si ibi esset, imaginor

6 Ad difficultatem respondendum est] Respondendum V2 | quia] quoniam V2 9 illud] aliud *a.c.* V1 11 parva] terra *add.* V2 19 ideo *om.* V2 22 sic est] sit V2 25 densa] dempsa *var.* V1 28 Dicendum quod] Ad quod dicendum V2 29 Si] ipsa *add.* V2 | imaginor] cum (?) *add. sed del.* V1

26 Cf. supra n. 26.

27 Cf. Blasius Pelacanus de Parma *Questiones disputate super octo libros Physicorum* VII q. 9 (2017) 128: Quinta difficultas: an magna terra et parva distantes a centro mundi per idem medium equaliter movebuntur.

28 Cf. supra n. 55.

enim quod A sit una parva terra summe gravis, | et gravitas medietatis infundatur in alia medietate; et clarum est quod illa medietas erit gravior in duplo quam primo esset. Sit igitur ista terra aequalis grano milii. Tunc probabile est dicere quod, si ista terra poneretur in media regione aëris, quod ipsa velocius moveretur deorsum quam si tota terra universi ibi esset. V1 137ra

[63] Octava difficultas: an eadem medicina sub maiori dosi, ut scamonia, maiorem effectum inducat in corpore quam sub minori quantitate.[29]

[64] Ad quod dicendum quod sic, propter ipsam ad corporis plures partes extendendi.

[65] Alia difficultas: utrum medicina quae ponitur calida in tertio possit inducere caliditatem quarti gradus quoad quantitatem ad auctam.

[66] Respondetur quod sic. Quoniam, ut dictum[30] fuit in quaestione, 'quantitas est de genere activorum'. Et haec de tertio articulo quaestionis.

[67] Quantum ad quartum, patet solutio ad rationes factas in contrarium a principio quaestionis. Et haec de articulo et tota quaestione. | V2 64ra

---

1 terra] quae est *add.* V2 5 si] sit V1 6 dosi] doxi *var.* V1 | scamonia] scamonoxa *var.* V1; scamonea *var.* V2 8 Ad quod dicendum] Respondendum V2 | ad] maiores *add. sed del.* V2 10 Alia] Ultima V2 13 articulo quaestionis *om.* V2 14–15 factas quaestione] patet solutio V2

---

29 Cf., e.g., Thomas Cantimpratensis *Liber de natura rerum* I, 52 (1973) 58: Qualiter scamonia aliquis purgandus sit: Si detur scamonia, colerico in pondere unius denarii detur, flegmatico in pondere duorum, melancolico in pondere trium. Colerici quidem de scamonia parum debent accipere; nam si multum acciperent, cum calidiores et fortiores sint aliis, scamonia eos magis calefaceret et potius parvam humiditatem consumeret quam humores nocivos purgaret.

30 Cf. supra n. 28.

APPENDIX

Blasius Pelacanus de Parma, *Quaestiones super VIII libros Physicorum* (redactio secunda)

Tabula quaestionum

V1 ms. Città del Vaticano, Biblioteca Apostolica Vaticana, Latin 2159.
V2 ms. Città del Vaticano, Biblioteca Apostolica Vaticana, Latin 3012.

Liber 1

Q. 1. Utrum scientifica notitia sit nobis de rebus naturalibus possibilis (V1 61ra–63ra | V2 5ra–6vb)

Q. 2. Utrum cognita causa totali alicuius rei cognoscatur statim illa res et non aliter (V1 63ra–65rb | V2 6vb–9ra)

Q. 3. Utrum in naturalibus ordine doctrine ab universalibus in singularia sit processus (V1 65rb–68rb | V2 9ra–11vb)

Q. 4. Utrum asserentes omnia esse unum possint probabiliter in hac opinione sustentari (V1 68rb–69va | V2 11vb–13ra)

Q. 5. Utrum totum sit omnes eius partes simul sumptae (V1 69va–71va | V2 13ra–15rb)

Q. 6. Utrum asserentes omnem rem extensam et suam extensionem non differre possint probabiliter positionem eorum substentare (V1 71va–74rb | V2 15rb–17vb)

Q. 7. Utrum in materia quantumcumque parva forma substantialis hora generationis producatur (V1 74rb–76vb | V2 17vb–20ra)

Q. 8. Utrum asserentes prima principia rerum naturalium esse tantum tria possint potenter impugnari (V1 76vb–79vb | V2 20ra–22vb)

Q. 9. Utrum per potentiam finitam vel infinitam possit aliquid fieri ex nihilo (V1 79vb–82vb | V2 22vb–25va)

Q. 10. Utrum quodlibet ens naturale appetat permanere (V1 82vb–84va | V2 25va–27ra)

 | DOI:10.1163/9789004734968_018

Liber II

Q. 1. Utrum domificator in faciendo domum faciat aliquid rebus naturalibus condistinctum (V1 84vb–86ra | V2 27rb–28ra)

Q. 2. Utrum quodlibet ens naturale habeat in se principium motus et quietis (V1 86ra–89rb | V2 28rb–30vb)

Q. 3. Utrum omnis forma in latitudine perfectionali entium sit perfectior quam sit materia (V1 89rb–91vb | V2 31ra–32vb)

Q. 4. Utrum diversae scientiae distinguantur secundum distinctionem obiectorum (V1 91vb–92va | V2 32vb–33rb)

Q. 5. Utrum diversae scientiae in perfectione essentiali secundum proportionem obiectorum proportionaliter excedant se (V1 92va–94ra | V2 33rb–34rb)

Q. 6. Utrum possit evidenter probari aliquid esse causa alterius (V1 94ra–96ra | V2 34rb–35vb)

Q. 7. Utrum ad cuiuscumque rei naturalis generationem praeter particulare requiratur influxus causae universalis, quae causa universalis dicitur sol (V1 96ra–98rb | 35vb–37rb)

Q. 8. Utrum inter agentia particularia, supposita semper generali influentia superiorum, possit qualitas una vel plures formam substantialem producere (V1 98rb–100ra | V2 37rb–38va)

Q. 9. Utrum asserentes omnia de necessitate evenire et nihil a casu vel a fortuna possint eorum positionem sustentare (V1 100ra–103rb | V2 38va–40va)

Liber III

Q. 1. Utrum in aliquo casu necesse sit ignorare materiam esse (V1 103va–104rb | V2 40va–41rb)

Q. 2. Utrum haec propositio 'motus est', significans motum esse et praecise sic, et non aliter, sit vera (V1 104rb–106rb | V2 41rb–42va)

Q. 3. Utrum motus sit ipsum mobile (V1 106rb–107vb | V2 42va–43vb)

Q. 4. Utrum contradictionem includat aliquam magnitudinem esse actu infinitam (V1 107vb–111va | V2 43vb–46vb)

Liber IV

Q. 1. Utrum possibile sit locum adequari locatum (Vi 111vb–114va | V2 47ra–49ra)
Q. 2. Utrum entia naturalia distantia ad eorum locis naturalibus moveantur ad illa impedimenta subtractis (Vi 114va–117rb | V2 49ra–51ra)
Q. 3. Utrum corpora naturalia appetant moveri ad eorum loca naturalia per lineas breviores (Vi 117rb–119ra | V2 51ra–52rb)
Q. 4. Utrum haec propositio 'vacuum est', significans vacuum esse et praecise sic, sit tanquam possibilis concedenda (Vi 119rb–122vb | V2 52rb–54vb)
Q. 5. Utrum in vacuo sit motus possibilis (Vi 122vb–126rb | V2 54vb–56vb)
Q. 6. Utrum penetratio corporum sit possibilis (Vi 126rb–127vb | V2 56vb–57va)
Q. 7. Utrum rarefactio sit possibilis (Vi 127vb–130rb | V2 57va–59rb)
Q. 8. Utum haec propositio sit concedenda 'nunc est tempus' (Vi 130rb–132rb | V2 59rb–60vb)
Q. 9. Utrum aliquid sit praecise per instans (Vi 132rb–135va | V2 60vb–62ra)

Liber V

Q. 1. Utrum agens naturale, hora qua agit in passum, agat in ipsum secundum se totum et quamlibet eius partem (Vi 134vb–137ra | V2 62rb–63vb)
Q. 2. Utrum generatio sit motus[1] (Vi 137ra–138vb | V2 64ra–65ra)
Q. 3. Utrum alteratio sit motus (Vi 138vb–140vb | V2 65ra–66rb)
Q. 4. Utrum augmentatio sit motus (Vi 140vb–143vb | V2 66rb–68ra)
Q. 5. Utrum contradictionem includat motum esse et non esse motum (Vi 143vb–145ra | V2 68ra–68va)
Q. 6. Utrum unitas motus sit magis attendenda penes unitatem temporis aut magis penes unitatem mobilis (Vi 145ra–147ra | V2 68va–69vb)
Q. 7. Utrum aliqui motus differant specie (Vi 147ra–149rb | V2 69vb–71ra)
Q. 8. Utrum in motibus sit penes contrarietatem terminorum ad quos contrarietas attendenda (Vi 149rb–151rb | V2 71ra–72ra)
Q. 9. Utrum possibile sit contraria in eodem simul complicari (Vi 151rb–154ra | V2 72ra–73va)
Q. 10. Utrum qualitas sit intensibilis et remissibilis (Vi 154ra–161rb | V2 73va–77va)

---

1 V2 Utrum in generatione sit servanda certa velocitas

Liber VI

Q. 1. Utrum per bonas rationes concludi possit continuum esse ex indivisibilibus compositum (V1 161va–164va | V2 77va–79ra)
Q. 2. Utrum continuum sit in infinitum divisibile (V1 164va–169vb | V2 79ra–82ra)
Q. 3. Utrum mobile velox per idem tempus vel equale plus pertranseat de spatio tardiori (V1 170ra–172ra | V2 82ra–83ra)
Q. 4. Utrum indivisibile moveri localiter vel alio modo rationibus obviet philosophorum (V1 172ra–173rb | V2 83ra–83vb)
Q. 5. Utrum sit possibile motum velocitari in infinitum (V1 173rb–174va | V2 83vb–84va)
Q. 6. Utrum omne quod movetur prius movebatur et post hoc movebitur (V1 174va–178va | V2 84va–87ra)
Q. 7. Utrum possibile sit magnitudinem infinitam transiri tempore finito et finitam transiri tempore infinito (V1 178vb–182ra | V2 87ra–89ra)
Q. 8. Utrum potenter possit improbari aliquod moveri localiter (V1 182ra–185rb | V2 89ra–91ra)

Liber VII

Q. 1. Utrum omne quod movetur moveatur ab alio (V1 185va–189ra | V2 91ra–93ra)
Q. 2. Utrum in motoribus et motis sit processus in infinitum aut potius sit venire ad primum motorem (V1 189ra–190ra | V2 93ra–94ra)
Q. 3. Utrum in omni motu movens et motum sint simul (V1 190rb–192va | V2 94ra–96ra)
Q. 4. Utrum motus rectus et circularis sint invicem comparabiles (V1 192va–193va | V2 96ra–96vb)
Q. 5. Utrum aequaliter gravia, quorum unum in brachio aequilibris est appensum, reliquum vero simpliciter suae naturae dimissum, possint aeque velociter in hac dispositione localiter moveri (V1 193va–194vb | V2 96vb–98ra)
Q. 6. Utrum in alteratione sit certa velocitas attendenda (V1 195ra–197va | V2 98ra–100ra)
Q. 7. Utrum in motu locali sit certa velocitas attendenda (V1 197va–200vb | V2 100ra–102vb)
Q. 8. Utrum in augmentatione sit certa velocitas attendenda (V1 200vb–203rb | V2 102vb–105ra

Q. 9. Utrum in motibus proportio velocitatum sit sicut proportio causarum (V1 203rb–209ra | V2 105ra–110ra)

Q. 10. Utrum omne agens naturale sit in eius actione limitatum (V1 209ra–210ra | V2 110ra–110vb)

Liber VIII

Q. 1. Utrum philosophicis rationibus patenter concludi possit motum fuisse ab aeterno (V1 210rb–213ra)

Q. 2. Utrum 'Deum non esse' contradictionem includat (V1 213ra–214ra)

Q. 3. Utrum contradictionem includat caelum fuisse aeternaliter productum (V1 214ra–214va)

Q. 4. Utrum caelum moveri in instanti contradictionem includat (V1 214va–215vb)

Q. 5. Utrum possibile sit primum motorem caelum movere in instanti (V1 215vb–218ra)

Q. 6. Utrum inanimate sive gravia sint sive levia ex se moveantur vel nata sint ex se moveri (V1 218ra–219vb)

Q. 7. Utrum motus localis sit primum motuum (V1 218vb–219va)

Q. 8. Utrum asserentes motos contrarios quiete medi interrumpi possint per rationes naturales improbari (V1 219va–221ra)

Q. 9. Utrum praecise motus circularis sit perpetuus (V1 221rb–222ra)

Q. 10. Utrum per rationes naturales rimari possit a quo proiecta moveantur contra inclinationes naturales cum ab impellente recesserunt (V1 222ra–223rb)

Q. 11. Utrum per rationes naturales concludi possit primum motorem qui est ipse Deus et vigore et duratione esse infinitum (V1 223rb–224va)

Q. 12. Utrum primus motor sit ubique, tamen magis in circumferentia quam in centro (V1 224va–225rb)

Bibliography

Manuscripts

Brugge, Stedelijke Openbare Bibliotheek 481
Milano, Biblioteca Ambrosiana, F. 145
Milano, Biblioteca Ambrosiana, P. 120 sup.
Vaticano, Biblioteca Apostolica Vaticana, Vat. lat. 1086
Vaticano, Biblioteca Apostolica Vaticana, Vat. lat. 2151
Vaticano, Biblioteca Apostolica Vaticana, Vat. lat. 2159
Vaticano, Biblioteca Apostolica Vaticana, Vat. lat. 2160
Vaticano, Biblioteca Apostolica Vaticana, Vat. lat. 3012
Vaticano, Biblioteca Apostolica Vaticana, Vat. lat. 4082
Vaticano, Biblioteca Apostolica Vaticana, Vat. lat. 4445
Vaticano, Biblioteca Apostolica Vaticana, Vat. Chig. O IV 41
Vaticano, Biblioteca Apostolica Vaticana, Vat. Pal. lat. 1056

Ancient and Medieval Authors

Aegidius Romanus. 1643. *In tertium librum Sententiarum quaestiones*. Romae.

Aegidius Romanus. 1646. *Quodlibeta*. Louvain.

Al-Fārābī. 2005. *De scientiis, "translatus a Girardo Cremonensi."* In Al-Fārābī. *Über die Wissenschaften / De scientiis*. Ed. Franz Schupp. Hamburg: Felix Meiner.

Albericus de Rosciate. 1601. *Dictionarium iuris quam civilis quam canonici*. Venetiis.

Albertus Magnus. 1890. *Analytica priora*. B. Alberti Magni *Opera omnia* II. Ed. A. Borgnet. Paris: Vivès.

Albertus Magnus. 1895. *Summa theologia*. B. Alberti Magni *Opera omnia* XXXI–III. Ed. A. Borgnet. Paris: Vivès.

Albertus Magnus. 1916–1921. *De animalibus* libri XXVI. 2 vols. Ed. H. Stabler (*Beiträge zur Geschichte der Philosophie des Mittelalters* XVI). Munich: Aschendorff.

Albertus Magnus. 1972. *Super Dionysium De divinis nominibus*. Alberti Magni *Opera omnia* XXXVII,1. Ed. Paulus Simon. Munich: Aschendorff.

Albertus Magnus. 1980. *De causis proprietatum elementorum*. Alberti Magni *Opera omnia* V,2. Ed. P. Hossfeld. Munich: Aschendorff.

Albertus Magnus. 1987. *Physica, libri 1–4*. Alberti Magni *Opera omnia* IV,1. Ed. P. Hossfeld. Munich: Aschendorff.

Albertus de Saxonia. 2008. *Quæstiones in Aristotelis De cælo*. Ed. Benoît Patar. *Philosophes médiévaux* t. 51. Louvain-la-Neuve / Louvain—Paris—Dudley, MA: Éditions de l'Institut supérieur de philosophie / Peeters.

Albertus de Saxonia. 2010. *Logica*. Ed. Harald Berger. Hamburg: Meiner.

Albertus de Saxonia (attrib.). 1999. *Expositio et quaestiones in Aristotelis "Physicam" ad Albertum de Saxonia attributae*. Ed. Benoît Patar. Louvain-la-Neuve: Ed. de l'Institut supérieur de philosophie.

Anonymus. *Sentencia probleumatum Aristotilis*. Cod. Brugge, Stedelijke Openbare Bibliotheek 481, ff. 105ra–110rb.

Archimedes. 1881. *De iis quae in humido vehuntur*. In *Opera omnia* II. Ed. J.L. Heiberg. Leipzig: Teubner. Reprinted Cambridge University Press, 2013.

ps.-Archimedes. 1952. *De insidentibus in humidum*. Ed. Ernest A. Moody and Marshall Clagett in *The Medieval Science of Weights (Scientia de Ponderibus) Treatises Ascribed to Euclid, Archimedes Thabit ibn Qurra, Jordanus de Nemore and Blasius of Parma*. Madison: University of Wisconsin Press.

Aristoteles Latinus. Corpus philosophorum medii aevi academiarum consociatarum auspiciis et consilio editum. Leiden: E.J. Brill, 1939 sqq.

pseudo-Aristoteles. 1562. *Physiognomia*, in *Aristotelis opera cum Averrois Commentariis*, vol. 7 (ff. 132va–139vb). Venetiis apud Iunctas. Reprinted Frankfort: Minerva, 1962.

Aristotle. 1984. *The Complete Works of Aristotle: The Revised Oxford Translation*. Edited by Jonathan Barnes. Princeton: Princeton University Press.

Auctoritates Aristotelis. 1974. Ed. Jacqueline Hamesse in *Les Auctoritates Aristotelis. Philosophes Médiévaux* XVII. Louvain: Publications Universitaires and Paris: Béatrice-Nauwelaerts.

Averroës. 1562. *Commentarium in libros Metaphysicorum*. In *Aristotelis opera cum Averrois Commentariis*, vol. 8. Venetiis apud Iunctas. Reprinted Frankfort: Minerva, 1962.

Averroës. 1562. *Commentarium in libros Physicorum*, in *Aristotelis opera cum Averrois Commentariis*, vol. 4. Venetiis apud Iunctas. Reprinted Frankfort: Minerva, 1962.

Averroës. 1562. *Commentarium in librum De caelo*, in *Aristotelis opera cum Averrois Commentariis*, vol. 4. Venetiis apud Iunctas. Reprinted Frankfort: Minerva, 1962.

Averroës. 1949. *Libri Aristotelis De sensu et sensato*. Ed. Aemilia Ledyard Shields, adiuvante Henrico Blumberg. In *Corpus commentariorum Averrois in Aristotelem*. Ed. Harry Austryn Wolfson. Cambridge, Mass: Mediaeval Academy of America.

Averroës. 1949. *Libri Aristotelis De sompno et vigilia*. Ed. Aemilia Ledyard Shields, adiuvante Henrico Blumberg. In *Corpus commentariorum Averrois in Aristotelem*. Ed. Harry Austryn Wolfson. Cambridge, Mass: Mediaeval Academy of America.

Avicenna. 1564. *Canon medicinae*. In Avicenna *Libri in re medica omnes, qui hactenus ad nos pervenere*. Venetiis: apud Vincentium Valgrisium.

Avicenna. 1595. *Canon medicinae*. Venetiis apud Iuntas.

Avicenna. 1987. *Liber tertius naturalium: De generatione et corruptione*. Ed. S. Van Riet. Leiden: Brill; Peeters: Louvain.

pseudo-Avicenna. 1949. *De diluviis dictis in T'imaeo Platonis*. Ed. M. Alonso Alonso in

"Homenaje a Avicena en su milenario: las traducciones de Juan Gonzalez de Burgosy Salomon." *Al-Andalus* 14, pp. 291–319 at 306–308.

Bartholomeus Coclitis. 1523. *Chyromantie ac physionomie anastasis*. Bononia.

Baverius de Baveriis. 1521. *Consilia de Baveriis*. Pavia.

Beda (attrib.). 1605. *Axiomata philosophica Venerabilis Bedae*. Coloniae.

Beda (attrib.). 1850. *Sententiae philosophicae. Patrologiae cursus completus, Series Latina* 90. Ed. J.P. Migne. Paris: Vivès.

Bero Magni de Ludosia. *Disputata super libros De anima*. Edited with an Introduction by Robert Andrews. Stockholm: Bibliotheca Philosophorum Suecorum Medii Aevi / Library of Swedish Philosophers in the Middle Ages, forthcoming.

Blasius Pelacanus de Parma. *Quaestiones circa libros De generatione et corruptione*. Cod. Vat. Chig. O IV 41.

Blasius Pelacanus de Parma. *Questiones circa libros Metaurorum Aristotelis*. Cod. Vat. lat. 2160.

Blasius Pelacanus de Parma. *Questiones circa libros Metaurorum Aristotelis* I q. 1. Ed. Joël Biard. https:https://bdeparme.hypotheses.org/quaestiones-in-meteorologica

Blasius Pelacanus de Parma. *Questiones disputate super octo libros Physicorum* prima redactio. Ed. Joël Biard. https://bdeparme.hypotheses.org/questions-sur-la-physique-premiere-redaction-padoue-1385-1388

Blasius Pelacanus de Parma. *Questiones disputate super octo libros Physicorum* secunda redactio. Ed. Joël Biard. https://bdeparme.hypotheses.org/questions-sur-la-physique.

Blasius Pelacanus de Parma. 1952. *De ponderibus*. Ed. Ernest A. Moody and Marshall Clagett in *The Medieval Science of Weights (Scientia de Ponderibus): Treatises Ascribed to Euclid, Archimedes, Thabit ibn Qurra, Jordanus de Nemore and Blasius of Parma*. Madison: University of Wisconsin Press.

Blasius Pelacanus de Parma. 1961. *Quaestiones de optica*. Ed. Franco Alessio in "Questioni inedite di ottica di Biagio Pelacani da Parma." *Rivista critica di storia della flosofa* 16: 79–110, 188–221.

Blasius Pelacanus de Parma. 1967. *Questiones super tractatum de ponderibus*. Ed. J.E. Brown in *The* Scientia de ponderibus *in the Latter Middle Ages*. PhD Dissertation, Madison, University of Wisconsin, pp. 348–569.

Blasius Pelacanus de Parma. 1974. *Quaestiones in Aristotelis De anima libri*. In *Le quaestiones de anima* di Biagio Pelacani da Parma. Ed. Graziella Federici Vescovini. Firenze: Olschki.

Blasius Pelacanus de Parma. 1976. *Quaestiones musicae*. Paris BN lat. 7372. Selections edited in John Murdoch, "Music and Natural Philosophy: Hitherto Unnoticed Questiones by Blasius of Parma(?)." *Manuscripta* 20, 3, pp. 119–136.

Blasius Pelacanus de Parma. 1994. *Quaestio de intensione et remissione formarum*. Ed. G. Federici Vescovini in "La Quaestio de intensione et remissione formarum de Biagio Pelacani da Parma." *Physis: Rivista internazionale di storia della scienza* 31: 433–535.

Blasius Pelacanus de Parma. 2001. *Questiones super tractatus logice magistri Petri Hispani.* Ed. Joël Biard and Graziella Federici-Vescovini. Paris: Vrin.

Blasius Pelacanus de Parma. 2005. *Questiones circa tractatum proportionum magistri Thome Braduardini.* Ed. Joël Biard and Sabine Rommevaux. Paris: Vrin.

Blasius Pelacanus de Parma. 2008. *Questiones circa tractatum proportionum magistri Thome Braduardini.* redactio prima, q. 7. Ed. Joël Biard et Sabine Rommevaux in "Annexe: transcription de la question 7 dans la première version des *Questions sur le traité des rapports de Thomas Bradwardine* par Blaise de Parme, d'après le manuscrit Milan, Ambrosiana, F. 145." In Joël Biard and S. Rommevaux, eds. *Mathématiques et théorie du mouvement. XIVe–XVIe siècles.* Villeneuve-d'Ascq: Presses Universitaires du Septentrion.

Blasius Pelacanus de Parma. 2009. *Questiones super perspectiva communi.* Ed. Graziella Federici-Vescovini et Joël Biard, avec la collab. d'Orsola Rignani et Valeria Sorge, et Riccardo Bellé pour les figures. Paris: Vrin.

Blasius Pelacanus de Parma. 2012. *Questiones disputate super octo libros Physicorum* secunda redactio IV q. 4–5. Ed. Joël Biard et Sabine Rommevaux in "Le vide secundum naturam dans les *Questions sur la physique* de Blaise de Parme," in *La Nature et le Vide dans la physique médiévale*, ed Joël Biard and Sabine Rommevaux. Studia Artistarum 22. Turnhout: Brepols, pp. 293–378.

Blasius Pelacanus de Parma. 2017. *Questiones disputate super octo libros Physicorum,* liber VII. Ed. Sabine Rommevaux. ⟨halshs-01663689⟩

Blasius Pelacanus de Parma. 2019. *Questiones circa libros Metaurorum Aristotelis* I q. 9. Ed. A. Panzica in "Les *Questions sur les Météorologiques* du manuscrit Vat. Lat. 4082: Blaise de Parme, Nicole Oresme et l'*Inter omnes impressiones*," in *Bulletin de philosophie médiévale* 61, pp. 153–182.

Blasius Pelacanus de Parma. 2019. *Quaestiones de anima* II q. 25. Ed. Tuomas Vaura. *Rationality in Perception: Transformations of Mind and Cognition 1250–1550*—project. URL = https://blogs.helsinki.fi/rationality-in-perception/texts/.

Buenaventura de San Agustín. 1738–1739. *Artium cursus: Libros de generatione, coelo mundo et anima appendicemque contra atomistas comprehendens* I–IV. Tertia editio. Cesar-Augustae: in Regali Monasterio Sanctae Engratiae.

Chartularium Universitatis Parisiensis. 1899. Ed. Heinrich Denifle et al. Parisiis: ex typis fratrum Delalain.

Conradus Gesneri. 1555. *Historiae animalium,* liber III qui est de auium natura. Tiguri: apud Christoph. Froschouerum.

Dominicus Gundissalinus. 1903. *De divisione philosophiae.* Ed. Ludwig Baur. Münster: Aschendorff.

Dominicus Gundissalinus. 1954. *De scientiis.* Ed. Manuel A. Alonso. Madrid/Granada: Consejo Superior de Investigaciones Científicas.

Euclides. 1983. *Elementa*. In *The First Latin Translation of Euclid's 'Elements', Commonly Ascribed to Adelard of Bath*. Ed. H.L.L. Busard. Toronto: Pontifical Institute of Mediaeval Studies.

Galenus. 1821. *Ars medica*. In *Claudi Galeni Opera Omnia* I. Ed. C.G. Kühn. Leipzig: Knobloch.

Giraldus Cambrensis. 1867. *Topographica Hibernica*. Ed. James F. Dimock in *Giraldi Cambrensis Topographica Hibernica et Expugnatio Hibernica*. Giraldi Cambrensis Opera 5. London: Longman, Green, Longman, and Roberts.

Gregorius Ariminensis. 1979. *Gregorii Ariminensis OESA Lectura super primum et secundum Sententiarum*. Ed. A. Damasus Trapp et al. Berlin: Walter de Gruyter.

Gualterus Burlaeus. *De somno*, q. 17, "Utrum cessante primo proiciente possit proiectum moveri." Cod. Vaticano, Biblioteca Apostolica Vaticana, Vat. Lat. 2151 ff. 105ra–106ra.

Gualterus Burlaeus. 1496. *De intensione et remissione formarum*, Venice: Octavianus Scotus.

Gualterus Burlaeus. 1955. *De puritate artis logicae tractatus longior, with a Revised Edition of the Tractatus brevior*. Ed. Philotheus Boehner. St. Bonaventure, N.Y.: The Franciscan Institute.

Gualterus Burlaeus. 1996. *Tractatus primus: de quattuor conclusionibus circa formas accidentales*. Partial edition in Lambert M. De Rijk, "Burley's so-called *Tractatus primus*, with an Edition of the Additional Quaestio '*Utrum contradictio sit maxima oppositio*'." *Vivarium* 34: 161–191.

Guillelmus Heytesbury. 1494. *Tractatus de sensu composito et diviso*. Ed. B. Locatellus. Venice.

Guillelmus Heytesbury. 2017. *Termini naturales*. Ed. Robert Andrews in "Tideman of Närke, the *Termini naturales*, and the *Liber naturarum* of the *Libellus sophistarum ad usum Cantabrigiensium*." In *Swedish Students at the Universities of Cambridge and Oxford in the Middle Ages*. Robert Andrews and Olle Ferm, eds. Stockholm: Sällskapet Runica et Mediaevalia, pp. 53–137.

Guillelmus de Ockham. 1944. *Tractatus de Successivis* (attributed to Ockham). Ed. P. Boehner. St. Bonaventure, N.Y.: The Franciscan Institute.

Guillelmus de Ockham. 1967. *Scriptum in librum primum Sententiarum, Ordinatio*: Prologus et distinctio prima. Guillelmi de Ockham *Opera Philosophica et Theologica. Opera Theologica* I. Ed. Gedeon Gál. St. Bonaventure, N.Y.: The Franciscan Institute.

Guillelmus de Ockham. 1974. *Summa logicae*. Guillelmi de Ockham *Opera Philosophica et Theologica. Opera Philosophica* I. Ed. P. Boehner et al. St. Bonaventure, N.Y.: The Franciscan Institute.

Guillelmus de Ockham. 1980. *Quodlibeta septem*. Guillelmi de Ockham *Opera Philosophica et Theologica. Opera Theologica* IX. Ed. Joseph C. Wey. St. Bonaventure, N.Y.: The Franciscan Institute.

Guillelmus de Ockham. 1981. *Quaestiones in librum secundum Sententiarum, Reportatio.* Guillelmi de Ockham *Opera Philosophica et Theologica. Opera Theologica* v. Ed. Gedeon Gál and Rega Wood. St. Bonaventure, N.Y.: The Franciscan Institute.

Guillelmus de Ockham. 1984. *Brevis Summa libri Physicorum, Summula Philosophiae Naturalis, et Quaestiones in libros Physicorum Aristotelis.* Guillelmi de Ockham *Opera Philosophica et Theologica. Opera Philosophica* VI. Ed. Stephen F. Brown. St. Bonaventure, N.Y.: The Franciscan Institute.

Guillelmus de Ockham. 1985. *Expositio in libros Physicorum Aristotelis* (Libri 4–8). Guillelmi de Ockham *Opera Philosophica et Theologica. Opera Philosophica* v. Ed. R. Wood, R. Green, G. Gal, J. Giermek, F. Kelley, G. Leibold, and G. Etzkorn. St. Bonaventure, N.Y.: The Franciscan Institute.

Guillelmus de Ockham. 1986. *Tractatus de corpore Christi.* Guillelmi de Ockham *Opera Philosophica et Theologica. Opera Theologica* X. Ed. C.A. Grassi. St. Bonaventure, N.Y.: The Franciscan Institute.

Iacobus de Ascoli. *Quodlibeta.* Cod. Vat. lat. 1086, ff. 53^a–78^b.

Iacobus Forliviensis. 1495. *De intensione et remissione formarum.* Venice: Octavianus Scotus.

Iacobus Forliviensis. 1495. *Expositio super primo Canonis Avicenne cum questionibus eiusdem.* Venetiis.

Iacobus Forliviensis. 1855. *Tituli quaestionum.* Ed. Francesco Puccinotti, "Ragguaglio del codice Vaticano 4445." *Storia della medicina* v. 2 pt. 1, "Questioni anonime" pp. CXL–CXLII.

Iacobus de Vitriaco. 1972. *Historia Occidentalis.* Ed. John Frederick Hinnesbusch, in *The Historia Occidentalis of Jacques de Vitry.* Spicilegium Friburgense 17. Fribourg, Switzerland: The University Press.

Iacobus Zabarella. 1602. *Commentarii in Aristotelis libros physicorum, item in libros De generatione et corruptione, item in Meteora.* Francofurti.

Ioannes Buridanus. 1518. *In Metaphysicen Aristotelis questiones*, Parisiis: Jodocus Badius [repr. Francfurt a.M.: Minerva, 1987].

Ioannes Buridanus. 1996. *Expositio et quaestiones in Aristoteles De caelo.* Ed. Benoît Patar. *Philosophes médiévaux* 33. Louvain-la-Neuve: Ed. de l'Inst. Supérieur de Philosophie.

Ioannes Buridanus. 1998. *Summulae De Suppositionibus.* Introduction, critical edition and indexes by Ria Van der Lecq. Nijmegen: Ingenium Publishers.

Ioannes Buridanus. 2004. *In Metaphysicam* VII q. 20. Excerpt in E.J. Ashworth, "Singular Terms and Singular Concepts." In *John Buridan and Beyond: Topics in the Language Sciences, 1300–1700.* Ed. Russell L. Friedman and Sten Ebblesen. Copenhagen: Det Kongelige Danske Videnskabernes Selskab.

Ioannes Buridanus. 2004. *Summulae De practica sophismatum.* Introduction, critical edition and indexes by Fabienne Pironet. Turnhout: Brepols.

Ioannes Buridanus. 2015. *Quaestiones super octo libros Physicorum Aristotelis (secundum ultimam lecturam) libri I–II*. Ed. Michiel Streijger, Paul J.J.M. Bakker, and Edith Dudley Sylla. Leiden: Brill.

Ioannes Buridanus. 2016. *Quaestiones super octo libros Physicorum Aristotelis (secundum ultimam lecturam) libri III–IV*. Ed. Michiel Streijger, Paul J.J.M. Bakker, and Edith Dudley Sylla. Leiden: Brill.

Iohannes Buridanus. 2023. *John Buridan's Questions on Aristotle's De Anima—Iohannis Buridani Quaestiones in Aristotelis De Anima*. Ed. and trans. Gyula Klima, Peter G. Sobol, Peter Hartman, and Jack Zupko. Historical-Analytical Studies on Nature, Mind and Action 9. Cham: Springer.

Ioannes de Celaya. 1517. *Expositio magistri Ioannis de Celaya Valentini in octo libros Physicorum*. Paris: Hémon Le Fèvre.

Ioannes Dumbleton. *Summa logica et philosophiae naturalis*. Cod. Vatican Pal. lat. 1056.

Ioannes Duns Scotus. 1895. *Quodlibet*. Ioannis Duns Scoti *Opera Omnia* XXV. Paris: Vivès.

Ioannes Duns Scotus. 1997–1998. *Doctoris Subtilis Ioannis Duns Scoti Quaestiones super libros Metaphysicorum Aristotelis*. Ed. R. Andrews, G. Etzkorn, G. Gál, R. Green, F. Kelley, T. Noone, and R. Wood. *Opera philosophica* vols. III–IV. St. Bonaventure, N.Y.: The Franciscan Institute.

Ioannes Pecham. 1970. *Perspectiva communis*. Ed. David Lindberg, in *John Pecham and the Science of Optics*. Milwaukee: University of Wisconsin Press.

Ioannes de Sacrobosco. c. 1478. *Sphaera*. Venice: Adam de Rottweil.

Ioannes de Sacrobosco. 1949. *Sphaera*. Ed. Lynn Thorndike in *The* Sphere *of Sacrobosco and Its Commentators*. Chicago: University of Chicago Press.

Iordanus Nemorarius. 1952. *Elementa Jordani super demonstrationem ponderum*. Ed. Ernest A. Moody and Marshall Clagett in *The Medieval Science of Weights* (Scientia de Ponderibus). *Treatises Ascribed to Euclid, Archimedes, Thabit ibn Qurra, Jordanus de Nemore, and Blasius of Parma*. Madison: University of Wisconsin Press.

Iordanus Nemorarius. 1952. *Liber Jordani de ponderibus*. Ed. Ernest A. Moody and Marshall Clagett in *The Medieval Science of Weights* (Scientia de Ponderibus). *Treatises Ascribed to Euclid, Archimedes, Thabit ibn Qurra, Jordanus de Nemore, and Blasius of Parma*. Madison: University of Wisconsin Press.

Leonardo da Vinci. 2000. *Il codice atlantico [della Biblioteca Ambrosiana di Milano]* III. Firenze: Giunti-Barberè.

Liber de causis. 1966. Ed. Adrien Pattin in "Le Liber de Causis. Édition établie à l'aide de 90 manuscrits." *Tijdschrift voor Filosofie* 28/1: 90–203.

Ludovicus Buccaferreus. 1590. *Lectiones super primum librum Meteorologicorum Aristotelis*. Venetiis.

Maier, Michael. 1619. *Tractatus de volucri arborea, absque patre et matre*. Francofurti.

Marsilio Ficino. 1956. *De amore: Commentaire sur le Banquet de Platon Marsile Ficin*. Ed. and trans. R. Marcel. Paris: Les Belles-Lettres.

Marsilius de Inghen. 1521. *Abbreviationes super octo libros Physicorum*. Venetiis.

Marsilius de Inghen. 1567. *Quaestiones super libros De generatione*. Venetiis.

ps. Marsilius de Inghen (pseudo-Ioannes Duns Scotus). 1891. *Expositio et Quaestiones in VIII libros Physicorum Aristotelis*. Ioannis Duns Scoti *Opera Omnia* III. Paris: Vivès.

Nicolaus de Argentina. 2009. *Summa* liber 2, 2; Tractatus 3–7. Ed. Kurt Flasch and Tiziana Suarez-Nani. *Corpus philosophorum teutonicorum medii aevi*, 5,2,2. Hamburg: Meiner.

Nicolaus de Autrecourt. 1939. *Tractatus utilis ad videndum an sermones peripateticorum fuerint demonstrativi* [otherwise called *Exigit ordo*]. Ed. J.R. O'Donnell in "Nicholas of Autrecourt." *Medieval Studies* 1: 179–266.

Nicolaus de Autrecourt. 1994. *His Correspondence with Master Giles and Bernard of Arezzo: A Critical Edition and English Translation* by L.M. de Rijk. Leiden: E.J. Brill.

Nicolaus Oresme. 1965. *Quaestiones super De caelo*. In *The "Quaestiones super De celo" of Nicole Oresme*. Ed. C. Kren. Unpublished doctoral thesis, The University of Wisconsin, Madison.

Nicolaus Oresme. 1966. *De proportionibus proportionum*. Ed. Edward Grant in *Nicole Oresme: "De proportionibus proportionum" and "Ad pauca respicientes."* Madison, Milwaukee, and London: University of Wisconsin Press.

Nicolaus Oresme. 1996. *Quaestiones super De generatione et corruptione*. Ed. Stefano Caroti. München: Beck.

Nicole Oresme. 2013. *Questiones super Physicam* (Books I–VII). Edited with Introduction and Indices by Stefano Caroti, Jean Celeyrette, Stefan Kirschner, and Edmond Mazet. Leiden: Brill.

Nicolaus Oresme. 2021. *Questiones in Meteorologica secundum ultimam lecturam, recensio Parisiensis. Study of the Manuscript Tradition and Critical Edition of Books I–II.10*. Ed. Aurora Panzica. Leiden; Boston: Brill.

Odo de Ceritona. 1896. *Fabulae*. Ed. Leopold Hervieux in *Les fabulistes latins* IV. Paris: Firmin-Didot.

Paracelsus, Aureolus Theophrastus. 1573. *Summi philosophi ac medici, De natura rerum: libri septem*. Perna.

Patricius, Episcopus (attrib.). 1955. *Versus sancti Patricii episcopi de mirabilibus Hibernie*. Ed. Aubrey Gwynn in *The Writings of Bishop Patrick 1074–1084*. Scriptores Latini Hiberniae, Volume I. Dublin: The Dublin Institute for Advanced Studies.

Paulus Veneti. 1978. *Logica magna*. P. 2. Fasc. 6, *Tractatus de veritate et falsitate propositionis* et *Tractatus de significato propositionis*. Ed. Francesco del Punta. Oxford: Oxford University Press for the British Academy.

Petrus de Abano. 1496. *Conciliator Differentiarum, quæ inter Philosophos et Medicos Versantur*. Venetiis.

Petrus de Bruxellis. 1510. *Argutissime subtiles et fecunde questiones phisicales*. Paris.

Petrus de Crescentius. 1471. *Ruralia commoda*. [Augsburg]: Johannes Schuessler.

Petrus Damianus. 1853. *De divina omnipotentia. Patrologiae cursus completus, Series Latina* 145. Ed. J.P. Migne. Paris: Vivès.

Petrus Pomponazzi. 1556. *De naturalium effectuum causis sive de incantationibus*. Basilea.

Plinius Secundus. 2002. *Historia Naturalis Libri III IV: Geografie Europa*. Ed. Roderich Winkler and Gerhard König. Berlin: Walter de Gruyter.

Porphyrius. 1966. *Isagoge. Aristoteles Latinus* I,6. Ed. L. Minio-Paluello. Leiden: E.J. Brill.

Proverbialia Dicteria: Ethicam et Moralem Doctrinam complectentia, Versibus veteribus Rhytmicis, ab antiquitate mutuatis, vna cum Germanica interpretatione, conscripta, & studiose collecta. 1575. Ed. Andreas Gartner. Frankfurt am Main.

Ptolemaeus. 1515. Claudius Ptolemaeus *Almagestum*. Venice: Petrus Lichtenstein.

pseudo-Ptolomeus.1501. *De nativitatibus*. Venezia: Giovanni Battista Sessa.

Quaestiones Cracovienses super octo libros Physicorum Aristotelis. 1969. Ed. Ryszard Palacz. *Studia mediewistyczne* 10, pp. v–xxviii and 1–265.

Razes (Rhases; Abii Bakr ar-Razi). 1893. *Almansor. Abubecri Rasis ad regem Mansorem De re medicina*. Ed. R. Förster in *Scriptores physiognomonici graeci et latini* II. Lipsiae.

Regiomontanus. *Epitome Almagesti*. Trans. Henry Zepeda, "Transcription of Regiomontanus, *Epitome Almagesti*" (update 25.05.2021). *Ptolemaeus Arabus et Latinus. Texts*, URL = https://ptolemaeus.badw.de/text/M597.

Renatus Cartesius. 1964–1974. *Oeuvres de Descartes*. 11 vols. Ed. Charles Adam and Paul Tannery. Paris: Vrin-CNRS.

Renatus Cartesius. 1984–1991. *The Philosophical Writings of Descartes*. 3 vols. Ed. and trans. John Cottingham, Robert Stoothoff, Dugald Murdoch, and (for vol. 3) Anthony Kenny. Cambridge: Cambridge University Press.

Richardus Divisensis. 1838. *De rebus gestis Ricardi Primi, regis Angliae*. Ed. Joseph Stevenson. English Historical Society Publications vol. 5. London: Sumptibus Societatis.

Richardus Swineshead. 1520. *Calculator: Subtilissimi Ricardi Suiseth Anglici Calculationes*. Venetiis: Scotus.

Richardus Swineshead. 1966. *Calculator* tr. XI. Ed. M.A. Hoskin and A.G. Molland in "Swineshead on Falling Bodies: An Example of Fourteenth-Century Physics." *The British Journal for the History of Science* 3: 150–182.

Robertus Alyngton. 1993. *Litteralis sententia super Praedicamenta Aristotelis*. Ed. Alessandro D. Conti in "Linguaggio e realtà nel commento alle Categorie di Robert Alyngton," *Documenti e studi sulla tradizione filosofica medievale* 4: 179–306.

Robertus Grosseteste. 1912. *De impressionibus aeris seu de prognosticatione*. Ed. L. Baur. *Beiträge zur Geschichte der Philosophie des Mittelalters* IX. Munich: Aschendorff.

Robertus Grosseteste. 1912. *De natura locorum*. Ed. L. Baur. *Beiträge zur Geschichte der Philosophie des Mittelalters* IX. Munich: Aschendorff.

Robertus Grosseteste. 2011. *De luce*. In *Roberto Grossatesta, La luce*. Introduzione, testo latino, traduzione e commento di Cecilia Panti, Prefazione di Pietro Bassiano Rossi. Pisa: Pisa University Press.

Robertus Grosseteste (attrib.). 1912. *Summa philosophia*. Ed. L. Baur. *Beiträge zur Geschichte der Philosophie des Mittelalters* IX. Munich: Aschendorff.

Robertus Kilwardby. 1976. *De ortu scientiarum*. Ed. Albert G. Judy. Oxford University Press: Oxford.

Rodolphus Goclenius. 1599. *Disquisitiones philosophicae*. Marpurgi Cattorum.

Rogerus Bacon. 1920. *Secretum secretorum cum glossis et notulis; tractatus brevis et utilis ad declarandum quedam obscure dicta Fratris Rogeri*. Ed. Robert Steele and A.S. Fulton. Oxonii: E Typographeo Clarendoniano.

Sadan (Shadhan b. Bahr). 1998. *Excerpta de secretis Albumazar*. Ed. G. Federici Vescovini in "La versio latina degli *Excerpta de secretis Albumasar* di Sadan." *Archives d'Histoire doctrinale et littéraire du Moyen Âge* 65: 297–330.

Simplicius. 1540. *In quatuor libros De celo Aristotelis*. Venice.

Thimo Judaei. 1518. *Thimo in Aristotelem de Meteoris*. Ed. Georgius Lokert. Paris.

Thomas de Aquino. 1852–1873. *Opera omnia*, I–XXV. Ed. Parma.

Thomas de Aquino. 1882–1948. *Opera omnia* I–XVI. Ed. Leoninae. Rome.

Thomas Bradwardine. 1961. *Tractatus de proportionibus velocitatum in motibus*. Ed. H. Lamar Crosby, Jr., in *Thomas of Bradwardine. His* Tractatus de proportionibus. *Its Significance for the Development of Mathematical Physics*. Madison: The University of Wisconsin Press.

Thomas Cantimpratensis. 1973. *Liber de Natura Rerum editio princeps secundum codices manuscriptos. Teil 1, Text*. Berlin: Walter De Gruyter.

Witelo. 1983. *Witelonis perspectivae liber quintus = Book V of Witelo's Perspectiva: an English translation, with introduction and commentary and Latin edition of the first catoptrical book of Witelo's Perspectiva*. Ed. A. Mark Smith. Studia Copernicana, 23. Wrocław: Ossolineum.

Modern Authors

Abattouy, Mohammed. 2002. "The Aristotelian Foundations of Arabic Mechanics: From the Ninth to the Twelfth Century." In *The Dynamics of Aristotelian Natural Philosophy from Antiquity to the Seventeenth Century*. Eds. Cees H. Leijenhorst, Christoph H. Lüthy, and Johannes M.M.H. Thijssen. Leiden / Boston / Cologne: Brill, pp. 109–140.

Abattouy, Mohammed. 2001. "Greek Mechanics in Arabic Context: Thābit ibn Qurra, al-Isfizārī and the Arabic Traditions of Aristotelian and Euclidean Mechanics," *Science in Context* 14 (1/2): 179–247.

Abattouy, Mohammed and Salim Al-Hassani. 2015. *The Corpus of Al-Isfizārī in the*

Sciences of Weights and Mechanical Devices. London: Al-Furqān Islamic Heritage Foundation.

Adams, Marilyn McCord. 1987. *William Ockham*. 2 volumes. Notre Dame, IN: University of Notre Dame Press.

Adams, Marilyn McCord. 2010. "Independent Accidents: Scotus and Ockham." In *Some Later Medieval Theories of the Eucharist: Thomas Aquinas, Giles of Rome, Duns Scotus, and William Ockham*. Ed. Marilyn McCord Adams. Oxford: Oxford University Press, pp. 197–226.

Affò, Ireneo. 1791. *Memorie degli scrittori et letterati parmigiani*. 7 vol. Parma: Stamperia reale.

Anstey, P.R. and Vanzo, A. 2023. *The Origins of the Experimental/Speculative Distinction*. Cambridge University Press: Cambridge.

Arlig, A.2015. "Medieval Mereology." In *The Stanford Encyclopedia of Philosophy*. Ed. Edward N. Zalta and Uri Nodelman. URL = https://plato.stanford.edu/entries/mereo logy-medieval/

Ashworth, E.J. 2004. "Singular Terms and Singular Concepts." In *John Buridan and Beyond: Topics in the Language Sciences, 1300–1700*. Ed. Russell L. Friedman and Sten Ebbesen. Copenhagen: Det Kongelige Danske Videnskabernes Selskab.

Bacon, Francis. 1857–1874. *The Works of Francis Bacon*. Ed. James Spedding, Robert Leslie Ellis, and Douglas Denon Heath. London: Longmans; reprinted New York: Garett, 1968.

Bacon, Roger. 1962. *Opus maius*. Vol. II. Translated by Robert Burke. New York: Russell and Russell.

Bakker, Paul J.J.M. 2024. "John Buridan on the Ontological Status of Artifacts: Interpreting His Commentaries on Aristotle's *Physics*." In *Interpreting Buridan: Critical Essays*. Ed. S. Johnston and H. Lagerlund. Cambridge: Cambridge University Press, pp. 101–134.

Bertolacci, A. 2013. "Averroes against Avicenna on Human Spontaneous Generation: The Starting-Point of a Lasting Debate." In *Renaissance Averroism and its Aftermath*. Eds. A. Akasoy / G. Giglioni. Springer, Dordrecht, pp. 37–54.

Biard, Joël and Graziella Federici Vescovini. 2001. "Introduction," in Blasius of Parma, *Questiones super Tractatus logice magistri Petri Hispani*. Ed. Joël Biard and Graziella Federici Vescovini. Paris: Vrin, pp. 7–36.

Biard, Joël. 2003. "Blaise de Parme et la théorie de la signification: '*Doctor Parisinus*'?" In *Ratio et superstitio. Essays in Honor of Graziella Federici Vescovini*. Ed. Giancarlo Marchetti, Orsola Rignani, and Valeria Sorge. Turnhout: Brepols, pp. 221–242.

Biard, Joël. 2003. "L'être et la mesure dans l'intension et la rémission des formes (Jean Buridan, Blaise de Parme)." *Medioevo* 27: 415–448.

Biard, Joël. 2003. "Le traité sur les *Catégories* de Blaise de Parme." In *La tradition médiévale des catégories, XIIe–XVe siècles. Actes du XIIIe Symposium européen de logique et de sémantique médiévales* (*Avignon, 6–10 juin 2000*). Ed. Joël Biard and Irène Rosier-Catach. Louvain-la Neuve / Leuven—Paris: Éditions de l'Institut supérieur de philosophie / Éditions Peeters, pp. 365–378.

Biard, Joël. 2009. "Blasius of Parma Facing Atomist Assumptions." In *Atomism in Late Medieval Philosophy and Theology*. Ed. Christophe Grellard and Aurélien Robert. Leiden / Boston: Brill, pp. 221–234.

Biard, Joël. 2017. "Matière, forme, qualité. Blaise de Parme et le statut de la matière." In *Materia. Nouvelles perspectives de recherche dans la pensée et la culture médiévales* (*XIIe–XVIe siècles*). Ed. Tiziana Suarez-Nani and Agostino Paravicini Bagliani. Florence: SISMEL—Edizioni del Galluzzo, pp. 191–215.

Biard, Joël. 2018. "Natures et degrés de perfection. L'échelle des étants chez Blaise de Parme." In *Edizioni, traduzioni e tradizioni filosofiche* (*secoli XII–XVII*). *Studi per Pietro B. Rossi*. Ed. Luca Bianchi, Onorato Grassi, and Cecilia Panti. Rome: Aracne, II, pp. 397–410.

Biard, Joël and Aurélien Robert, eds. 2019. *La Philosophie de Blaise de Parme. Physique, psychologie, éthique*. Florence: SISMEL—Edizioni del Galluzzo.

Biard, Joël. 2022. "Blasius of Parma on the Calculation of the Variation of Qualities and Aristotelian Physics." In *Quantifying Aristotle. The Impact, Spread and Decline of the Calculatores Tradition*. Ed. Daniel A. Di Liscia and Edith D. Sylla. Leiden / Boston: Brill, pp. 232–254.

Biard, Joël. 2022. *Hypothèse matérialiste et pensée radicale. La philosophie de la nature de Blaise de Parme*. Paris: Vrin.

Biard, Joël. 2022. "La matière est-elle le sujet des accidents? Un débat au XIVe siècle." *Acta Philosophica. Rivista internazionale di filosofia* 31/1: 15–31.

Bodnár, István. 2023. "Aristotle's Natural Philosophy." In *The Stanford Encyclopedia of Philosophy* (Summer 2023 Edition). Ed. Edward N. Zalta and Uri Nodelman. URL = https://plato.stanford.edu/entries/aristotle-natphil/

Boler, John. 1985. "Ockham's Cleaver." *Franciscan Studies* 45: 119–144.

Bos, Egbert P., ed. 2013. *Medieval Supposition Theory Revisited: Studies in Memory of L.M. de Rijk*. Leiden: Brill.

Brown, Deborah J. and Calvin Normore. 2019. *Descartes and the Ontology of Everyday Life*. Oxford: Oxford University Press.

Brown, Deborah J. 2023. "Nature, Artifice, and Discovery in Descartes' Mechanical Philosophy." *Philosophies* 8, no. 5: 85. https://www.mdpi.com/2409-9287/8/5/85

Brown, Joseph E. 1968. *The* Scientia de ponderibus *in the Later Middle Ages*. Unpublished doctoral thesis. Madison: University of Wisconsin.

Bruno, Francesca. 2019. *The Quantity Debate in Late Medieval and Early Modern Europe*. PhD diss., Cornell University.

Buzzetti, Dino, Roberto Lambertini, and Andrea Tabarroni. 1997. "Tradizione testuale e insegnamento nell'Università di Medicina e Arti di Bologna dei secoli XIV e XV." In *Annali di storia dell'università italiane*, vol. 1. Bologna: CLUEB, pp. 77–96.

Caroti, Stefano. 1989. "Da Buridano a Marsilio di Inghen: la tradizione parigina della discussione 'De reactione'." *Medioevo* 15: 173–233.

Caroti, Stefano. 1995. "Da Walter Burley al 'Tractatus de sex inconvenientibus'. La tradizione inglese della discussione medievale 'De reactione'." *Medioevo* 21: 257–374.

Caroti, Stefano. 2002. "*Generatio/Generare*: Ontological Problems in John Buridan's Natural Philosophy." *Medioevo* 27: 373–413.

Caroti, Stefano. 2004. "Some Remarks on Buridan's Discussion on Intension and Remission." *Vivarium* 42/1: 58–85.

Caroti, Stefano. 2006. "La 'reactio' in Italia: Jacopo da Forlì." In *"Ad ingenii acuitionem": Studies in Honour of Alfonso Maierù*. Ed. Stefano Caroti. Louvain-La Neuve: Brepols, pp. 13–38.

Ceccarelli, Marco and Stephanos A. Paipetis, eds. 2014. *The Genius of Archimedes—23 Centuries of Influence on Mathematics, Science and Engineering. Proceedings of an International Conference Held at Syracuse, Italy, June 8–10, 2010*. Dordrecht: Springer.

Celeyrette, Jean. 2004. "La problématique du point chez Jean Buridan." *Vivarium* 42: 86–108.

Chandelier, Joël and Andrea Tabarroni. 2023. "Philosophie, médecine et frères mendiants à Bologne dans la première moitié du XIVe siècle." In *Savoirs profanes dans les ordres mendiants en Italie, XIIIe–XVe siècle*. Eds. Joël Chandelier and Aurélien Robert. Rome: Publications de l'École française de Rome, 2023, pp. 199–233.

Clagett, Marshall. 1964–1984. *Archimedes in the Middle Ages*. 5 vols. Madison: The University of Wisconsin Press.

Clagett, Marshall. 1941. *Giovanni Marliani and Late Medieval Physics*. New York: Columbia University Press.

Clagett, Marshall and Ernest A. Moody. 1952. *The Medieval Science of Weights* (Scientia de Ponderibus). *Treatises Ascribed to Euclid, Archimedes, Thabit ibn Qurra, Jordanus de Nemore and Blasius of Parma*. Madison: The University of Wisconsin Press.

Corkum, Phil. 2023. "Aristotle on Artifactual Substances." *Metaphysics* 6 (1): 24–36.

Crisciani, Chiara and Gianfranco Fioravanti. 2016. "I filosofi e i medici come gruppo: autorappresentazione e autopromozione." In *La filosofia in Italia al tempo di Dante*. Eds. Carla Casagrande and Gianfranco Fioravanti. Bologna: Il Mulino, pp. 77–90.

Cross, Richard. 1998. "Infinity, Continuity, and Composition: The Contribution of Gregory of Rimini." *Medieval Philosophy and Theology* 7: 89–110.

De Libera, Alain. 1990. "Le développement des nouveaux instruments conceptuels et leur utilisation dans la philosophie de la nature au XIIIe siècle." In *Knowledge and the Sciences in Medieval Philosophy. Proceedings of the Eighth International Congress of Medieval Philosophy (S.I.E.P.M.), Helsinki 24–29 August 1987*. Vol. 1. Ed. Monika Asztalos, John E. Murdoch, and Ilkka Niiniluoto. Helsinki: Yliopistopaino, pp. 159–196.

Descartes, René. 1984–1985. *The Philosophical Writings of Descartes*. Translated by John Cottingham, Robert Stoothoff, and Dugald Murdoch. 2 vol. Cambridge: Cambridge University Press.

Des Chene, Dennis. 1996. *Physiologia. Natural Philosophy in Late Aristotelian and Cartesian Thought*. Ithaca, NY: Cornell University Press.

Di Liscia, Daniel A. 2007. "Kalkulierte Ethik: Vives und die 'Zerstörer' der Moralphilosophie (Le Maistre, Cranston und Almain)." In *Ethik: Wissenschaft oder Lebenskunst? Modelle der Normenbegründung von der Antike bis zur frühen Neuzeit*. Ed. Sabrina Ebbersmeyer and Eckhard Kessler. Münster: LIT Verlag, pp. 75–105.

Di Liscia, Daniel A. 2019. "Biagio Pelacani da Parma's Geometrisation of Latitudes and the Problems of the Mean Degree Theorem." In *La philosophie de Blaise de Parme. Physique, psychologie, éthique*. Ed. Joël Biard and Aurélien Robert. Florence: SISMEL—Edizioni del Galluzzo, pp. 129–152.

Di Liscia, Daniel A. 2022. "Perfections and Latitudes: The Development of the Calculators' Tradition and the Geometrisation of Metaphysics and Theology." In *Quantifying Aristotle: The Impact, Spread and Decline of the* Calculatores *Tradition*. Ed. Daniel A. Di Liscia and Edith D. Sylla. Leiden / Boston: Brill, pp. 278–327.

Di Liscia, Daniel A. and Edith D. Sylla, eds. 2022. *Quantifying Aristotle: The Impact, Spread and Decline of the* Calculatores *Tradition*. Leiden / Boston: Brill.

DuCange, Charles Du Fresne. 1845. *Glossarium mediae et infimae Latinitatis*. Paris: Didot.

Duhem, Pierre. 1905–1906. *Les origines de la statique*, 2 vols. Paris: Hermann.

Duhem, Pierre. 1913–1959. *Le système du monde. Histoire des doctrines cosmologiques de Platon à Copernic*. 10 vols. Paris: Hermann.

Dutton, Blake D. 1996. "Nicholas of Autrecourt and William of Ockham on Atomism, Nominalism, and the Ontology of Motion." *Medieval Philosophy and Theology* 5: 63–85.

Ebbesen, Sten. 2009. "Early Supposition Theory (12th–13th Centuries)." In Sten Ebbesen, *Topics in Latin Philosophy from the 12th–14th Centuries*, vol. 2. Aldershot: Ashgate, pp. 1–14.

Jung, E. 2016 "Mathematics and the *Secundum Imaginationem* Procedure in Richard Kilvington," *Przegląd Tomistyczny* 22: 109–120.

Encyclopaedia Britannica. 1926. 13th ed. London: Encyclopædia Britannica Company.

Federici Vescovini, Graziella. 1961. "Le questioni di 'Perspectiva' di Biagio Pelacani da Parma." *Rinascimentoi* 12: 163–243.

Federici Vescovini, Graziella. 1976. "A propos de la diffusion des œuvres de Jean Buridan en Italie du XIVe au XVIe siècle." In *The Logic of John Buridan*. Ed. Jan Pinborg. Copenhagen: Museum Tusculanum Press, pp. 21–45.

Federici Vescovini, Graziella. 1979. *Astrologia e scienza: la crisi dell'aristotelismo sul cadere del Trecento e Biago Pelacani de Parma*. Florence: E. Vallecchi.

Federici Vescovini, Graziella. 1981. "Su alcune testimonianze dell'influenza di Alberto Magno come 'metafisico', scienziato e 'astrologo' nella filosofia padovana del cadere del secolo XIV: Angelo di Fossombrone e Biagio Pelacani da Parma." In *Albert der Große: Seine Zeit, sein Werk, seine Wirkung*. Ed. Albert Zimmermann. Berlin: De Gruyter, pp. 155–176.

Federici Vescovini, Graziella. 1991. "Notes sur la circulation en Italie du commentaire d'Albert de Saxe sur le *De caelo*." In *Itinéraires d'Albert de Saxe: Paris-Vienne au XIVe siècle*. Ed. Joël Biard. Paris: Vrin, pp. 235–251.

Federici Vescovini, Graziella. 1994. "La '*Quaestio de intensione et remissione formarum*' di Biagio Pelacani da Parma." *Physis* 31: 433–535.

Fioravanti, Gianfranco. 2016. "I filosofi e la medicina: una progressiva autonomia." In *La filosofia in Italia al tempo di Dante*. Eds. Carla Casagrande and Gianfranco Fioravanti. Bologna: Il Mulino, 65–76.

Geach, Peter. 1967. "Infinity in Scholastic Philosophy." In *Problems in the Philosophy of Mathematics*. Ed. Imre Lakatos. Amsterdam: North Holland, pp. 41–42.

Grellard, Christophe. 2005. *Croire et savoir. Les principes de la connaissance selon Nicolas d'Autrécourt*. Paris: Vrin.

Grellard, Christophe. 2019. "Une histoire naturelle des religions. Blaise de Parme, les astres et les sectes". In *La philosophie de Blaise de Parme. Physique, psychologie, éthique*. Ed. Joël Biard and Aurélien Robert. Florence: SISMEL—Edizioni del Galuzzo, pp. 59–82.

Hackett, Jeremiah. 1997. "Roger Bacon on scientia experimentalis." In J. Hackett (ed.) *Roger Bacon and the Sciences*. Brill: Leiden, pp. 277–315.

Harrison, Anne. 1983. "Blasius of Parma's Critique of Bradwardine's *Tractatus de proportionibus*." In *Scienza e filosofia all'Università di Padova nel Quattrocento*. Ed. Antonino Poppi. Padua / Trieste: Edizioni Lint, pp. 19–69.

Hilpinen, Risto. 1992. "Artifacts and Works of Art." *Theoria*, 58(1): 58–82.

Hoskin, Michael A. and A. George Molland. 1966. "Swineshead on Falling Bodies: An Example of Fourteenth-Century Physics." *The British Journal for the History of Science* 3: 150–182.

Høyrup, Jens. 1988. "Jordanus de Nemore, 13th Century Mathematical Innovator: An Essay on Intellectual Context, Achievement, and Failure," *Archive for History of Exact Sciences* 38/4: 307–363.

Hugonnard-Roche, Henri. 1988. "Modalités et argumentation chez Nicole Oresme." In *Nicole Oresme. Tradition et innovation chez un intellectuel du XIVe siècle*. Eds. Pierre Souffrin and Alain-Philippe Segonds. Paris / Padua: Les Belles Lettres / Programma e 1+1, pp. 145–163.

Hugonnard-Roche, Henri. 1991. "Le possible et l'imaginaire dans la physique d'Albert de Saxe," In *Itinéraires d'Albert de Saxe. Paris-Vienne au XIVe siècle*. Ed. Joël Biard. Paris: Vrin, pp. 161–173.

Hugonnard-Roche, Henri. 1997. "L'hypothétique et la nature dans la physique parisienne du XIVe siècle." In *La nouvelle physique du XIVe siècle*. Eds. Stefano Caroti and Pierre Souffrin. Florence: Olschki, pp. 161–177.

Jung, Elżbieta. 2011. "Intension and Remission of Forms." In *Encyclopedia of Medieval Philosophy*. Ed. Henrik Lagerlund. Berlin: Springer, pp. 551–555.

Kaluza, Zénon. 1995. *Nicolas d'Autrécourt, ami de la vérité*. Histoire littéraire de la France 42–41. Paris: Imprimerie nationale.

King, Peter. 1985. "Introduction," in John Buridan, *Jean Buridan's Logic: The Treatise on Supposition, The Treatise on Consequences*, trans. P. King. Dordrecht / Boston: D. Reidel, pp. 3–82.

Klima, Gyula. 1993. "'Debeo tibi equum': A Reconstruction of the Theoretical Framework of Buridan's Treatment of the Sophisma." In *Sophisms in Medieval Logic and Grammar. Acts of the Ninth European Symposium for Medieval Logic and Semantics, Held at St Andrews, June 1990*. Ed. Stephen Read. Dordrecht / Boston: Springer, pp. 333–347.

Klima, Gyula. 2005. "The Essentialist Nominalism of John Buridan." *Review of Metaphysics* 58(4): 739–754.

Kretzmann, Norman, ed. 1982. *Infinity and Continuity in Ancient and Medieval Thought*. Ithaca / London: Cornell University Press.

Lagerlund, Henrik. 2012. "Material Substance." In *The Oxford Handbook of Medieval Philosophy*. Ed. J. Marenbon. Oxford: Oxford University Press, pp. 468–485.

Lagerlund, Henrik. 2020. *Skepticism in Philosophy: A Comprehensive, Historical Introduction*. New York: Routledge.

van der Lugt, Maikke. 2004. *Le Ver, le Démon et la Vierge. Les théories médiévales de la génération extraordinaire*. Paris: Les Belles Lettres.

Mahoney, Edward P. 1982. "Metaphysical Foundations of the Hierarchy of Being According to Some Late-Medieval and Renaissance Philosophers." In *Philosophies of Existence, Ancient and Medieval*. Ed. Parviz Morewedge. New York: Fordham University Press, pp. 165–267.

Maier, Anneliese. 1958. *Zwischen Philosophie und Mechanik*. Studien zur Naturphilosophie der Spätscholastik, Bd. V. Roma: Storia e Letteratura.

Maier, Anneliese and Marco Vattasso. 1961. *Codices Vaticani Latini*, Codices 2118–2192. Bibliothecae Apostolicae Vaticanae codices manu scripti recensiti iussu Ioannis XXIII Pontificis Maximi. Vaticana: Biblioteca Apostolica Vaticana.

Maier, Anneliese. 1968. *Zwei Grundprobleme der scholastischen Naturphilosophie*. Rome: Edizioni di storia et letteratura.

Maierú, Alfonso. 2013 "Robert Kilwardby on the Division of the Sciences." In H. Lagerlund and Paul Thom (eds.) *A Companion to the Philosophy of Robert Kilwardby*. Brill: Leiden, pp. 353–389.

Majcherek, Kamil. 2020. "Paul of Venice's Metaphysics of Artefacts." *British Journal for the History of Philosophy* 28(1): 29–48. DOI: 10.1080/09608788.2019.1588703.

Majcherek, Kamil. 2022. *Medieval Metaphysics of Artifacts 1250–1500*. PhD dissertation, University of Toronto.

Miglio, Massimo et al., eds. 1986. *Un pontificato ed una città: Sisto IV (1471–1484): atti del Convegno, Roma, 3–7 dicembre 1984*. Rome: Istituto storico italiano per il medioevo.

Murdoch, John E. 1969. "*Mathesis in philosophiam scholasticam introducta*. The Rise and Development of the Application of Mathematics in Fourteenth-Century Philosophy and Theology." In *Arts libéraux et philosophie au Moyen Âge. Actes du quatrième Congrès international de philosophie médiévale*. Montreal: Institut d'Études médiévales, pp. 215–254.

Murdoch, John E. 1974. "Philosophy and the Enterprise of Science in the Late Middle Ages." In *The Interaction between Science and Philosophy*. Ed. Yehuda Elkana and Samuel Sambursky. Atlantic Highlands: Humanities Press, pp. 51–74.

Murdoch, John E. 1975. "From Social into Intellectual Factors: An Aspect of the Unitary Character of Late Medieval Learning." In *The Cultural Context of Medieval Learning. Proceedings of the First International Colloquium on Philosophy, Science, and Theology in the Middle Ages*. Ed. John E. Murdoch and Edith D. Sylla. Dordrecht / Boston: D. Reidel Publishing Company, pp. 271–339.

Murdoch, John E. 1978. "*Subtilitates Anglicanae* in Fourteenth-century Paris: John of Mirecourt and Peter Ceffons." In *Machaut's World. Science and Art in the Fourteenth Century*. Ed. Madeleine Pelner Cosman and Bruce Chandler. New York: New York Academy of Sciences, pp. 51–86.

Murdoch, John E. 1979. "Propositional Analysis in Fourteenth-Century Natural Philosophy: A Case Study." *Synthese* 40: 117–146.

Murdoch, John E. 1981. "*Scientia mediantibus vocibus*: Metalinguistic Analysis in Late Medieval Natural Philosophy." In *Sprache und Erkenntnis im Mittelalter*. Ed. Wolfgang Kluxen. Berlin: De Gruyter, pp. 73–106.

Murdoch, John E. 1982. "Infinity and Continuity." In *The Cambridge History of Later Medieval Philosophy*. Ed. Norman Kretzmann, Anthony Kenny, and Jan Pinborg. Cambridge: Cambridge University Press, pp. 564–591.

Murdoch, John E. 1984. *Album of Science: Antiquity and Middle Ages*. New York: Charles Scribner's Sons.

Murdoch, John E. 1987. "Thomas Bradwardine: Mathematics and Continuity in the Fourteenth Century." In *Mathematics and Its Applications to Science and Natural Philosophy in the Middle Ages: Essays in Honour of Marshall Clagett*. Ed. Edward Grant and John E. Murdoch. Cambridge: Cambridge University Press, pp. 103–137.

Murdoch, John E. 1991. "Pierre Duhem and the History of Late Medieval Science and Philosophy in the Latin West." In *Gli studi di filosofia medievale fra otto e novecento*. Ed. Ruedi Imbach and Alfonso Maieru. Storia e letteratura 179. Rome: Edizioni di Storia e Letteratura.

Neilson, George. 1895. *Caudatus Anglicus: A Mediæval Slander*. Edinburgh: George P. Johnston.

Normore, Calvin. 2006. "Ockham's Metaphysics of Parts." *Journal of Philosophy* 103(12): 737–754.

O'Brien, Denis. 1981–1984. *Theories of Weight in the Ancient World. Four Essays on Democritus, Plato, and Aristotle. A Study in the Development of Ideas*. 2 vols. Leiden: Brill.

Palmerino, Carla Rita. 2011. "Galileo's Use of Medieval Thought Experiments." In *Experiments in Methodological and Historical Contexts*. Ed. Katerina Ierodiakonou and Sophie Roux. Leiden / Boston: Brill, pp. 101–125.

Panaccio, Claude. 2012. "Ockham and Buridan on Epistemic Sentences: Appellation of the Form and Appellation of Reason." *Vivarium* 50(2): 139–160.

Pasnau, Robert. 2011. *Metaphysical Themes 1274–1671*. Oxford: Oxford University Press.

Pelletier, Jenny. 2022. "Getting Real: Ockham on the Human Contribution to the Nature and Production of Artifacts." *Philosophies* 7 (90). https://doi.org/10.3390/philosophies7050090.

Piacentini, Paola. 1986. "Ricerche sugli antichi inventari della Biblioteca Vaticana." In *Un pontificato ed una città: Atti del convegno, Roma, 3–7 dicembre 1984*. Ed. Massimo Miglio and Francesca Nuttia. Roma: Nella sede dell'Istituto, pp. 115–178.

Preston, Beth. 2022. "Artifact." In *The Stanford Encyclopedia of Philosophy* (Winter 2022 Edition). Ed. Edward N. Zalta and Uri Nodelman. URL = https://plato.stanford.edu/entries/artifact/

Puccinotti, Francesco. 1855. *Storia della medicina* v. 2 pt. 1. Livorno: Presso Massimiliano Wagner Editore.

Queiroz, L. 2018. "On the Relation between Natural Philosophy, Mathematics and Logic in the Investigation on Atomism in the Middle Ages." *Enrahonar*, Supplement Issue, pp. 135–149.

Rashed, Marwan. 2009. "Contre le mouvement rectiligne naturel: trois adversaires (Xénarque, Ptolémée, Plotin) pour une thèse." In *Physics and Philosophy of Nature in Greek Neoplatonism. Proceedings of the European Science Foundation Exploratory Workshop (Il Ciocco, Castelvecchio Pascoli, June 22–24, 2006)*. Eds. Riccardo Chiaradonna and Franco Trabattoni. Leiden: Brill, pp. 17–42.

Robert, Aurélien. 2009. "William Crathorn's Mereotopological Atomism." In *Atomism in Late Medieval Philosophy and Theology*. Ed. Christophe Grellard and Aurélien Robert. Leiden: Brill, pp. 127–162.

Robert, Aurélien. 2019. "Désir de persévérer dans l'être et mort volontaire chez Nicole Oresme." *Quaderni di Noctua* 5: 199–239.

Robert, Aurélien. 2019. "L'éthique de Blaise de Parme dans ses *Questions sur la Physique*." In *La Philosophie de Blaise de Parme. Physique, psychologie, éthique*. Ed. J. Biard et A. Robert. Florence: SISMEL—Edizioni del Galluzzo, pp. 33–58.

Rommevaux, Sabine. 2008. "Les règles du mouvement de Blaise de Parme dans les deux versions de ses '*Questions sur le traité des rapports de Thomas Bradwardine*'." In *Mathématiques et théorie du mouvement, XIVe–XVIe siècles*. Ed. J. Biard and S. Rommevaux. Villeneuve d'Ascq: Presses Universitaires du Septentrion, pp. 31–52.

Rommevaux, Sabine. 2010. "Magnetism and Bradwardine's Rule of Motion in Fourteenth- and Fifteenth-Century Treatises." *Early Science and Medicine* 15: 618–647.

Rommevaux, Sabine. 2012. "Le vide *secundum naturam* dans les *Questions sur la Physique* de Blaise de Parme." In *La nature et le vide dans la physique médiévale. Études dédiées à Edward Grant*. Ed. Joël Biard and Sabine Rommevaux. Turnhout: Brepols, pp. 293–316.

Rommevaux, Sabine. 2019. "Ajout de conditions dans les règles du mouvement formulées par Blaise de Parme dans la seconde rédaction de ses Questions sur la Physique." In *La philosophie de Blaise de Parme: Physique, psychologie, éthique*. Ed. Joël Biard and Aurélien Robert. Florence: SISMEL—Edizioni del Galluzzo, pp. 153–174.

Roques, Magali. 2019. "Blaise de Parme et la Quantité." In *La philosophie de Blaise de Parme: Physique, psychologie, éthique*. Ed. Joël Biard and Aurélien Robert. Florence: SISMEL—Edizioni del Galluzzo, pp. 175–196.

Rota, M. 2004. "Substance and Artifact in Thomas Aquinas." *History of Philosophy Quarterly* 21(3): 241–259.

Roudaut, Sylvain. 2021. *La mesure de l'être. Le problème de la quantification des formes au Moyen Âge (ca. 1250–1370)*. Leiden / Boston: Brill.

Roudaut, Sylvain. 2022. "Paul of Venice's Theory of Quantification and Measurement of Properties." *Noctua* 9(2): 104–158.

Roudaut, Sylvain. Forthcoming. "The Substrate of Change in Buridan's Theory of Nature: A Semantic Reinterpretation of Hylomorphism." In *The Elusive Substrate: Prime Matter and Hylomorphism from Ancient Rome to Early Qing China*, vol. 2. Ed. Nicola Polloni. Cambridge: Cambridge University Press.

Sarnowsky, Jürgen. 1997. "God's Absolute Power, Thought Experiments and the Concept of Nature in the 'New Physics'." In *La nouvelle physique du XIVe siècle*. Eds. Stefano Caroti and Pierre Souffrin. Florence: Olschki, pp. 179–201.

Shields, Christopher. 2004. *Aristotle*. Second edition. London: Routledge.

Sinkler, Georgette. 1989. "William Heytesbury's Word-Order Theory of Propositional Sense." *Journal of the History of Philosophy* 27 (3): 365–377.

Solère, Jean-Luc. 2000. "Plus ou moins: le vocabulaire de la latitude des formes." In *L'élaboration du vocabulaire philosophique au Moyen Âge*. Ed. Jacqueline Hamesse and Carlos Steel. Turnhout: Brepols, pp. 437–488.

Solère, Jean-Luc. 2007. "Tension et Intention. Esquisse de l'histoire d'une notion." In *Questions sur l'intentionnalité*. Ed. Lambros Couloubaritsis and Antonino Mazzù. Brussels: Ousia, pp. 59–124.

Sylla, Edith D. 1971–1972. “Medieval Quantifications of Qualities: The ‘Merton School’.” *Archive for History of Exact Sciences* 8: 9–39.

Sylla, Edith D. 1973. “Medieval Concepts of the Latitude of Forms: The Oxford Calculators.” *Archives d'histoire doctrinale et littéraire du Moyen Âge* 40: 223–283.

Sylla, Edith D. 1991. *The Oxford Calculators and the Mathematics of Motion, 1320–1350. Physics and Measurement by Latitudes*. New York / London: Garland.

Sylla, Edith D. 1993. “Aristotelian Commentaries and Scientific Change: The Parisian Nominalists on the Cause of the Natural Motion of Inanimate Bodies.” *Vivarium* 31(1): 37–83.

Sylla, Edith D. 2010. “The Oxford Calculators' Middle Degree Theorem in Context.” *Early Science and Medicine* 15: 338–370.

Sylla, Edith D. 2020. “Infinity and Continuity: Thomas Bradwardine and His Contemporaries.” In *The History of Continua. Philosophical and Mathematical Perspectives*. Ed. Stewart Shapiro and Geoffrey Hellman. Oxford: Oxford University Press, pp. 49–81.

Tabarroni, Andrea. 2016. “Ambienti culturali prossimi a Dante nell'esilio: lo Studio bolognese di Arti e Medicina.” In *Dante fra il settecentocinquantenario della nascita (2015) e il settecentenario della morte (2021)*. Eds. Enrico Malato and Andrea Mazzucchi. Rome: Antenore, pp. 327–349.

Tabarroni, Andrea. 2016. “La nascita dello Studio di Medicina e Arti a Bologna.” In *La filosofia in Italia al tempo di Dante*. Eds. Carla Casagrande and Gianfranco Fioravanti. Bologna: Il Mulino, pp. 25–36.

Thijssen, Johannes M.M.H. 1990. “The Book-collection of Bernardus a Campanea de Verona (fl. end 14th/beginning 15th century): The Dissemination of English and French Natural Science in Italy.” *Scriptorium* 44(2): 299–312.

Thijssen, Johannes M.M.H. 2004. “The Buridan School Reassessed. John Buridan and Albert of Saxony.” *Vivarium* 42/1: 18–42.

Thijssen, Johannes M.M.H. 2009. “The Debate over the Nature of Motion: John Buridan, Nicole Oresme and Albert of Saxony. With an Edition of John Buridan's ‘*Quaestiones super libros Physicorum, secundum ultimam lecturam*’, Book III, Q. 7.” *Early Science and Medicine*, vol. 14, no. 1/3. *Evidence and Interpretation: Studies on Early Science and Medicine in Honor of John E. Murdoch*, pp. 186–210.

Thorndike, Lynn. 1967. *Science and Thought in the Fifteenth Century: Studies in the History of Medicine and Surgery, Natural and Mathematical Science, Philosophy and Politics*. New York: Hafner.

Trifogli, Cecilia. 2000. *Oxford Physics in the Thirteenth Century (ca. 1250–1270). Motion, Infinity, Time, and Place*. Leiden: Brill.

Trifogli, Cecilia. 2002. “Matter and Form in Thirteenth-Century Discussions of Infinity and Continuity.” In *The Dynamics of Aristotelian Natural Philosophy from Antiquity to the Seventeenth Century*. Eds. Cees H. Leijenhorst, Christoph H. Lüthy, and Johannes M.M.H. Thijssen. Leiden / Boston / Cologne: Brill, pp. 169–187.

Trifogli, Cecilia. 2015. "The Reception of Averroes' View on Motion in the Latin West. The Case of Walter Burley." In *Averroes' Natural Philosophy and Its Reception in the Latin West*. Ed. Paul J.J.M. Bakker. Leuven: Leuven University Press, pp. 127–139.

Uckelman, Sara L. 2015. "The Logic of Categorematic and Syncategorematic Infinity." *Synthese* 192(8): 2361–2377.

Wang, Xiaona. 2022. "Rarefaction and Condensation." In *Encyclopedia of Renaissance Philosophy*. Ed. Marco Sgarbi. Springer eBook, pp. 2764–2770.

Weill-Parot, Nicolas. 2019. "Une position originale? Le contact entre le moteur et le mû dans les *Questiones disputatae super octo libros Physicorum* de Blaise Parme." In *La Philosophie de Blaise de Parme. Physique, psychologie, éthique*. Ed. Joël Biard and Aurélien Robert. Florence: SISMEL—Edizioni del Galluzzo, pp. 197–217.

Zoubov, Vassili P. 1963. "Blaise de Parme et le principe d'inertie." *Physis* 5: 39–42.

Zupko, Jack. 2018. "'Nothing in Nature Is Naturally a Statue': William of Ockham on Artifacts." *Metaphysics* 1: 88–95. DOI: https://metaphysicsjournal.com/articles/10.5334/met.7

Index

Åkerlund, Erik 9, 72, 93
Albert of Saxony 1, 3, 33, 38, 46, 72, 121, 122, 148
Analytical languages 10, 115, 116, 117, 128, 142, 158, 159
Andrews, Robert 8, 163
Appellation 119, 126, 128
Aquinas, Thomas 80, 106, 107, 111
Archimedes (also Pseudo-) 7, 19, 21, 31, 35, 37, 38, 229
Aristotle 2, 5, 6, 9, 13, 15, 17, 18, 22, 26, 27, 32, 33, 35, 36, 37, 38, 40, 41, 45, 46, 47, 51, 53, 57, 62, 63, 70, 73, 74, 75, 76, 77, 79, 81, 82, 83, 88, 89, 93, 104, 105, 106, 116, 119, 120, 143, 147
Artifact 6, 7, 8, 10, 13, 21, 29, 102–114, 128
Averroes 73, 76, 77, 78, 83
Avicenna 66, 67, 192, 196, 203, 208, 211

Bacon, Francis 102, 103, 114
Bacon, Roger 6, 7, 8
Bayle, Pierre 8
Biard, Joël 7, 9, 39, 43, 45, 142, 150, 151, 166
Body 5, 19, 20, 22, 23, 24, 25, 26, 27, 28, 29, 30, 31, 33, 34, 38, 39, 58, 64, 67, 68, 73, 75, 78, 79, 81, 82, 83, 84, 85, 86, 87, 90, 91, 95, 96, 98, 100, 124, 125, 126, 127, 128, 129, 130, 131, 132, 133, 134, 135, 136, 137, 138, 139, 144, 147, 148, 150, 152, 153, 154, 155, 157
Bradwardine, Thomas 2, 3, 40, 139, 140, 141, 144
Brown, Joseph 14, 15, 43
Buridan, John 3, 9, 10, 33, 34, 45, 46, 50, 61, 65, 72, 73, 77, 79, 80, 83, 93, 94, 96, 99, 100, 105, 108, 109, 110, 111, 114, 118, 121, 122, 126, 128, 148
Burley, Walter 145, 146

Carneades 8
Category 74, 77
Causation 9, 46, 52, 61, 87
Centueri, William 1
Cicero 8
Clagett, Marshall 14, 15, 43
Condencation 8, 9, 10, 93–101, 138, 139, 144, 157
Continuity 10, 62, 64, 68, 115, 116, 118, 128, 129, 130, 131, 132, 135, 136, 141, 142, 144, 150, 159
Crathorn, William 93, 94, 95, 100, 101

Des Chene, Dennis 93
Descartes, René 4, 102, 103, 104, 114
Doubt (skeptical) 46, 49, 51, 52, 53, 54, 55, 71
Duhem, Pierre 72
Dumbleton, John 3, 134, 148
Duns Scotus, John 106, 107

Efficient causality 33, 45, 46, 56, 103, 114, 206
Euclid 116
Evidence 47, 50, 53, 54, 66, 69, 71, 126, 127, 155
Experimental philosophy 6, 7, 8, 260, 281

Federici Vescovini, Grazziella 15, 163, 164, 166
Final Causality 45, 47, 114, 263

Galileo 4, 7, 31, 80
Gassendi, Pierre 8
Godfrey of Fontaine 145
Gravity 19, 31, 42

Heaviness and lightness 4, 18, 21, 22, 33, 35, 36–40, 82, 157
Heytesbury, William 3, 127, 128, 148
Hydrostatics 14, 15, 31, 37, 39

Imaginary physics 4, 27, 31, 32, 125, 130, 133, 138, 141
Impetus 73, 79, 80, 87, 89, 196
Infinity 10, 115, 116, 118, 128, 129, 130, 131, 132, 133, 134, 135, 136, 137, 138, 139, 141, 142, 144, 147, 150, 159
Intensive properties 2, 39, 137, 142–158, 159, 272, 273, 278, 280

Jordanus de Nemore 7, 14, 15, 18, 19, 23

Kilvington, Richard 3
Kilwardby, Robert 5, 6

Lagerlund, Henrik 10, 72, 102
Laws of nature 16, 21, 22, 23, 24, 28, 30, 40, 42, 45, 92, 114, 142, 159
Leonardo da Vinci 2

Mair, John 72
Marsilius of Inghen 1, 3, 72, 148
Materialism 1, 8, 33, 43
Mathematics 2, 5, 6, 7, 8, 13, 17, 18, 130, 136, 157
Matter 4, 15, 32, 33, 34, 35, 44, 47, 58, 59, 60, 62, 65, 66, 68, 69, 70, 71, 80, 81, 83, 93, 94, 95, 97, 99, 100, 106, 107, 122, 124, 129, 130, 131, 138, 145, 150, 151, 155
Mechanical philosophy 4, 5, 6, 7, 16, 17, 44, 103, 104
Mechanics 4, 5, 6, 7, 8, 9, 13–44, 72, 104
Moody, Ernest 43
Motion 4, 5, 8, 9, 14, 18, 22, 25, 26, 28, 30, 31, 32, 33, 34, 35, 37, 39, 40, 41, 45, 60, 72–92, 105, 106, 108, 109, 111, 120, 121, 124, 127, 131, 134, 138, 149
  Natural 5, 10, 34, 73, 81, 84, 85, 88, 89, 128
  Local 4, 9, 35, 40, 43, 45, 75, 78, 79, 80, 82, 85, 86, 89, 90, 92, 95, 111, 112, 142, 144, 151, 158
  Violent 5, 10, 72, 73, 81, 84, 85, 86, 87, 88, 89
Murdoch, John 10, 115, 116, 117, 118, 128, 142, 144, 158, 159, 165

Nicholas of Autrecourt 8, 50, 53, 54
Nominalism 3, 10, 33, 104, 105, 108, 109, 110, 111, 114, 117, 119, 120, 122, 123, 137

Oresme, Nicole 1, 3, 38, 46, 61, 72, 141, 148
Oxford Calculators 3, 4, 7, 28, 32, 45, 72, 117, 125, 132, 134, 135, 136, 142, 144, 152, 154, 157

Parmenides 74
Peckham, John 2
Peter of Abano 3
Peter of Spain 2, 119
Plato 116, 248
Power 9, 40, 41, 45–71, 85, 87, 103, 135, 139, 147, 150, 151, 152
Probabilism 7, 8, 122, 155, 227, 245, 268, 271
Problemata mechanica 16, 18, 19
Proportions 2, 10, 17, 18, 20, 22, 23, 24, 25, 26, 27, 29, 31, 57, 58, 60, 64, 65, 66, 69, 70, 115, 116, 129, 132, 133, 134, 135, 136, 139, 140, 141, 142, 144, 147, 148, 149, 150, 152, 153, 155, 156, 157, 159, 166, 180, 181, 182, 183, 191, 193, 194, 195, 197, 199, 209, 212, 230, 231, 232, 233, 234, 245, 247, 253, 256, 257, 259, 262, 268, 279, 281, 282

Quality 25, 58, 59, 60, 61, 69, 74, 75, 77, 80, 83, 84, 86, 87, 94, 134, 137, 138, 145, 147, 148, 149, 150, 151, 153, 155
Quantity 10, 41, 74, 75, 77, 80, 93, 94, 95, 96, 97, 99, 100, 107, 136, 137, 138, 144, 145

Rarefaction 8, 9, 10, 93–101, 131, 132, 136, 137, 144, 157
Realism 109, 111, 120
Reductionism 3, 32
Roudaut, Sylvain 10, 13, 72, 115

Salisbury, John 8
Science of weights 2, 4, 6, 7, 14, 15, 16–35, 36, 40, 41, 42, 43, 157
Shape 29, 30, 31, 42, 104, 108, 109, 110, 111, 112
Statics 2, 4, 6, 7, 8, 14, 16, 18, 19, 21, 22, 28, 30, 42, 43, 44
Substantial forms 4, 8, 34, 35, 38, 43, 44, 57, 58, 59, 60, 61, 64, 66, 83, 90, 97, 107, 121, 130
Successive entity 78, 79, 87, 193
Supposition theory 10, 51, 115, 116, 118–128, 144, 158
Swineshead, Richard 3, 28, 133, 136, 148, 149

Themon Judaeus 72
Thought-Experiment 4, 7, 8, 28, 29, 31, 39, 40, 41, 42, 94, 100, 125, 130–134, 139, 141, 152, 158

Vittorino de Feltre 2
Void 29, 30, 32, 40, 41, 42, 90, 93, 123, 124, 125, 158

William Ockham 1, 9, 10, 46, 73, 77, 78, 79, 80, 93, 94, 95, 96, 99, 100, 105, 106, 107, 108, 109, 110, 111, 126

Zeno of Elea 74

www.ingramcontent.com/pod-product-compliance
Lightning Source LLC
Chambersburg PA
CBHW071333230926
51849CB00002B/389
9789004734272